Studies in Systems, Decision and Control

Volume 256

Series Editor

Janusz Kacprzyk, Systems Research Institute, Polish Academy of Sciences, Warsaw, Poland

The series "Studies in Systems, Decision and Control" (SSDC) covers both new developments and advances, as well as the state of the art, in the various areas of broadly perceived systems, decision making and control–quickly, up to date and with a high quality. The intent is to cover the theory, applications, and perspectives on the state of the art and future developments relevant to systems, decision making, control, complex processes and related areas, as embedded in the fields of engineering, computer science, physics, economics, social and life sciences, as well as the paradigms and methodologies behind them. The series contains monographs, textbooks, lecture notes and edited volumes in systems, decision making and control spanning the areas of Cyber-Physical Systems, Autonomous Systems, Sensor Networks, Control Systems, Energy Systems, Automotive Systems, Biological Systems, Vehicular Networking and Connected Vehicles, Aerospace Systems, Automation, Manufacturing, Smart Grids, Nonlinear Systems, Power Systems, Robotics, Social Systems, Economic Systems and other. Of particular value to both the contributors and the readership are the short publication timeframe and the world-wide distribution and exposure which enable both a wide and rapid dissemination of research output.

** Indexing: The books of this series are submitted to ISI, SCOPUS, DBLP, Ulrichs, MathSciNet, Current Mathematical Publications, Mathematical Reviews, Zentralblatt Math: MetaPress and Springerlink.

More information about this series at http://www.springer.com/series/13304

Nishchal K. Verma · Al Salour

Intelligent Condition Based Monitoring

For Turbines, Compressors, and Other Rotating Machines

Springer

Nishchal K. Verma
Department of Electrical Engineering and
Inter-disciplinary Program in Cognitive Science
Indian Institute of Technology Kanpur
Kanpur, Uttar Pradesh, India

Al Salour
Boeing Research and Technology
Saint Louis, MO, USA

ISSN 2198-4182 ISSN 2198-4190 (electronic)
Studies in Systems, Decision and Control
ISBN 978-981-15-0511-9 ISBN 978-981-15-0512-6 (eBook)
https://doi.org/10.1007/978-981-15-0512-6

This Springer imprint is published by the registered company Springer Nature Singapore Pte Ltd.
The registered company address is: 152 Beach Road, #21-01/04 Gateway East, Singapore 189721, Singapore

Preface

Intelligent condition-based monitoring of machines for early recognition of faults saves industry from heavy losses occurring due to machine breakdowns. It is the process of monitoring machine parameters associated with various conditions, such that a significant change can be indicated for a developing failure. It allows continuous monitoring of machines to avoid the consequences of failure before the failure occurs. The process includes details of data acquisition, sensitive position analysis for deciding suitable sensor locations, signal pre-processing, feature extraction, feature selection, and classification.

Intelligent condition-based monitoring increases the reliability of a process using artificially intelligent agents like fuzzy systems, artificial neural networks, support vector machines, etc. The main objective of this book is to introduce the fault diagnosis framework for intelligent condition-based monitoring of machines, their applications, and case studies.

With the rapid development of signal processing techniques, the parameters such as vibration, displacement, or sound emitted from the machine can be used in intelligent condition-based monitoring because they always carry the dynamic information of the machine. To understand the fault dynamics of a machine, the parameters such as vibration, displacement, and acoustic emissions are considered in this book. Fault diagnosis is an integral component of condition-based monitoring. It involves mainly four stages:

(1) Capturing the signal of machine in both states, healthy as well as faulty.
(2) Removing the noise content from the monotonic function known as pre-processing of data.
(3) Extracting the features from pre-processed data.
(4) Finally, a relevant set of features are used for classification.

The contents of this book are briefly explained below:

- Chapter 1 gives a brief introduction to the general framework for fault diagnosis, sequence of operations, and various techniques used in fault diagnosis of air compressor.

- Chapter 2 begins with a review of the working principle of air compressor, followed by an introduction to the faults being investigated in the work. Further, there is a brief introduction to the data acquisition setup, and finally, it ends up describing the data collection procedure of the healthy and faulty compressor.
- Chapter 3 gives a review of the various signal pre-processing methods.
- Chapter 4 is dedicated to feature extraction in time, frequency, and time–frequency domain.
- Chapter 5 starts with a brief introduction to the feature selection techniques and dwells into the details of feature selection procedures based on principal component analysis (PCA), mutual information (MI), Bhattacharyya distance (BD), and independent component analysis (ICA).
- Chapter 6 highlights different classifiers and explains support vector machine in detail.
- Chapter 7 highlights a brief introduction to quick fault diagnosis system using Palmtop.
- Chapter 8 highlights an improved fault diagnosis model.
- Chapter 9 introduces Android app development for fault diagnosis.

Kanpur, India Nishchal K. Verma
Saint Louis, USA Al Salour

Contents

About the Authors

Dr. Nishchal K. Verma (SM'13) is a Professor in Department of Electrical Engineering and Inter-disciplinary Program in Cognitive Science at Indian Institute of Technology Kanpur, India. He obtained PhD in Electrical Engineering from Indian Institute of Technology Delhi, India. He is an awardee of Devendra Shukla Young Faculty Research Fellowship by Indian Institute of Technology Kanpur, India for year 2013-16.

His research interests include intelligent fault diagnosis systems, prognosis and health management, big data analysis, deep learning of neural and fuzzy networks, machine learning algorithms, computational intelligence, computer vision, brain computer/machine interface, intelligent informatics, soft-computing in modelling and control, internet of things/cyber physical systems, and cognitive science. He has authored more than 200 research papers.

Dr. Verma is an IETE Fellow. He is currently serving as a Guest Editor of the IEEE *Access*: special section on "Advance in Prognostics and System Health Management," an Editor of the IETE Technical Review Journal, an Associate Editor of the IEEE Transactions on Neural Networks and Learning Systems, an Associate Editor of the IEEE Computational Intelligence Magazine, an Associate Editor of the Transactions of the Institute of Measurement and Control, U.K. and editorial board member for several journals and conferences.

Dr. Al Salour is a Boeing Technical Fellow and the enterprise leader for the Network Enabled Manufacturing technologies. He is responsible for systems approach to develop, integrate, and implement affordable sensor based manufacturing strategies and plans to provide real time data for factory systems and supplier networks. He is building a model for the current and future Boeing factories by streamlining and automating data management to reduce factory direct labour and overhead support and promote manufacturing as a competitive advantage.

Dr. Salour's accomplishments include machine health monitoring integrations, asset tracking and RFID system installations; and safety systems for automated guided vehicles. Dr. Salour is the research investigator with national and international premiere universities and research labs. He serves as a committee vice chair for the ASME's prognostics and health management national society. He is also a member of Industrial wireless technical working group with the National Institute of Standards and Technology (NIST).

Dr. Salour has 31 invention disclosures, 22 patents and 1 trade secret in manufacturing technologies.

Abbreviations

AC	Alternating Current
ACF	Autocorrelation Function
ACR	Accumulative Contribution Rate
ADT	Android Developer Tools
AE	Auto-Encoder
AI	Analog Input
BD	Bhattacharyya Distance
BJD	Born–Jordan Distribution
CBM	Condition-Based Monitoring
cDAQ	Chassis for CompactDAQ
CM	Corrective Maintenance
CWD	Choi–William Distribution
CWT	Continuous Wavelet Transform
DAQ	Data Acquisition
DC	Direct Current
DCT	Discrete Cosine Transform
DDAG	Decision-Directed Acyclic Graph Method
DHPC	Dynamic Host Configuration Protocol
DWT	Discrete Wavelet Transform
EMD	Empirical Mode Decomposition
emf	Electromagnetic Force
EVs	Eigenvectors
FDF	Fuzzy Decision Function Method
FT	Fourier Transform
FVs	Feature Vectors
GA	Graphical Analysis
GMFD	Generalized Model for Fault Detection
HP	High Pressure
ICA	Independent Component Analysis
IDE	Integrated Development Environment

IEPE	Integrated Electronic Piezoelectric
IFDM	Improved Fault Detection Model
IMFs	Intrinsic Mode Functions
IMU	Inertial Measurement Units
k-NN	*k*-Nearest Neighbour
LB	Lower Bound
LEDs	Light-Emitting Diode
LIV	Leakage Inlet Valve
LOV	Leakage Outlet Valve
LP	Low Pressure
MI	Mutual Information
MIFS	Mutual Information Feature Selector
MIFS-U	Mutual Information Feature Selection Under Uniform Information Distribution
mRMR	Minimum Redundancy, Maximum Relevance
NI	National Instruments
NMIFS	Normalized Mutual Information Feature Selection
NRV	Non-Returning Valve
NRVF	Non-Returning Valve Fault
OAA	One-Against-All
OAO	One-Against-One
PCA	Principal Component Analysis
PCs	Principal Components
PM	Preventive Maintenance
PWD	Pseudo-Wigner–Ville Distribution
QMF	Quadrature Mirror Filter
RIO	Reconfigurable IO Modules
RMS	Root Mean Square
SD	Secure Digital
SDK	Software Development Kit
SPA	Sensitive Position Analysis
STFT	Short-Time Fourier Transform
SVM	Support Vector Machine
UB	Upper Bound
UMT	Updated Morlet Transformei
VIs	Virtual Instruments
WP	Windows Phone
WPD	Wavelet Packet Decomposition
WPT	Wavelet Packet Transform
WVD	Wigner–Ville Distribution
ZCR	Zero-Crossing Rate

List of Figures

List of Tables

Chapter 1
Introduction

The cost-effective maintenance of machines is the primary and utmost concern of any industrial plant [1]. With rapid advancement in automation, the usage of number of machines in industries is also increasing. Design of these machines is very compact with the advancement of technology which makes the task of maintenance more difficult. To resolve these issues, the development of intelligent fault diagnosis systems will be needed. In this book, intelligent refers to automated, quick, and reliable system that uses advanced machine learning and signal processing approaches for decision making of its own in an adaptive fashion. It is especially required in an industry where the breakdown of machines can have a significant impact on production capacity and safety of employees. For example, in the case of single-stage reciprocating air compressor, certain valves are subject to frequent wear and tear, where damage can lead to dangerous consequences. Therefore, diagnosis of faults is necessary to overcome the damage in machines and to maintain a healthy production environment. Three widely practised approaches for machine maintenance are Preventive Maintenance (PM), Corrective Maintenance (CM), and Condition-Based Maintenance (CBM) [2].

CM is performed once machine component fails and operation gets interrupted. As a part of maintenance strategy, it is restored to the normal condition. It is also known as breakdown maintenance. The second most popular approach is PM where machine maintenance is carried out at predetermined intervals and if needed corrective actions are taken. It reduces the failure risk. The third most effective and commonly used approach is CBM where machine parameters like vibration, acoustics, temperature, pressure, etc. are continuously monitored [3–5]. Deviation of these parameters from standard values reflects that there is some fault in the machine. The selection of parameters to be monitored also depends on the machine to be monitored. For example, in the case of induction motor faults such as damage in bearings, rotor, and stator bars are diagnosed by analysing motor's current and vibration data. For rotating machines, the most commonly used measurements for fault recognition are acoustic and vibration measurements. Some advantages of Auto-Encoder (AE) signals have over vibration signals are being relatively insensitive to structural resonances and mechanical background noises, more sensitive to disturbances caused

N. K. Verma and A. Salour, *Intelligent Condition Based Monitoring*, Studies in Systems, Decision and Control 256, https://doi.org/10.1007/978-981-15-0512-6_1

by faults, and provide good tending parameters. The selection of an appropriate maintenance strategy is another very interesting research problem.

According to past studies, CBM has proven to be the most cost-effective and reliable solution among three approaches [5]. Fault diagnosis is an integral component of CBM systems. It generally involves two aspects, namely detection and isolation of faults. The CBM solutions are expected to do both at a much earlier stage, i.e., before breakdown occurs in machines. Intelligent fault diagnosis approach has been implemented for both, i.e., machine fault detection and manufacturing material fault detection. There are two types of approaches used for fault diagnosis, i.e., physics-based approach and data-driven based approach. In this book, the data-driven based approach has been used where firstly the data is collected from the machine and then supervised machine learning approaches are used to train the model. The detailed description of the methodology is described in the later chapters.

This book mainly covers the design and implementation details of intelligent condition-based monitoring along with the results, discussions, and case studies of each component. It details the significance of each module in the fault diagnosis framework as well as presents how these can be implemented.

1.1 Fault Diagnosis System

The fault diagnosis system is a supervision system used to detect and isolate faults and identify their types and characteristics. It is a search or reasoning process for detection, isolation, analysis, and recovery of occurring faults. A fault is an unexpected change or malfunction in a system, although it may not lead to physical failure or breakdown. Fault detection implies the decision that a fault is present in a system and fault isolation implies the determination of source of the fault. After a fault is detected and isolated, actions should be taken to eliminate the fault to prevent further damage or propagation.

The data generated from rotating machines plays a very crucial role in the fault diagnosis procedure because it measures parameters of machines. Figure 1.1 shows the general framework used in fault diagnosis of various industrial machines. The sensor network collects the data from rotating machines. The data processing and decision making are done using different signal processing and machine learning algorithms.

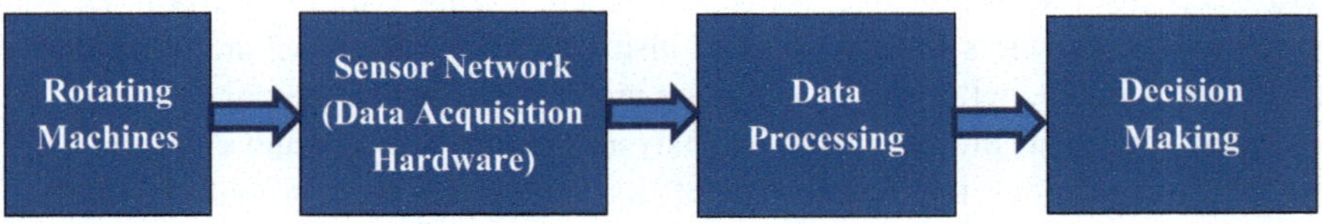

Fig. 1.1 General framework of the fault diagnosis system

1.2 Sequence of Operations Involved in Fault Diagnosis

The fault diagnosis procedure involves the following sequence of operations:

- Collect data samples from rotating machines in a different health condition.
- Pre-process the collected data.
- Extract features from the pre-processed data.
- Select salient features.
- Develop a classification algorithm.

1.3 Brief Introduction to the Techniques Adopted

The sequence of operations involved in any fault diagnosis procedure may be the same but techniques involved in performing each operation might be different. A brief introduction of techniques adopted and instrumentations used in accomplishing the tasks are briefly explained in next subsections.

1.3.1 Data Acquisition

Data acquisition (DAQ) begins by collecting the machine-generated raw data from appropriate locations. As expected, there exist many locations on a machine from where the data can be acquired. For finding appropriate location(s), sensitive position analysis (SPA) is discussed in future chapters. The collected raw data can be very noisy and subjected to various environmental contaminants. The acquisition process involves data collection using sensors and transducers (acoustic or vibration) deployed on or around air compressors and motors in various conditions like healthy and faulty. It includes a set of four accelerometers, mics, a CCD laser, and a DAQ hardware (NI cDAQ 9172) manufactured by National Instruments (NI). The software code has been developed using LabVIEW, proprietary software supplied by NI, which accomplishes the task of forming the DAQ setup. With the help of DAQ hardware, samples of healthy and faulty compressors are recorded.

Pre-processing The recorded samples are corrupted by surrounding noise in an industrial plant. It is of high importance that we remove the noise content from the data before using it for further analysis. It thereby improves quality of the data and brings it in right format. This procedure is termed as “data pre-processing”.

The four major pre-processing steps applied on datasets are:

- **Clipping:** reducing size of the sample for reducing time in decision making.
- **Filtering:** removing unwanted signals from the signal acquired.
- **Smoothing:** removing outliers from the data.
- **Normalization:** scaling of the data into specified numerical limits.

1.3.2 Feature Extraction

To perform fault diagnosis, the information from acoustic signals needs to be extracted in a smaller number of variables so that it can be easily processed. Features are characteristics of the signal which represent the entire signal in a much lower dimension. The feature extraction is carried out after transforming the signal into different domains. Three types of signal processing features can be extracted as mentioned below:

- **Time Domain:** The pre-processed data is in time domain. Several parameters/features like absolute mean, maximum peak, Root Mean Square (RMS) value, variance, kurtosis, crest factor, shape factor, and skewness are calculated. The time domain representation is a very basic way of representing the data and very little inference can be drawn from it.
- **Frequency Domain:** The frequency domain analysis is based on transforming the time domain signal to frequency domain signal. The advantage of frequency domain analysis over time domain is its ability to easily identify and isolate the contribution of different frequencies using spectral analysis. Total spectral energy of the signal is calculated and is divided into eight equal segments, each one is called as a bin. The ratio of individual bin energy to the total energy is considered as a feature.
- **Time-Frequency Domain:**

  - **Continuous Wavelet Transform:** A signal is first transformed from time domain to time-frequency domain by convolving it with Morlet wavelet. Certain features like wavelet entropy, sum of peaks, kurtosis, zero-crossing rate, variance, and skewness are calculated from resultant wavelet coefficients.
  - **Discrete Wavelet Transform:** In this method, the pre-processed signal is decomposed up to nth level using low pass and high pass filters. The features like mean and variance of coefficients at different decomposed levels are calculated. Mallat's fast wavelet transform is used for signal decomposition.
  - **Wavelet Packet Transform:** This transform is similar to the discrete wavelet transform, except the hierarchy of signal decomposition. The low pass and high pass filters are also used here in the signal decomposition process. The wavelet packet node energy is calculated as a feature.

1.3.3 *Feature Selection Algorithms*

Having more features may seem to improve performance but generally when the number of features is more and performance may deteriorate due to the curse of dimensionality. For this purpose, dimensionality reduction is done where a small subset of relatively good features is selected and used for further processes. Four feature selection algorithms have been applied as mentioned below:

- Principal Component Analysis (PCA)
- Mutual Information (MI)
- Bhattacharyya Distance (BD)
- Independent Component Analysis (ICA)
- Graphical Analysis (GA)

1.3.4 *Classification*

Classification is the process of learning about the relationship between training samples and their class. This learning is later used for predicting the class of the test data. Some popular algorithms for classification are as follows:

- *k*-Means Clustering,
- *k*-Nearest Neighbour (*k*-NN) Classifier,
- Support Vector Machine, and
- *z*-Score.

References

1. Verma, N.K., Ghosh, A., Dixit, S., Salour, A.: Cost-benefit and reliability analysis of prognostic health management systems using fuzzy rules. In: IEEE Workshop on Computational Intelligence Theories, Applications and Future Directions, IIT Kanpur, India, pp. 1–9 (2015)
2. Verma, N.K., Subramanian, T.: Cost benefit analysis of intelligent condition based maintenance of rotating machinery. In: Proceedings of the 7th IEEE Conference Industrial Electronics and Applications, Singapore, pp. 1390–1394 (2012)
3. Verma, N.K., Sreevidya: Cost benefit analysis for condition based monitoring. In: IEEE International Conference on Prognostics and Health Management, Maryland, USA, pp. 1–6 (2013)
4. Verma, N.K., Sreevidya: Study on multi unit models for machine maintenance. In: IEEE Workshop on Computational Intelligence: Theories, Applications and Future Directions, IIT Kanpur, India, pp. 181–184 (2013)
5. Verma, N.K., Sreevidya: Cost benefit analysis for maintenance of rotating machines. In: IEEE International Conference on Prognostics and Health Management, Austin, USA, pp. 1–7 (2015)

Chapter 2
Faults and Data Acquisition

Abstract This chapter provides detail about the data acquisition process. It is the foremost step of a fault diagnosis framework where machine characteristics are measured and recorded for further analysis. The chapter starts with an introduction of air compressor, its parts, working principle, and common occurring faults. It further presents a methodology of placing the sensors on a machine known for sensitive position analysis. A case study on air compressor and motor has been presented at the end of the chapter.

2.1 Air Compressor

An air compressor works by transforming electrical energy into kinetic energy in the form of air pressure, which is held in a highly compressed space before sudden release. The compressor increases the pressure of air by reducing its volume. In the compressed form, the air is under pressure greater than of the normal atmospheric pressure. It characteristically attempts to return to its normal state. Since energy is required to compress the air, energy is released as the air expands and returns to its atmospheric pressure levels. This energy is released when the compressed air is let out and can be used for inflation to push a piston, cleaning under pressure, and turning, generating torque, driving, or other similar movements with the help of force. The compressed air provides torque and rotation power to pneumatic tools, such as drills, brushes, nut runners, riveting guns, and screwdrivers.

The components of an air compressor are cylinders, piston, crankshaft, valves, and housing blocks. Its principle is similar to that of an internal combustion engine but follows it reversely. A crankshaft moves a piston through a connecting rod, in a linear fashion along the length of the cylinder. The air is drawn in through an inlet valve as the piston moves downwards, providing ample space for a high volume of air. In the return stroke, the inlet valve closes, and the piston moves up, which compresses the air. This compressed air has excessive pressure because of the restricted space into which it has been forced. Due to the excess pressure, the outlet valve opens confining the compressed air into the cylinder for storage. The cylinders may be placed in series, so that compressed air is fed to one cylinder after another, which compresses

N. K. Verma and A. Salour, *Intelligent Condition Based Monitoring*, Studies in Systems, Decision and Control 256, https://doi.org/10.1007/978-981-15-0512-6_2

it further. Oil is used to lubricate the piston as it moves inside the cylinder. Rings act as seals to reduce the amount of oil that can mix with compressed air. Reciprocating or piston based compressor is a positive displacement-type compressor and is most commonly used. The positive displacement-type compressor fills an air chamber and then reduces it to increase the pressure.

2.1.1 Types of Air Compressor

There are two common types of air compressor:

- Reciprocating air compressors.
- Screw air compressors.

2.1.1.1 Reciprocating Air Compressors

Reciprocating air compressors are most widely used in industrial plant air systems. It is basically a piston and cylinder device with (automatic) spring-controlled inlet and exhaust valves as shown in Fig. 2.1. It uses the piston inside a cylinder to compress air. The piston performs an upward and downward motion inside the cylinder, which is also known as the reciprocating motion. As the piston moves downwards, a vacuum is created inside the cylinder. Because the pressure above the intake valve is greater than the pressure below it, the intake valve is open forcefully, and the air is sucked into the cylinder. Once the piston reaches the bottom of the cylinder, it begins to move upwards. The intake valve closes and traps air inside the cylinder. As the piston continues to move upwards, it compresses the air and consequently increasing the air pressure. At a certain point, the pressure exerted by the air forces the exhaust valve to open, and it flows out of the cylinder towards the receiver. Once the piston reaches its top-most position in the cylinder, it starts moving downwards again, and the cycle is repeated again and again as shown in Fig. 2.2.

Structure and Working Principle Reciprocating air compressors utilize crankshaft-driven pistons to compress gases which are used in various processes. Like a small internal combustion engine, a conventional piston compressor has a crankshaft, a connecting rod, a piston, a cylinder, and a valve head. The crankshaft is driven by either an electric motor or a gas engine. While there are small models which comprise pump and motor, most of compressors also have an air tank to hold a quantity of air within a pre-set pressure range. At the top of the cylinder, there is a valve head which has inlet and discharge valves. Both are simply thin metal flaps–one mounted underneath and the other on top of the valve head.

As the piston moves down, a vacuum develops above it, which allows outside air at atmospheric pressure to fill the area above it by opening the inlet valve. Once the air

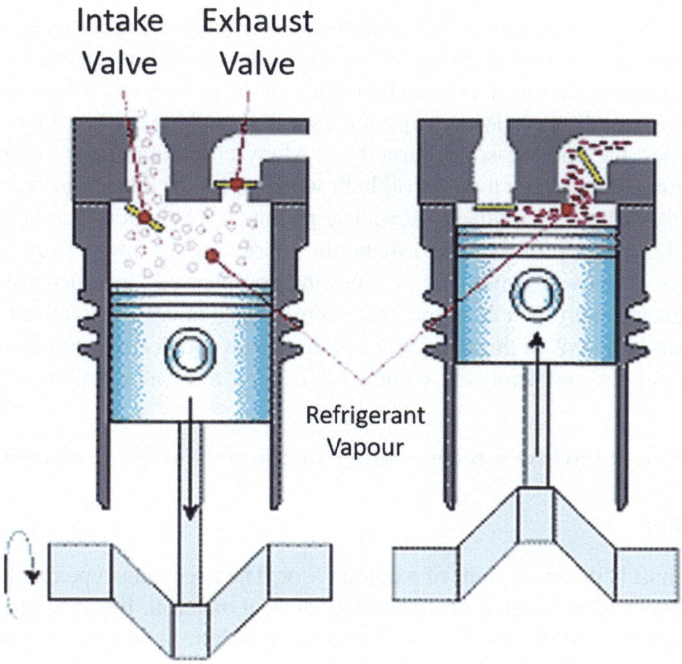

Fig. 2.1 Working principle of the air compressor. Image Courtesy: HVAC Classes and Labs

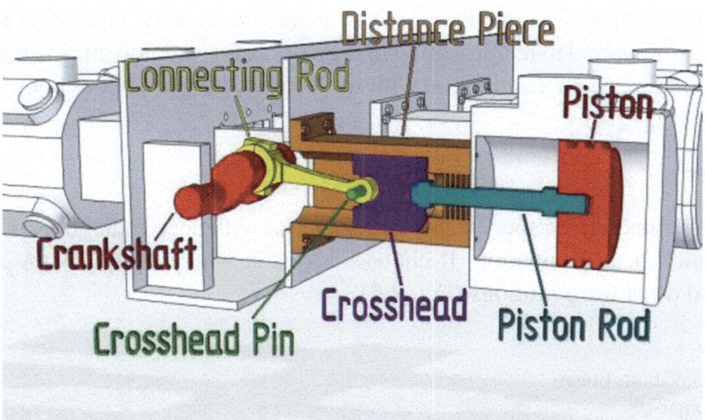

Fig. 2.2 Cross-sectional view of the air compressor. Image Courtesy: HVAC Classes and Labs

is filled in the vacuum, the inlet valve is shut, and the piston moves up. As the piston moves up, the air above it compresses and pushes the discharge valve to open. The air moves from the discharge port to the tank. With each stroke, more air enters the tank and the pressure rises. The compressor has a pressure switch to stop the motor when tank pressure reaches a pre-set limit-of about 125 psi (for many single-stage models).

A regulator can be used if this much pressure is not required. A gauge placed before the regulator monitors tank pressure and a gauge after it monitors airline pressure. In addition, the tank has a safety valve, which opens if the pressure switch malfunctions. The pressure switch may also incorporate an unload valve which reduces the tank pressure when the compressor is turned off. Many articulated piston compressors are oil lubricated, i.e., they have an oil bath which splashes lubricated bearings and cylinder walls where the crank rotates. The pistons have rings which help to keep compressed air on top of them and make the lubricating oil away from air. These rings are not effective. Some oil may enter the compressed air in aerosol form. In practical design, there is a clearance between the piston crown and the top of the cylinder. Air "trapped" in this clearance volume is never delivered. It expands as the piston moves back and limits the volume of fresh air to a value less than the swept volume.

i. **The different parts of a reciprocating air compressor are explained below**.

- **Crankshaft**

The crankshaft is the main shaft of a compressor. The overhung-type crankshaft has a replaceable crankpin bush made from special steel material. It is precision-ground and runs with minimum vibration. A crankshaft is balanced and rotates over two or three ball bearings as shown in Fig. 2.3.

- **Connecting rod**

A connecting rod acts as a link between piston and crankshaft. The rotary motion of crankshaft is converted to reciprocating motion of piston by connecting rod. Precisely machined, one-piece connecting rod with solid end construction does not require any adjustment (Fig. 2.4).

- **Cylinder**

Cylinders are made from special cast-iron material with deep radial fins on the outer side for quick heat dissipation. It chooses a piston which performs the repetitive upward and downward motions (Fig. 2.5).

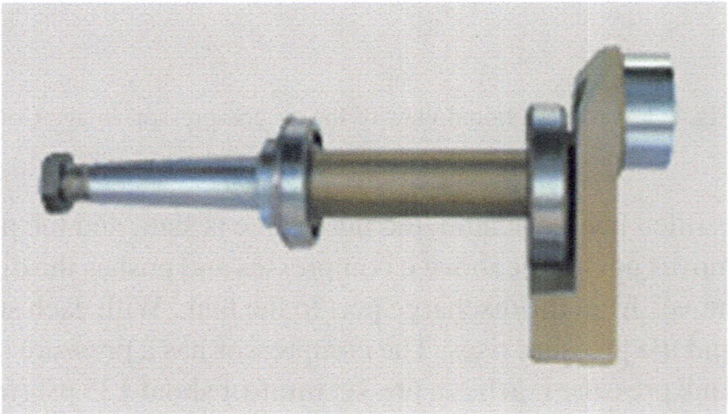

Fig. 2.3 Crankshaft. Image Courtesy: Gajjar Compressors Pvt. Ltd.

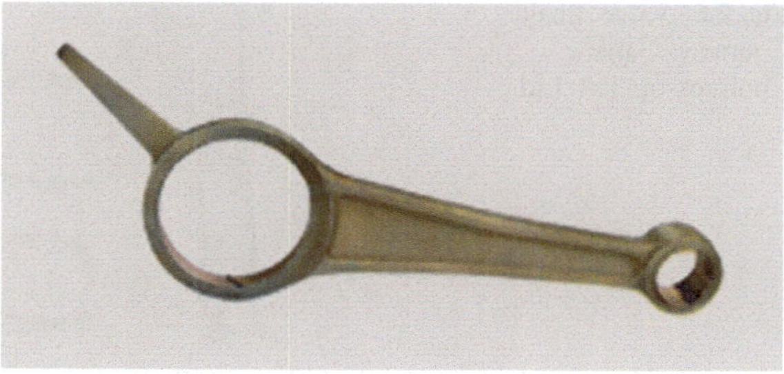

Fig. 2.4 Connecting rod. Image Courtesy: Gajjar Compressors Pvt. Ltd.

Fig. 2.5 Cylinder. Image Courtesy: Gajjar Compressors Pvt. Ltd.

- **Valve**

There are two types of valve: suction valve and discharge valve. Through the suction valve, low-pressure air is sucked inside the cylinder, whereas through the discharge valve, compressed high-pressure air is discharged. The operation of the suction valve is such that it opens when the piston moves downwards and closes when air is discharged. The discharge valve opens only when the piston reaches to a certain level inside the cylinder and air reaches the desired level of pressure level. The discharge valve closes once the air is delivered from the cylinder. Finger and concentric rings are two widely used types of valve (Fig. 2.6).

- **Piston and piston rings**

The piston performs a reciprocating motion, i.e., upward and downward motions inside the cylinder. During its motion, it enables suction and compression of air. During reciprocating motion, piston rings come in contact with the walls of the cylinder. The two types of equalized and balanced pistons are made from aluminium

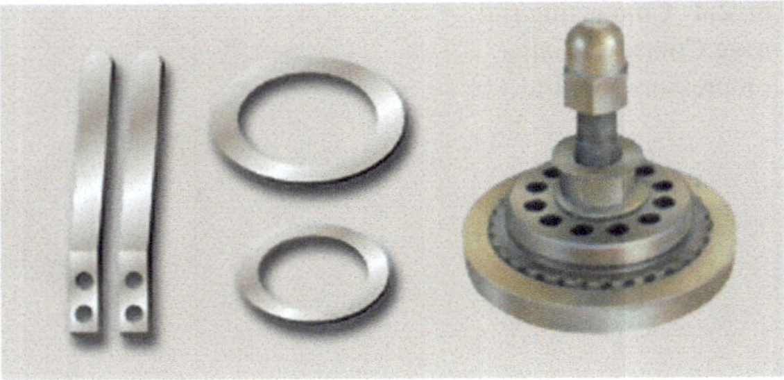

Fig. 2.6 Valve. Image Courtesy: Gajjar Compressors Pvt. Ltd.

Fig. 2.7 Piston and piston rings. Image Courtesy: Gajjar Compressors Pvt. Ltd.

or cast-iron material. Both have compression and oil control rings for optimum efficiency. A piston ring circles the piston. There is a lot of friction between cylinder walls and piston rings; therefore, it is replaced regularly for the proper functioning of the compressor (Fig. 2.7).

- **Ball bearing**

A bearing consists of several hard steel balls rolling between a metal sleeve fitted over the rotating shaft and an outer sleeve held in the bearing housing. It reduces friction between moving parts while providing support to the shaft (Fig. 2.8).

- **Cooling fan**

The fan provides ventilation to a system. It generates pressure to move air (or gases) against the resistance caused by ducts, dampers, or other components in the system (Fig. 2.9).

Fig. 2.8 Ball bearing. Image Courtesy: Gajjar Compressors Pvt. Ltd.

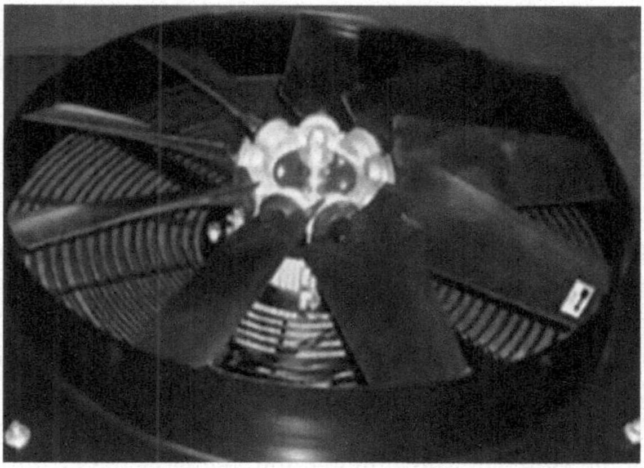

Fig. 2.9 Cooling fan. Image Courtesy: Oil-Injected Rotary Screw Compressors, Atlas Copco

ii. **Types of reciprocating air compressor**

- **Single-stage reciprocating air compressor**

A single-stage compressor has one/two/three piston(s) and a cylinder of same bore size. The compressor is ideal for a multitude of small business uses including pneumatic tools, panel spray painting, nailers, staplers, blow guns, liquid transfer, granting, and cleaning. They are a reliable source of low-pressure air for numerous applications like instrumentation, process and boiler fuel oil automation, filtration plant, blow moulding, etc. (Fig. 2.10).

Specifications of single-stage air compressor

Air pressure range	Type	Range	DIFF	Induction motor	Pressure switch
0–500 lb/m^2 0–35 kg/cm^2	PR-15	100–213 psi 7–15 kg/cm^2	24–100 psi 1.7–7 kg/cm^2	5 HP, 415 V, 5 amp, 3 ph, 50 Hz, 1440 rpm	Delta compressor India 6 A, 500 V AC

Fig. 2.10 Single-stage reciprocating air compressor. Image Courtesy: ELGI

In a single-stage reciprocating air compressor, entire compression is carried out in a single cylinder. It is known as single acting, if the compression is affected at one end of the piston and cylinder, whereas it is a doubleacting compressor if the compression is affected in both ends of a piston and cylinder. The opening and closing of a simple check-valve (plate or spring valve) depends upon the difference in pressure. The volume of air in the cylinder will be zero when the piston is at top neglecting the clearance volume. As the piston starts moving downwards, the pressure inside the cylinder falls below atmospheric level, and the suction inlet valve opens. Air then draws into the cylinder through the suction valve. This operation is known as the suction stroke. As the piston moves upwards, the inlet valve closes. The pressure inside the inlet valve reaches an atmospheric level and compresses air in the cylinder. The compression increases the piston to move toward the top of its stroke until pressure in the cylinder exceeds that in the receiver. This is known as the compression stroke. At the end of this stroke, the discharge or delivery valve opens, and the air is delivered to the receiver. At the end of the delivery stroke, a small quantity of air at high pressure is left in the clearance space. As the piston starts, its suction stroke air contained in the clearance space expands until its pressure falls below atmospheric pressure. At this stage, the inlet valve opens as a result of which fresh air is sucked into the cylinder. This cycle is repeated every time (Figs. 2.11 and 2.12).

Specifications of single-stage reciprocating air compressor: LG 02100 model

Motor power	Piston displacement	Free air delivery	Comp. speed	Max. pressure	Air receiver	Net weight	Dimensional data
2 HP	7.06 cfm	3.89 cfm	870 rpm	145 psi	100 L	167 kg	$1350 \times 530 \times 1000\ mm^3$

Fig. 2.11 Single-stage reciprocating air compressor. Image Courtesy: ELGI

Fig. 2.12 Two-stage reciprocating air compressors. Image Courtesy: ELGI

- **Two-stage reciprocating air compressors**

Two-stage air compressors are used when the air is required at high pressure. In such case, two or more cylinders in a series are used to compress the air. Employing single-stage compression for generating high pressure air leads to many drawbacks.

Two-stage air compressors consist of two or more cylinders. In the first stage, the low-pressure (LP) cylinder is always larger in diameter than the second stage, high-pressure (HP) cylinder. The sizes of two cylinders (i.e., HP and LP) are to be adjusted to suit the volume and pressure of the air. The atmospheric air enters LP cylinders through the inlet filter and valves. The air after compression in LP cylinder passes to the HP cylinder through the intercooler and HP inlet valves. The air acquires higher pressure by the HP cylinder and is delivered to the destination. The highly efficient intercooler provides maximum heat dissipation between stages. The efficiency of the intercooler plays an important role in the working of a two-stage reciprocating air compressor. The work done per kg of air is reduced in multistage compression with an intercooler as compared to a single-stage compression with the same pressure. These compressors are suitable for applications where air pressure is required up to 17.5 (250 psi). They are widely used in textile industries, plastic industries, paper industries, spray paintings, cleaning, chemical plants, garages, pneumatic operations, shot blasting, cement plants, mining, pharmaceuticals, pesticides, foundries, oil industries, etc.

Specifications of two-stage reciprocating air compressor (TS05HN, ELGI equipment)

Air pressure range	Capacity	Max. W/Pr	Test Pr	Dispt Vol	Motor hp	Unit rpm
0–25 Kgf/cm^2	220 L	12 kgf/cm^2	19.8 Kgf/cm	500 lpm	5	925

2.1.1.2 Screw Air Compressors

Screw air compressors use a pair of helical rotors to compress the gas. Rotors turn in opposite directions with very little clearance between them. As the rotors rotate, they intermesh, alternately exposing and closing off inter-lobe spaces at the ends of the rotors. When an inter-lobe space at the intake end opens, the air is sucked into it. As the rotors continue to rotate, the air becomes trapped inside the inter-lobe space, and is forced along the length of the rotors. The volume of the inter-lobe space decreases, and the air is compressed. The compressed air exists when the inter-lobe space reaches the other end inside a sealed chamber. The compressor screw element is also called air-end. They are constant flow (volume) with variable pressure compressors, which

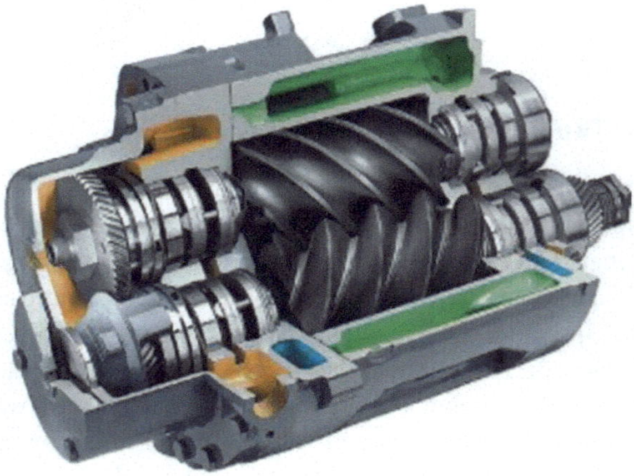

Fig. 2.13 Compressor element (oil-free type). Image Courtesy: The Workshop Compressor

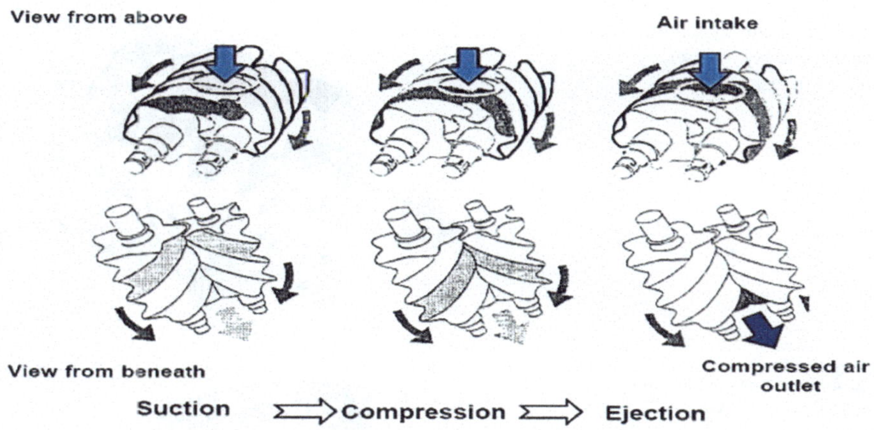

Fig. 2.14 Working of screw air compressor. Image Courtesy: Kaeser Compressors, Inc.

means that at a given speed (rpm), they always supply the same amount of air (e.g. in litres per second) but can do so at different pressures (Figs. 2.13 and 2.14).

Structure and Working Principle A rotary screw air compressor has two interlocking helical rotors contained in the housing. Air comes in through the inlet valve and is taken into the space between the rotors. As the screws turn, they reduce the volume of the air, thus increasing the pressure. The screw element is the most important part of any screw-type compressor. It is that part of the machine where the actual compression takes place. It is the heart of the rotary screw air compressor. Different parts of screw air compressors are mentioned below:

- **Rotatory Screw**: When the inlet valve is open, the air enters the compressor screw element. The screw element works like a pump, and it compresses the air as shown in Fig. 2.15.
- **Air/Oil Separators**: During the compression process, oil is injected in the screw element to cool the air, as the air gets very hot during compression. Now we need to separate the oil and the air. The oil is separated by the separator element, which looks like a big filter as shown in Fig. 2.16.
- **Air/Oil Filters**: The oil filters remove all the dirt and dust that has been collected in the oil as shown in Fig. 2.17.
- **Heat Exchangers**: Heat exchangers are used to reduce final discharge air temperature so that air bottle size can be reduced and to reduce air volume after it has been compressed to the final pressure (Fig. 2.18).

i. **Types of screw air compressors**

The rotary screw compressors are available in oil-injected and oil-free versions. The basic principle is the same (the rotors "push" the air to one side), but they are quite different in construction. The oil-injected version needs oil to operate properly but the oil-free version does not need oil. Due to this, the rotors used in oil-free screw

Fig. 2.15 Rotatory screw. Image Courtesy: Genser Tecnica Industrial

Fig. 2.16 Air/oil separators. Image Courtesy: Air Engineering

compressors are of superior quality with very little space in between them. They do not touch each other, otherwise, they would wear-down too quickly. For this reason, they are a lot more expensive.

Oil-free compressors are used in medical research and semiconductor manufacturing where entrained oil carryover is not acceptable. Oil-injected compressors are used in machinery, workshops, and factories where the minor oil carryover of the compressor is not problematic.

- **Oil-injected screw compressors**

In oil-injected screw compressor, the air is sucked in by the screw element through the air inlet filter. The inlet filter makes sure that all the dust and dirt stays outside. It protects the screw element (which is very expensive and can be damaged). It is also

Fig. 2.17 Air/oil filters. Image Courtesy: Air Engineering

Fig. 2.18 Heat exchangers. Image Courtesy: L. G. Steels

the first step in making sure that whether the compressed air is clean. All the dust that is sucked in will eventually end up in the compressed air system (Fig. 2.19).

The air is passed through the inlet valve before it enters the screw element. The inlet valve opens and closes the air supply to the screw element. When the valve is open, the compressor is in "loaded" condition: it is compressing air and pumping it into the compressed air system. When the valve is closed, it shuts off the air supply to the compressor element, the motor and screw element are turning, but the compressor is not sucking any air in and is not pumping any air to the system. During the process of air compressing, oil is injected into the element. The oil serves five purposes. It cleans, cools, lubricates, seals, and protects. As the air gets very hot during compression, oil is there to cool the air. It is also there for lubrication and sealing off the clearances between the screws. Now, as output, there is a mixture of compressed air and compressor oil. This mixture leaves the screw element through a one-way valve and flows into a separator tank. This valve makes sure that the

Fig. 2.19 Oil-injected screw compressor. Image Courtesy: Kasi Sales and Services Pvt. Ltd.

oil cannot flow back into the compressor element through the exit pipe (this could happen when the compressor stops). Now we need to separate the oil and the air. Most of the oil is separated from the compressed air by centrifugal force. In the process of centrifugal force, rapid spinning allows the heavier oil particles to drop to the bottom, while the lighter air spins around on top (just like what happens in a clothes tumble dryer). The remaining oil is separated using the separator element (such as a big filter). The air with oil flows through the separator element. The element separates the oil from the compressed air. The separated oil is collected at the bottom of the separator and is removed by the scavenge line. It sucks the collected oil back to the compressor element. Now the clean compressed air is ready to leave the compressor by first passing through the minimum pressure valve and the aftercooler. The minimum pressure valve is a spring-loaded valve that opens at a certain pressure, about 2.5 bars. The minimum pressure valve makes sure that there is always a minimum pressure inside the compressor. This pressure is needed for correct operation of the air compressor. The compressed air is still very hot at this point, about 80 °C. The compressed air is now cooled by the aftercooler before it leaves the compressor. The air temperature after the cooler is around 25–40 °C. The compressed air and oil mixture was successfully separated by the separator but the separated oil is very hot. It absorbs the heat of the compression and can be as hot as 120 °C (anything more and the compressor will shut down). The oil is cooled by the oil cooler. The amount of cooling is controlled by a thermostatic valve. If the oil is still cold, the oil cooler is completely bypassed. If the oil is very hot, all the oil is led through the oil cooler. The thermostatic valve regulates the oil temperature. Finally, the oil flows through the oil filter. The oil filters remove all the dirt and dust that has collected in the oil. Too much dirt in the oil will damage the screw element. The oil filter has an internal bypass valve which opens when the pressure difference over the filter becomes too high (when the filter is very dirty or when the oil is still very cold). Oil-injected screw compressors, therefore, do not produce oil-free air, and they cannot be used in places where oil-free air is needed such as medical research and semiconductor manufacturing.

Specifications of oil-injected screw air compressor (MAS GA 110)

Motor power	Outlet air temperature	Max. pressure	Net weight	Dimensional data
168 HP	Cooling water temperature + 5 °C	14 bar	1938 kg	3000 × 799 × 1850 mm^3

- **Oil-free screw compressors**

In the oil-free screw compressor, the air is sucked in through the unloader valve and inlet air filter. The filter protects the compressor elements from damage by keeping all dust and dirt outside of the compressor. The unloader valve is opened and closed by the control system. When the valve is open, the compressor is in loaded condition (it is pumping air). When the valve is closed, the compressor is in an unloaded condition. When the compressor is in loaded condition and the unloader (inlet) valve is open, the air is sucked into the first (low pressure) compressor element. In the low-pressure element, the air is compressed to about 2–2.5 bars. The compression is done without oil, only air (as opposed to oil-injected rotary screw compressors). Because of this, the air becomes hot.

Normal temperature for low-pressure element outlet in oil-free element is between 160 and 180 °C, whereas oil-injected screw elements have an outlet temperature of about 80 °C. Also, the (low-pressure) oil-free element only compresses it to about 2.5 bars, compared to 7–13 bars for oil-injected screw elements. The air is cooled by the intercooler to about 25–30 °C. There is a moisture trap installed after the intercooler to remove the water from the air.

The air is further compressed by the high-pressure element to the final pressure, which depends on the compressor specifications and is normally anywhere between 7 and 13 bars. Because of the compression, the air is, again, very hot around 140–175 °C. So, it is cooled again, by the aftercooler. Before it enters the aftercooler, it is passed through a pulsation damper and a check-valve. The check-valve makes sure that compressed air does not flow back into the compressor when it is stopped. After going through the aftercooler, the air reaches its outlet temperature of about 25 °C. There is another moisture trap installed to remove any water that may have formed inside the aftercooler. All the air that is compressed by the low-pressure elements need to be sucked in by the high-pressure element. If there is no balance, the pressure in the intercooler will rise or fall. The elements are designed for a certain pressure ratio. That is the outlet pressure divided by the inlet pressure. If the pressure ratio over a compressor element becomes too large, it will eventually break down. If one of the elements wears down or breaks down, it disturbs the balance and can take the other element down with it (Fig. 2.20).

Fig. 2.20 Oil-free screw compressor. Image Courtesy: HTDS

2.1.2 Types of Probable Faults in Air Compressor

i. **Leakage exhaust and inlet valves**

During the air intake stroke, when the air inlet valve in a healthy air compressor gets closed, the fresh air in the cylinder is compressed to around 50–60 bars. The exhaust and inlet valves are meant to be closed during this period. They maintain a perfect air seal by virtue of the metal to the metal seating between the valve lid and valve seat. The air confined within the piston rings against the liner through the complex action of the air pressures acting both downwards along the vertical axis of the cylinder liner, pushing the piston rings against the liner surface. Any loss of pressure will directly affect air pressure oscillations within the cylinder. The patterns of these pressure oscillations cause most of the vibration in a normally aspirated compressor. Due to this, the air leakage gets detected. To rectify, the leaking exhaust and air inlet valves, the original components were replaced by valves which got semi or completely opened. This produced a unique signature for vibration and acoustic emission.

ii. **Blocked exhaust and inlet valves**

The air gets sucked into the cylinder during the intake stroke, while the outlet valve is closed. Due to the mechanical action, the outlet valve gets opened, and the inlet valve is closed. The compressor suction end was induced to blockage partially or completely. However, the outlet valve gets neither partially nor completely blocked as the piston will exert very high pressure on both the valves. This may prove to be fatal, leading to an explosion of the cylinder. Hence, only the blockage of the inlet valve is considered for the simulation of the fault. Owing to the blockage of the suction valve, the air is not able to enter the cylinder, and the compression cycle stopped. The piston in the cylinder only runs in oscillation during both strokes but the air is cycled to compression. This momentary run of the piston exhibits significant variations in the audio and vibration as well. The acoustic emission from the compressor reduces considerably, whereas the vibration of the cylinder increases.

iii. **Rider belt faults**

A rider belt is a mechanism for the conversion of rotary motion into linear. Rider belt in the compressor is like a clutch in an automobile. It connects the shaft of the prime mover and the flywheel of the compressor. The distance between the flywheel and the prime mover shaft will remain fixed. The rider belt is made from nylon rubber. As the pressure inside the compressor increases, it resists the air being pumped through the outlet, leading to slippery in the rider belt on the shaft. The pulley of the prime mover and the flywheel of the compressor must be positioned in the straight line. If the alignment in the rider belt contains any deviation from the actual horizontal line, the slippery action arises at the shaft.

iv. **Slaps in the piston/screw**

The primary movement of the piston is to move inside the cylinder parallel to the bore. Excessive piston slaps occur when the clearance between the piston and the cylinder bore is increased beyond the actual dimensions due to which the piston movement deviates from its primary movement. The change in clearance occurs due to either wear and tear, incorrect piston, and cylinder dimension at manufacturing.

vi. **Bearing faults**

Air compressor has two main bearings, one at each end of the crankshaft. They are responsible for smoothing the rotation of the crankshaft and reduce the friction with a connecting rod. Wearing out of bearing leads to an increase in heat and pressure, and the metal-to-metal contact can distort the crankshaft also, due to which connecting rod results in catastrophic failure.

2.1.3 Induction Motor

Electrical power is converted into mechanical power in the rotating part of an induction motor. In DC motors, electric power is directly transferred to the armature (i.e., rotating part) through brushes and commutator. Hence, it is called a conduction motor. In AC motors, the rotor does not receive electric power by conduction. The power is transferred by induction in the same way as the secondary of a two-winding

Fig. 2.21 Squirrel cage induction motor. Image Courtesy: ELGI

transformer, which receives its power from the primary winding. Due to this reason, these motors are known as induction motors. An induction motor can be treated as a rotating transformer, i.e., one in which primary winding is stationary, but the secondary is free to rotate. Figure 2.21 shows the squirrel cage induction motor.

The polyphase induction motor is extensively used for various kinds of industrial drives. It has the following advantage:

- Very simple and extremely rugged, almost unbreakable construction (especially squirrel cage type).
- Low cost and very reliable.
- Requires minimum maintenance.

The disadvantage of the polyphase induction motor is that speed cannot be varied without sacrificing some of its efficiency and its speed decreases with an increase in load. An induction motor consists essentially of two main parts, i.e., stator and rotor.

i. **Stator**

The stator carries a balanced three-phase winding, which means that the number of turns in each phase, connected in star/delta, is equal to and fed from a three-phase supply. The windings of the three phases are placed 120° (electrical) apart. It is wound for a definite number of poles (the exact number of poles being determined by the requirement of speed). The greater the number of poles, the lesser will be the speed and vice versa. When the stator windings are supplied with the three-phase currents, they produce a magnetic flux of constant magnitude, which rotates at a synchronous speed (given by $N_s = 120$ f/P). This revolving flux induces an electromagnetic force (emf) in the rotor by mutual induction (Fig. 2.22).

ii. **Rotor**

There are mainly two types of rotor:

a. Squirrel cage rotor.
b. Wound (slip ring) rotor.

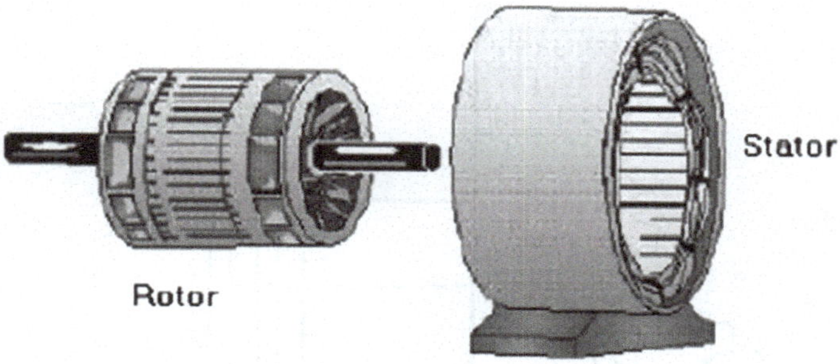

Fig. 2.22 Main parts of an induction motor. Image Courtesy: HLQ Induction Equipment Co., Ltd.

Fig. 2.23 Squirrel cage rotor

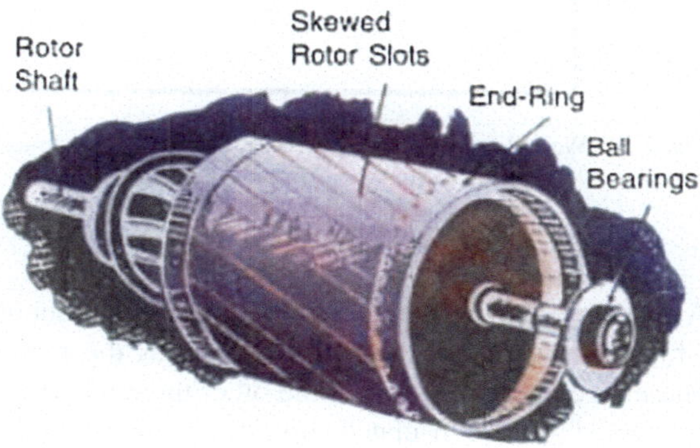

a. **Squirrel cage rotor**

The squirrel cage induction motor is preferred for compressors, as it is cheap and robust. The rotor consists of a cylindrical laminated core with parallel slots for carrying the rotor conductors, which are made of heavy bars of copper, aluminium, or alloys. One bar is placed in each slot. The rotor bars are short-circuited at both ends by the end rings. This rotor has an advantage of the flexibility that this may be used for the stator with the different number of poles. The current in the bars of a cage rotor follows the pattern of current in the stator winding. In case the number of poles in the stator is changed, still, there is no need to change rotor as the current pattern in the rotor bars gets adapted it automatically. However, the equivalent resistance of the rotor is constant since external resistance cannot be added in series with the rotor circuit. The rotor bars are skewed (as shown in Fig. 2.23), which provides us with two advantages:

(1) It makes motor run quietly by reducing the magnetic hum.
(2) It reduces the locking tendency of the rotor, i.e., tendency of the rotor teeth to remain under stator teeth due to direct magnetic attraction between the two.

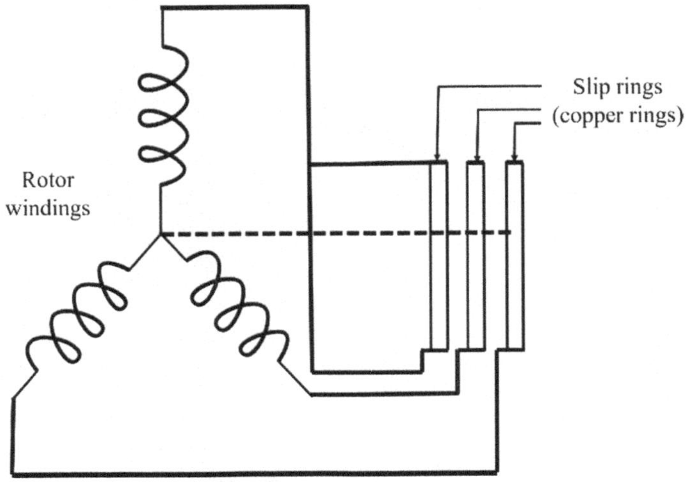

Fig. 2.24 Wound rotor (slip ring) of an induction motor. Image Courtesy: IDEA Lab, IIT Kanpur

b. **Wound (slip ring) rotor**

A slip ring rotor is provided with three-phase, double-layer, distributed windings, which are wounded for as many poles as the number of stator poles. The three phases are internally started; the other three winding terminals are brought out and connected to three insulated slip-rings mounted on the shaft with brushes resting on them. At these points, external resistance can be added to increase the starting torque requirement.

The other parts of the induction motor are:

(1) Frame: Outer structure made of close-grained alloy cast iron.
(2) Rotor core: built from high-quality silicon steel laminations to prevent eddy current.
(3) Stator and rotor windings: These windings have moisture-proof insulation embodying mica and high-quality varnishes. These are also carefully spaced for most effective air circulation and are rigidly braced to withstand centrifugal forces and any short-circuit stresses.
(4) Shaft and bearings.
(5) Fans: Light aluminium fans are used for adequate circulation of cooling air and are securely placed onto the rotor shaft.
(6) Terminal box: The current supply given to the stator winding is given through the terminal box (Figs. 2.24 and 2.25).

When the balanced stator windings are fed with three-phase supply, they produce a rotating magnetic field of constant magnitude with synchronous speed (given by $N_s = 120$ f/P). The magnetic flux lines in the air gap (Fig. 2.26a) cut both stator and rotor (being stationary, initially as the motor speed is zero) conductors at the same speed. The emf in both stator and rotor is induced at the same frequency, i.e., line

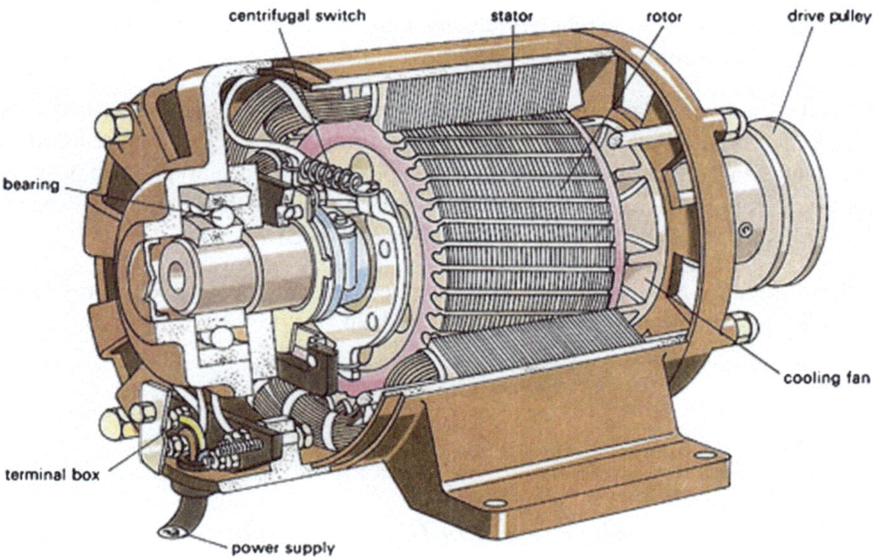

Fig. 2.25 Parts of an induction motor. Image Courtesy: Magnetics

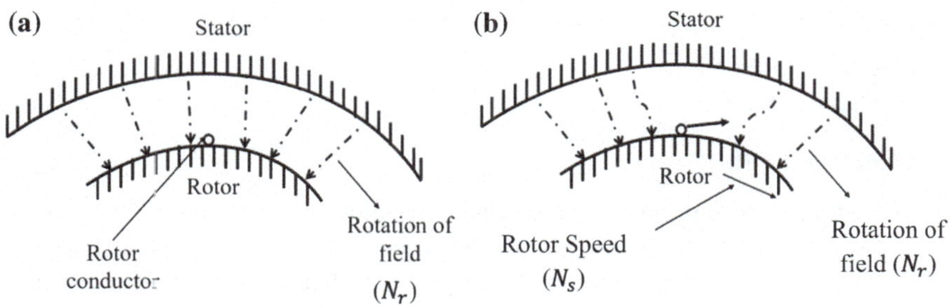

Fig. 2.26 Flux linkage with rotor bars under **a** stationary and **b** rotating conditions. Image Courtesy: IDEA Lab, IIT Kanpur

or supply frequency (f) with the number of poles for both stator and rotor windings being the same. As the rotor windings are short-circuited, current flows within the rotor windings. The electromagnetic torque in the motor produced is in the direction as that of the rotating flux produced in the air gap based on Lenz's law. The developed torque is in such direction that it will oppose the cause, which results in the current flowing in the rotor winding. The current in the rotor bars interacts with the air-gap flux to develop the torque, irrespective of the number of poles for which the winding in the stator is designed. The induced emf and the current in the rotor are due to the relative velocity between the rotor conductors and the rotating flux in the air gap, which is maximum when the rotor is stationary. In Fig. 2.26b, as the rotor starts rotating in the same direction as that of the rotating magnetic field, the production of the torque decreases the relative velocity along with lower values of induced emf and current in the rotor.

Note If the rotor speed is same as synchronous speed, then the relative speed becomes zero, which causes both the induced emf and current in the rotor to be reduced to zero. Under this condition, torque will not be produced. Hence, to produce positive (motoring) torque, the rotor speed must always be lower than the synchronous speed. The rotor speed can never be equal to synchronous speed.

The difference between the synchronous speed and the rotor speed, expressed as a ratio of the synchronous speed, is termed as "slip."

$$\text{Slip} = (N_s - N_r)/N_s.$$

2.1.4 Probable Faults Present in Induction Motor

Any deviation from the normal functioning of a machine, which can cause subpar performance, is termed as faults. The types of faults, which can be possible in an induction motor are broadly divided into two categories:

i. Electrical faults.
ii. Mechanical faults.

i. **Electrical faults**

   (a) **Stator winding faults**: Approximately stator faults account for 38% of all the faults. The faulty and asymmetric winding may produce spatial harmonics of any wave number into the air-gap field. However, all these harmonics vary at a single frequency, i.e., the supply frequency of the sinusoidal voltage source. The stator harmonics induce currents in the rotor cage and detect from the rotor as new air-gap field harmonics.

      (1) Turn-to-turn fault: This occurs when the insulation between two turns in the same coil breaks down and reduces the coil's ability to produce a balanced magnetic field. Unbalanced magnetic fields result in vibration, which can then cause degradation of the insulation as well as bearing failures. Localized heating can also spread to other coils, resulting in a coil-to-coil short. Excessive heating will eventually not only destroy the motor windings, but will also damage the insulation between the laminations of the stator core.

      (2) Phase-to-Phase fault: Another fault that can occur with motor windings is a phase-to-phase fault. This results from the insulation breaking down between two separate phases, usually lying adjacent to each other in the same slot. A higher difference in voltage potential tends to make this fault accelerate very quickly.

(3) Phase-to-Earth Short: This fault occurs at low voltage.

(b) **Loose connections**: Overheating and arcing of loose connections cause further corrosion and in time the failure of the connection. This can lead to increased levels of heat at the connections, which eventually leads to combustion and fire.

(c) **Rotor bar damage**: The other reason that can cause faults in squirrel cage induction motors is broken or cracked rotor bars and end rings damage. There are many reasons for such faults, which include thermal stress due to overload, non-uniform heat distribution, hot spot and arc, magnetic stresses due to the electromotive force, unsymmetrical magnetic force, electromagnetic vibrations, residual stress due to axial torque, materials wearing by chemical materials and humidity, mechanical stress due to mechanical fatigue of different parts, faulty ball bearings and loosening of laminations. Figures 2.27 and 2.28 show the distributed form of rotor cage for a healthy motor and for a faulty motor with one broken rotor bar, respectively (arrows representing the interbar current).

ii. **Mechanical faults**

(a) **Air-gap eccentricity**: When rotational axes of a motor do not coincide with the axes of the rotor and stator, the air gap becomes non-uniform. Approximately 80% of the mechanical faults lead to the stator and rotor eccentricity. This fault may occur during the process of manufacturing and fixing the rotor and can lead to a rub between the rotor and stator, causing

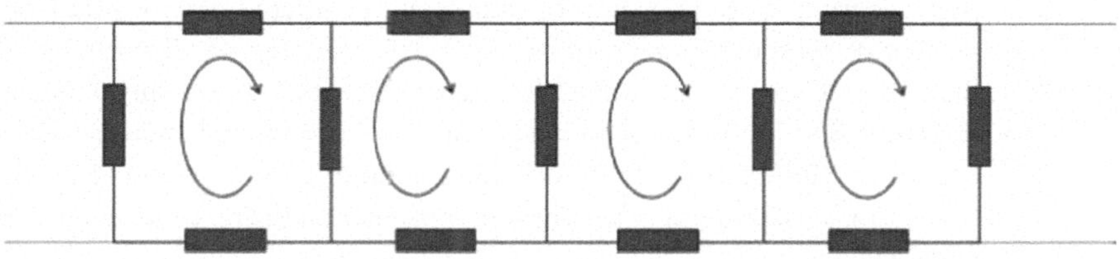

Fig. 2.27 Distributed form of rotor cage for the healthy motor. Image Courtesy: IDEA Lab, IIT Kanpur

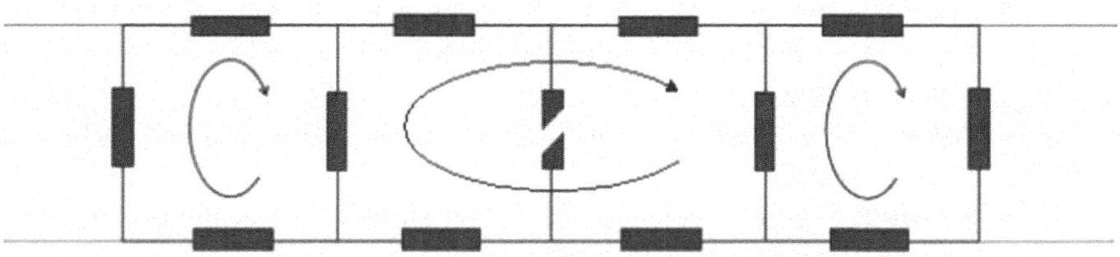

Fig. 2.28 Distributed form of rotor cage for the faulty motor. Image Courtesy: IDEA Lab, IIT Kanpur

serious damage to the machine. There are two types of eccentricity, **static** and **dynamic**.

- Static eccentricity is when the rotor rotational axis coincides with its symmetry axis but has a displacement with the symmetrical axis. In this case, the air gap around the rotor misses its uniformity, but it is invariant with time. This can be caused by stator core ovality or incorrect positioning of the rotor or stator at the commissioning stage. At the position of the minimum air gap, there is an unbalanced magnetic pull, which tries to deflect the rotor, thus increasing the amount of air-gap eccentricity.
- Dynamic eccentricity describes the condition when the minimum air gap revolves with the rotor. It can be caused by a bent shaft, mechanical resonances at critical speeds, or bearing wear.

(b) **Belts**: When the output of the motor is transmitted through belts, then belt or pulley mounted on the motor can be the source of the fault such as their misalignment.

(c) **Shaft misalignment**: The inherent problems of the rotating machine dysfunction are often caused by shaft misalignment. This defect generates some heavy loads and vibrations and can lead to premature failure of the bearing, the shaft, or the coupling.

(d) **Motor bearings**: Common bearing failures are as follows:

- **Flaking or surface fatigue**: Flaking occurs when small pieces of bearing material are split off from the smooth surface of the raceway or the rolling elements. This causes regions with a rough and coarse texture.
- **Peeling**: Dull or cloudy spots appear on the raceway surface along with light wearing. Tiny microscopic cracks are generated downwards from these cloudy spots to a depth of 5–10 μm. Peel the small particles of the material from the surface with areas of minor flaking starting to occur.
- **Scoring**: Scoring is surface damage due to accumulated small seizures caused by sliding under improper lubrication or severe operating conditions. Linear damage appears circumferentially on the raceway and roller surfaces.
- **Smearing**: Smearing is surface damage, which occurs from a collection of small seizures between bearing components caused by oil film rupture and sliding. Surface roughening occurs along with melting.
- **Fracture**: Fracture refers to small pieces, which broken off due to excessive load or shock were loading acting locally on a roller corner or rib of a raceway ring.
- **Pitting**: It has a dull lustre and appears on the rolling element surface or raceway surface.
- **Fretting**: Fretting occurs at the fitting surface and at the contact area between raceway ring and rolling elements.
- **False brinelling**: False brinelling is the occurrence of hollow spots that resemble brinell dents and are due to wear caused by vibration and swaying at the contact points between the rolling elements and raceway.

- **Electrical corrosion**: When an electric current passes through a bearing, arcing and burning occur through the thin oil film at points of contact between the raceway and rolling elements. The points of contact melted locally.
- **Discoloration**: Discoloration of cages, rolling elements, and raceway rings occur due to their reacting with lubricant at high temperature.

2.2 Data Acquisition

The process of measuring real-world physical variables and converting it into digital values that can be manipulated by a computer is known as data acquisition (DAQ) [1]. Data is collected using transducers/sensors placed on or around the machine in many positions. Most commonly used sensors in this application are microphones and accelerometers. Microphones capture the sound emanating from the machine, whereas accelerometers measure the compressor vibrations under running condition. Sound emission and vibration signal of engine often give much dynamic information of mechanical system condition. The positioning of a sensor is also an important step for capturing the condition of the machine. The complete data acquisition system comprises of data acquisition hardware, sensors to sense the machine condition, and a software interface to work with hardware comfortably. In this subsection, the sensors, along with specifications and their usages are described. Following subsection details the procedure of data acquisition, interface details, and wireless data acquisition. The detailed procedure for finding a sensitive position on the machine is also covered in this section.

2.2.1 Microphones

Acoustic data is collected using microphones. These are transducers responsible for the conversion of the sound signal to an electrical signal, where thin membrane vibrates in accordance with sound pressure. According to the requirement, we place unidirectional and omnidirectional microphones. Both receive acoustics from outside sources and convert them to electronic impulses but unidirectional microphones only pick up sounds aimed directly into their centres. This is ideal for one speaker or solo instrument but not as useful when sounds arrive from all directions. Omnidirectional microphones can pick up sounds from virtually any direction. Therefore, for our

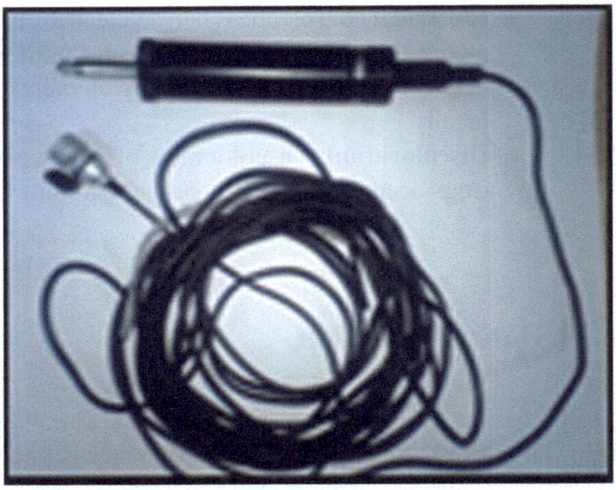

Fig. 2.29 UTP-30 Electret condenser-type unidirectional microphone. Image Courtesy: Ahuja Radios

Fig. 2.30 CTP-10DX electret condenser-type omnidirectional microphone. Image Courtesy: Ahuja Radios

application of recording sounds from the air compressor and motors, we have chosen a unidirectional microphone UTP-30, which is a cardioid-type microphone.

i. **Microphone type**

See Figs. 2.29 and 2.30.

Parameter	UTP-30	CTP-10DX
Direction	Unidirectional	Omnidirectional
Sensitivity	7.0 mv/pa	5.0 mv/pa
Frequency response	100–15,000 Hz	100–15,000 Hz
Impedance	1000 Ω	1000 Ω
Power supply	1.5 V DC	1.5 V DC

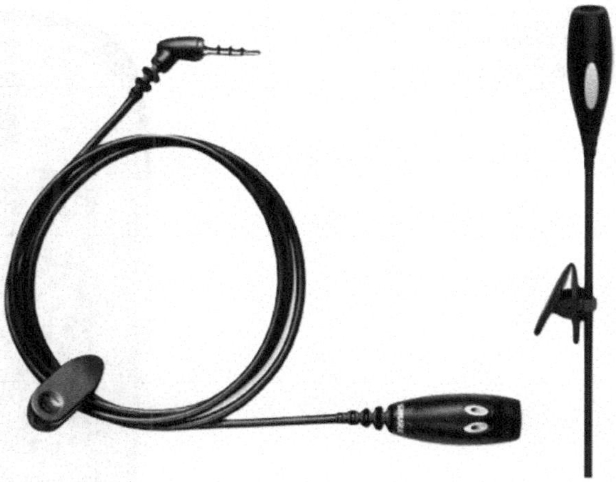

Fig. 2.31 Shure omnidirectional microphone. Image Courtesy: HIBINO

ii. **Shure omnidirectional microphone specifications**

See Figs. 2.31 and 2.32.

Model	WL183
Type	Omnidirectional (condenser)
Sensitivity	−40.0 dbv/pa
Dynamic range	102.5 db
Frequency response	50–17,000 Hz
Output impedance	1800 Ω
Operates on	1 × 1.5 v (UM-3)
Power requirements	1.5–10 V DC

iii. **PCB130D20 omnidirectional microphone specifications**

See Fig. 2.33.

Type	Omnidirectional
Frequency response	20–15,000 Hz
Sensitivity	45 mv/pa
Constant current excitation	2–20 Ma
Output impedance	70 Ω
Dynamic range	>122
Excitation voltage	18–30 V DC

Fig. 2.32 LG smartphone model P690. Image Courtesy: GSMArena

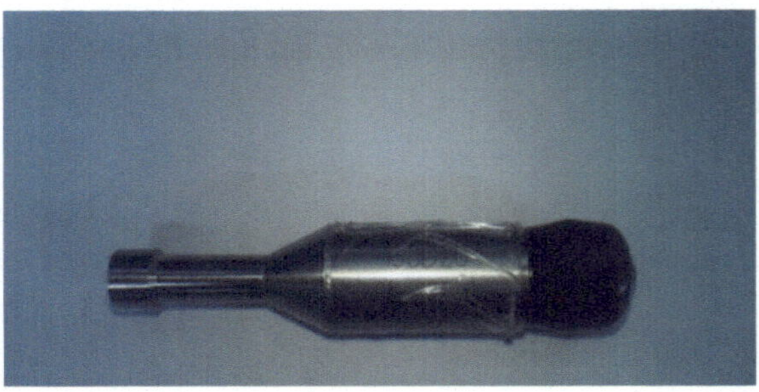

Fig. 2.33 PCB 130D20 omnidirectional microphone. Image Courtesy: PCB 130D20, pre-polarized condenser microphone

Fig. 2.34 LG smartphone model P690. Image Courtesy: GSMArena

iv. **Smartphone (L.G P690 model) specifications**

See Fig. 2.34.

Media player	Audio formats: MP4/H.264/H.263
Alert tone	72 polyphonic, MP3, WAV
Speakerphone	Yes
Audio connector	3.5 mm
OS	Android OS, v2.3.4 (Gingerbread)
Chipset	Qualcomm MSM7227T
CPU	800 MHz ARMv6
GPU	Adreno 200

2.2.2 Accelerometer

These are transducers for vibration signal-to-electric signal conversion. Normally, they use piezoelectric effect of certain material to measure dynamic changes in mechanical vibrations. The piezoelectric sensor is much more accurate and less prone to noise. For condition-based monitoring of machines, these accelerometers are the best choice.

i. **Integrated Circuit Piezoelectric (ICP) piezoelectric uniaxial accelerometer model: 603C01, ICP piezoelectric triaxial accelerometer model: 356A15 and triaxial accelerometer model: 352C33**

   See Figs. 2.35, 2.36, and 2.37.

Fig. 2.35 PCB 603C01 (uniaxial) accelerometer. Image Courtesy: Direct Industry

Fig. 2.36 PCB 356A15 (triaxial) accelerometer. Image Courtesy: SINUS

Fig. 2.37 PCB 352C33 (triaxial) accelerometer. Image Courtesy SINUS

Parameter	603C01	356A15	352C33
Sensitivity (± 10%)	100 mV/g	100 mV/g	100 mV/g
Frequency range	0.4 Hz–10 kHz	1.4 Hz–6.5 kHz	0.3–15,000 Hz
Temperature range	−54 to +121 °C	−54 to +121 °C	−54 to +93 °C
Excitation voltage	18–28 V DC	20–30 V DC	18–30 V DC

2.2.3 *Acoustic Data Acquisition Using Smartphone*

Here, our data acquisition process is performed by recording sound generated by a machine (acoustic data) using inbuilt the microphone of the smartphone [2, 3] as shown in Fig. 2.34. Acoustic data carries real-time dynamic information of machine characteristics. There are some key points, which need to be set while recording data.

i. **Sampling rate**—To record digital sound, an analog signal is sampled. Samples are measures of the intensity of a sound at the moment in time. The number of samples received per second is called the sampling rate, which is measured in hertz. There are many sampling rates supported by Android devices such as 22,050, 16,000, 11,025, and 44,100 Hz. Out of these, only 44.1 kHz is the only rate that is guaranteed to work on all devices. Higher the sampling rate, greater is the audio quality and ensures greater precision in your high notes and low notes.
ii. **Encoding scheme**—Most common technique used for encoding acoustic data (analog one) is ENCODING_PCM_16 BIT or ENCODING_PCM_8 BIT. In this technique, the magnitude of the analog signal is sampled regularly at the uniform interval. Then each sample is quantized, and Pulse Code Modulation (PCM) pulses are obtained. These pulses are encoded to produce a bit of streams.

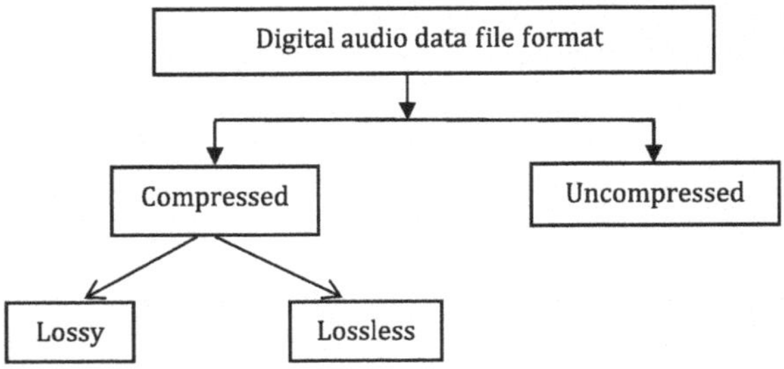

Fig. 2.38 Types of file format. Image Courtesy: IDEA Lab, IIT Kanpur

iii. **Channel**—Mono and stereo are two classifications of reproduced sound. Mono is the term used to describe sound that is only from one channel, while stereo uses two or more channels to provide a natural experience.
iv. **Audio file format**—To store digital data on a system, the file format must be followed. Audio file format is classified into number of categories as shown in Fig. 2.38. It can be saved either in compressed audio format or uncompressed audio format.

(a) **Compressed audio format**—A digital audio data is compressed, resulting in smaller files. It has two subgroups.

- **Lossless compressed audio formats**—These audio formats compress digital audio data but there is no loss of data or degradation of sound quality during the compression process, e.g. Free Lossless Audio Codec (FLAC) file.
- **Lossy compressed audio formats**—These audio formats compress digital audio data but are known to eliminate certain information and frequencies to reduce the file size. Lossy compressed audio formats cause degradation in audio quality. The difference in audio quality can be large or small, depending upon how much data has been removed. Generally, the quality loss is such that it is not noticeable by the user. The classic example of lossy compression is MP3.

(b) **Uncompressed audio format**—Uncompressed audio files are digital representations of sound wave, which are most suitable for archiving and delivering audio at high resolution due to their accuracy, i.e., WAV, AIFF, and BWF. Some of these audio file types are "wrapper" formats that use PCM audio and add additional data to enable compatibility with specific codecs and operating systems. In this application, we are using .wav format, which is most widely used and records data at standard compact disc quality. Chosen specifications for recording machine acoustic are as follows:
- Sampling rate: 44.1 kHz per second
- Channel: Stereo
- Encoding scheme: ENCODING_PCM_16BIT
- Format: *.wav* format (Fig. 2.39).

Fig. 2.39 Data acquisition using smartphone [2]

Fig. 2.40 CCD laser and controller (KEYENCE model: LK-031). Image Courtesy: Keyence

2.2.4 Laser Displacement Sensor

The Charged Coupled Device (CCD) laser displacement sensor uses a triangulation measurement system. The LK Series uses a CCD as the light-receiving element. The light reflected by a target passes through the receiver lens and is focused on the Position Sensitive Device (PSD) or CCD.

The CCD detects the peak value of the light quantity distribution of the beam spot for each pixel and identifies this as the target position. Therefore, it enables stable highly accurate displacement measurement, regardless of the light quantity distribution of the beam spot (Figs. 2.40 and 2.41).

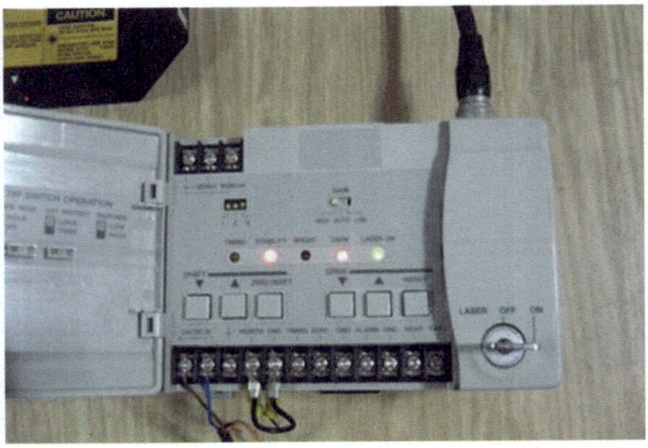

Fig. 2.41 Laser controller in working condition. Image Courtesy: Keyence

Type	High precision
Model	Sensor head: LK-031
Controller	LK-2001
Light source	Visible red semiconductor Laser ($\lambda = 670$ nm
Pulse duration	FDA—3 to 482 μs IEC—3 to 482 μs

Reference distance	30 mm
Measuring range	−5 mm to +5 mm
Sampling time	512 μs
Resolution	1 μm
Power supply	24 V DC
Current consumption	400 mA (max)

2.2.5 Data Acquisition Hardware

Microphone and accelerometer convert the signal into electric form but that must be converted into digital form, and to store in computer. DAQ module and its graphic user interface (LabVIEW) serve this purpose. DAQ module used was cDAQ-9172 made by National Instruments (Fig. 2.42).

i. **NI cDAQ-9172 8-slot USB 2.0 chassis for Compact DAQ**

Manufacturer: National Instruments
Model: NI cDAQ-9172

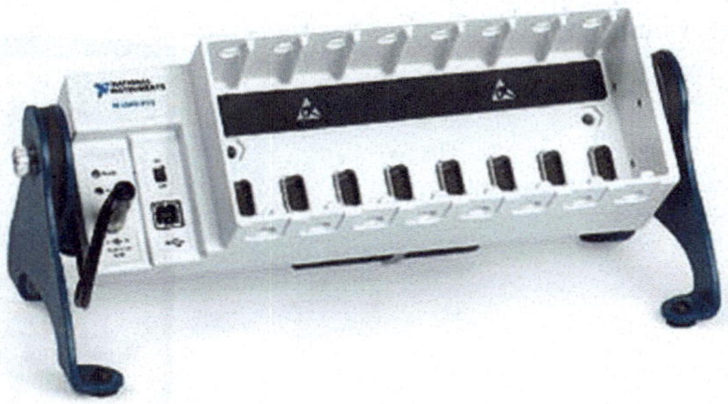

Fig. 2.42 DAQ chassis (NI cDAQ 9172)

Specifications NI cDAQ-9172

More than 5 ms/s streaming acquisition per chassis
Hi-speed Universal Serial Bus (USB) connectivity to Personal Computer (PC)
Power input: 11–30 V DC at 15 W
Accepts up to 8 C series Input-Output (I/O) modules
NI C series modules (NI-9234 module).

ii. **DAQ NI-9234 (Digital-to-Analog (A/D) converter)**

cDAQ—9172 need to have signal acquisition modules. Four-channel dynamic signal acquisition module NI 9234 is used for this purpose. The NI 9234 is a four-channel C series dynamic signal acquisition module for making high-accuracy audio frequency measurements from integrated electronic piezoelectric (IEPE) and non-IEPE sensors with NI Compact DAQ or Compact Reduced Input Output (RIO) systems. The NI 9234 delivers 102 dB of dynamic range and incorporates software-selectable AC/DC coupling and IEPE signal conditioning for accelerometers and microphones. The four input channels simultaneously digitize signals at rates up to 51.2 kHz per channel with built-in anti-aliasing filters that automatically adjust to our sampling rate (Fig. 2.43).

Specifications of NI 9234

Software-selectable IEPE signals conditioning (0 or 2 *mA*).
51.2 kS/s per-channel maximum sampling rate.
±5 V input.
Software-selectable AC/DC coupling.
AC-coupled (0.5 Hz).
NIST-traceable calibration.
Smart Transducer Electronic Data Sheet (TEDS) sensor compatibility.
24-bit resolution.
102 dB dynamic range.
Anti-aliasing filter.

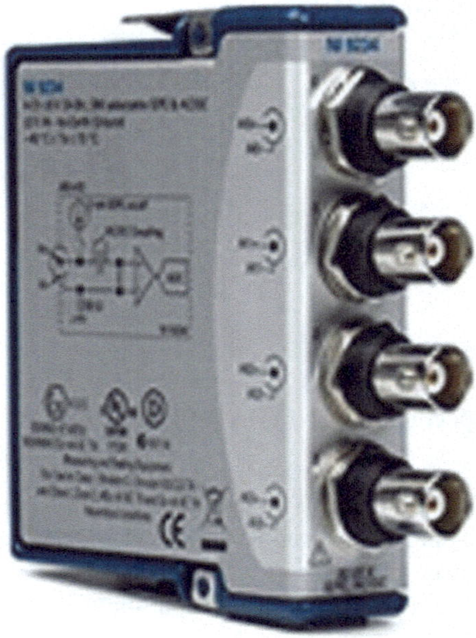

Fig. 2.43 DAQ (NI 9234). Image Courtesy: Artisan

iii. **Recommended software for DAQ**

NI sound and vibration analysis software include the NI sound and vibration measurement suite and the NI sound and vibration toolkit. It provides signal processing functionality for performing audio measurements, fractional-octave analysis, frequency analysis, transient analysis, and order tracking. NI analysis software features NI sound and vibration assistant interactive software for quickly acquiring, analyzing, and logging acoustic, noise, and vibration data. With a configuration-based flexible measurement library and open-analysis capability, the sound and vibration assistant is designed for quick data capture through a unique software-based measurement approach to create the customized applications.

2.2.6 Data Acquisition Software

LabVIEW is a graphical programming environment used by millions of engineers and scientists to develop sophisticated measurement, test, and control systems using intuitive graphical icons and wires that resemble a flow chart. LabVIEW offers unrivaled integration with thousands of hardware devices and provides hundreds of built-in libraries for advanced analysis and data visualization. The LabVIEW platform is scalable across multiple targets and operating systems, and since its introduction in 1986, it has become an industry leader.

LabVIEW VI contains three components—the front panel, the block diagram, and the icon and connector pane. In LabVIEW, we build a user interface, or front panel, with controls and indicators. Controls are knobs, push buttons, dials, and other feed devices. Indicators are graphs, LEDs, and other displays. After we build the user interface, we add code using VIs and structures to control the front panel objects. LabVIEW ties the creation of user interfaces (called front panels) into the development cycle. LabVIEW programs/subroutines are called virtual instruments (VIs). Each VI has three components: a block diagram, a front panel, and a connector panel. The last is used to represent the VI in the block diagrams of other, calling VIs. Controls and indicators on the front panel allow an operator to input data into or extract data from a running virtual instrument. However, the front panel can also serve as a programmatic interface. Thus, a virtual instrument can either be run as a program, with the front panel serving as a user interface, or, when dropped as a node onto the block diagram, the front panel defines the inputs and outputs for the given node through the connector pane. This implies each VI can be easily tested before being embedded as a subroutine into a larger program. The graphical approach also allows non-programmers to build programs simply by dragging and dropping virtual representations of laboratory equipment with which they are already familiar. The LabVIEW programming environment, with the included examples and the documentation, makes it simple to create small applications. This is a benefit on one side but there is also a specific danger of underestimating the expertise needed for good quality "G" programming. For complex algorithms or large-scale code, the programmer must possesses an extensive knowledge of the special functions in LabVIEW. Using LabVIEW, we can create test and measurement, data acquisitions, instrument control, data logging, measurement analysis, and report generation applications.

2.2.7 *Data Acquisition Toolbox*

Data Acquisition Toolbox provides a complete set of tools for analog input, analog output, and digital I/O from a variety of PC-compatible data acquisition hardware. The toolbox lets you configure your external hardware devices, read data into LabVIEW and Simulink environments for immediate analysis, and send out data. Data Acquisition Toolbox enables you to customize your acquisitions, access the built-in features of hardware devices, and incorporate the analysis and visualization features of LabVIEW and related toolboxes into your design. You can analyze or visualize your data, save it for post-processing. Data Acquisition Toolbox also supports Simulink with blocks that enable you to incorporate live data or hardware configuration directly into Simulink models. Acquired data is displayed, analyzed, and stored on a computer specialized programming language used for data acquisition include LabVIEW, which offers a graphical programming environment optimized for data acquisition.

2.2.7.1 Working with Data Acquisition Toolbox

The toolbox provides functions for creating device objects that are directly associated with your hardware. These objects include base properties that apply to all supported hardware, such as sample rate, trigger settings, and channel properties. They also include device-specific properties that let you access the specific features and capabilities of your hardware.

It supports three device objects: analog input, analog output, and digital I/O. Data Acquisition Toolbox automatically performs Analog-to-digital (A/D) and D/A data conversions for receiving or sending data.

i. Analog input: The analog input functions let you acquire signals from your hardware. You can create an analog input object, add channels to the object, acquire data to memory, read data into the workspace, and preview the most recently acquired data.
ii. Analog output: Analog output functions let you send signals out to your hardware. You can create an analog output object, add channels, queue datasets to the output, and generate analog signals.
iii. Digital I/O: Digital I/O functions enable you to generate or read digital signals using your hardware. You can create digital I/O objects, add lines, send data to the hardware, and read data into the workspace.
iv. Channels and lines: Data Acquisition Toolbox channels and lines are mapped to your hardware's channels and lines. The toolbox supports number of channels/lines, enabling you to use as many as your hardware permits.

2.2.7.2 Controlling Your Acquisition

Data Acquisition Toolbox supports a wide range of functions for controlling your acquisition. For example, you can set event information, evaluate the acquisition status, preview data while the device is running, and perform analysis on-the-fly. The toolbox also supports several hardware-specific properties that can be displayed and customized to your specifications.

2.2.7.3 Performing Data Acquisition in Simulink

Data Acquisition Toolbox provides Simulink blocks for acquiring live or measured data directly into your models, or configuring hardware interfaced to data acquisition devices.

These blocks enable us to quickly evaluate the response of our Simulink models and algorithms with real-world data, instead of designing systems against static datasets, such as those saved in files. We can also use these blocks to verify and validate our models against live, measured data as part of the system development process. Data Acquisition Toolbox provides four Simulink blocks.

i. Analog input—Acquire data from analog channels
ii. Analog output—Output data to analog channels
iii. Digital input—Acquire the latest set of values from digital lines
iv. Digital output—Output data to digital lines

The output blocks let you configure hardware from our Simulink models, including instructing the hardware to send data. The input blocks enable us to acquire live data from hardware and incorporate that data directly into our models. Each block provides a dialogue box that enables you to configure parameters including, the device type, channels, and lines. The analog blocks also allow us to configure other relevant parameters such as asynchronous versus synchronous acquisition, the sample rate, block size, and data type. We can connect our Simulink model to a broad range of data acquisition hardware, and later change devices with minimal changes to your model. Using Simulink with Data Acquisition Toolbox enables system simulation and design with live data in a single software environment. Apart from this, communication between information systems and shop floor equipment is essential for real-time monitoring in the industrial scenario. For this purpose, there are some protocols such as MTConnect [4], Open Platform Communication [5] to achieve better performance in various applications.

2.2.8 Wireless Data Acquisition

For wireless data acquisition, there are two different wireless data acquisition modules, NI 9163 and NI 9191, which can be used in conjunction with NI 9234. Both DAQ modules support Ethernet network and wireless network, so both modes can be used for data acquisition.

Latest DAQ devices from National Instruments offer connectivity over wireless and cabled Ethernet. NI Wi-Fi DAQ devices combine IEEE 802.11g wireless or Ethernet communication, direct sensor connectivity. Here, we have chosen NIWLS-9163 and NI WLS 9191DAQ modules for acquiring the data through NI cDAQ 9234. Specification details of each hardware are as follows.

NI WLS-9163: It is a IEEE 802.11B/G wireless communication interface, which gives 30 m indoor, 100 m line of sight wireless signal range. It can send data to a desktop PC over Ethernet or Wi-Fi. It also provides advanced security with 128-bit AES data encryption (Fig. 2.44).

NI WLS-9191: It is also a Wi-Fi DAQ, which holds a single C Series module and is capable of streaming continuous waveform data at more than 50 kS/s sampling rate per channel with 24 bits of resolution. In addition, built-in NIST-approved 128-bit AES encryption advanced IEEE 802.1X network authentication offer the highest commercially available network security (WPA2 Enterprise) (Fig. 2.45).

NI cDAQ 9234: It is a four-channel dynamic signal acquisition module specifically used for sound and vibration measurement. It has four BNC connectors that provide connections to four simultaneously sampled analog input channels. It also provides

Fig. 2.44 NI WLS 9163

Fig. 2.45 NI WLS 9191.
Image Courtesy: NI

the flexibility of IEPE signal conditioning. The NI 9234 uses a combination of analog and digital filtering to provide an accurate representation of in-band signals while rejecting out-of-band signals. It captures signal at the sampling rate of 51.2 kS/s per channel.

2.2.8.1 Data Acquisition via Ethernet Cable on Laptop Using 9163 Chassis

For this case of data acquisition, we have used laptop with LabVIEW toolbox Measurement and Automation Explorer installed. Detailed procedure for setup is as follows:

Step 1 Connect the chassis to the laptop using the Ethernet cable and detect the device using LabVIEW MAX (Fig. 2.46).

Step 2 The device is now visible also with the connection type to be wired. Select the detected device (Fig. 2.47).

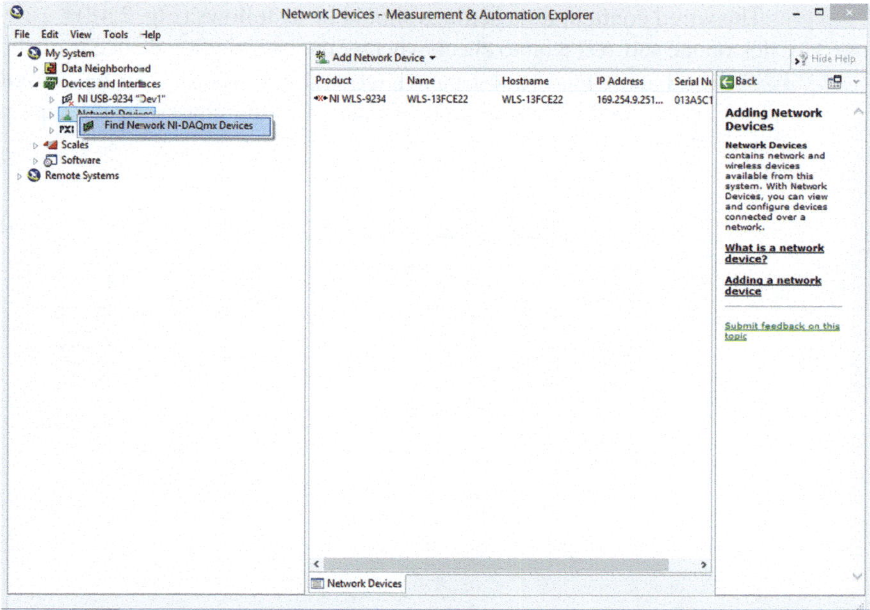

Fig. 2.46 Data acquisition via Ethernet cable using 9163 (Step 1). Image Courtesy: NI

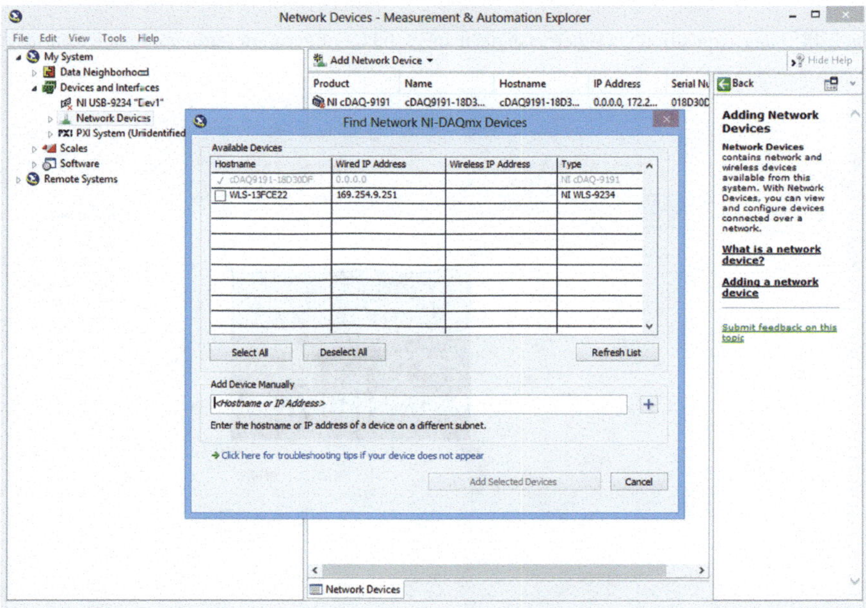

Fig. 2.47 Data acquisition via Ethernet cable using 9163 (Step 2). Image Courtesy: NI

The wired configuration settings should be as follows (Fig. 2.48):

Step 3 Click on the self-test icon to check whether data acquisition device is ready. To see the data, click the test panels tab and a waveform can be viewed indicating successful transmission of data from the chassis to the laptop (Fig. 2.49).

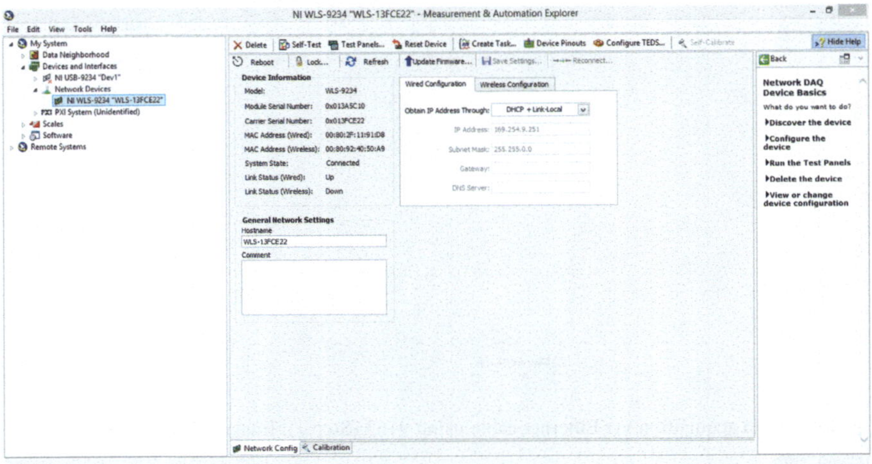

Fig. 2.48 Data acquisition via Ethernet cable using 9163 (configuration settings). Image Courtesy: NI

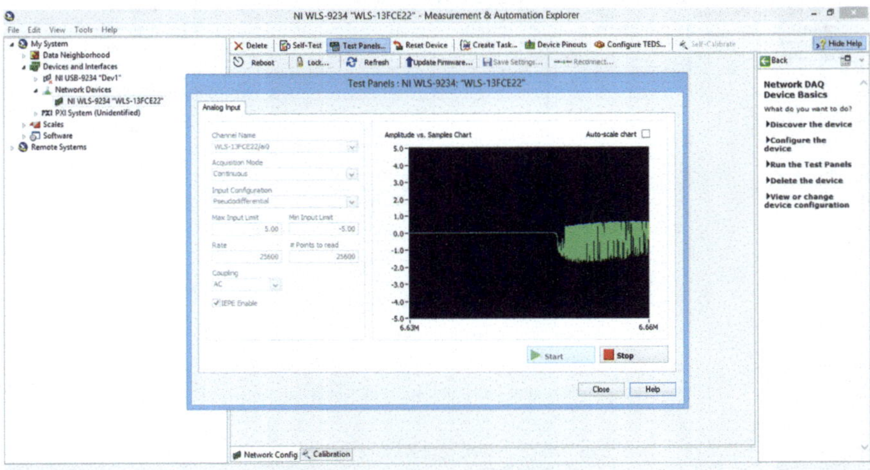

Fig. 2.49 Data acquisition via Ethernet cable using 9163 (Step 3). Image Courtesy: NI

2.2.8.2 Data Acquisition via Ethernet Cable on Laptop Using 9191 Chassis

CDAQ-9191 has four counters and 5 Watts of power. NI WLS/ENET-9163 has no counters and 4.5 Watts of power. The physical dimensions seem comparable but do differ slightly. Though the two devices differ structurally, the functionality of these two devices remains the same. The setup for this DAQ remains the same as that of 9163. The only difference is that one needs to reserve the chassis in 9191. This choice is not available in case of 9163. Screenshot for the chassis reservation is shown in Fig. 2.50. Since the chassis is reserved for a particular laptop, only one laptop can access the data at one time (Fig. 2.50).

Data Acquisition Via Wi-Fi on Laptop Using 9163 Chassis

There are two types of network, which can be used for wireless data acquisition ad hoc network and infrastructure network. On wireless computer networks, ad hoc mode is a method for wireless devices to directly communicate with each other. Operating in ad hoc mode allows all wireless devices within range of each other to discover and communicate in peer-to-peer fashion without involving central access points (including those built into broadband wireless routers). The chassis is first configured to the laptop using Ethernet cable.

Step 1 Click on the chassis icon, go to network settings, and under the "wireless mode", select "Create wireless network" option.

Add a suitable name (like "9163" here), and let the IPv4 address be "DHPC or Link Local". This will create the ad hoc network. Connect to the network, and note down the IP address and the subnet mask of the network that shows up in the control panel under the Wi-Fi settings. Now uncheck

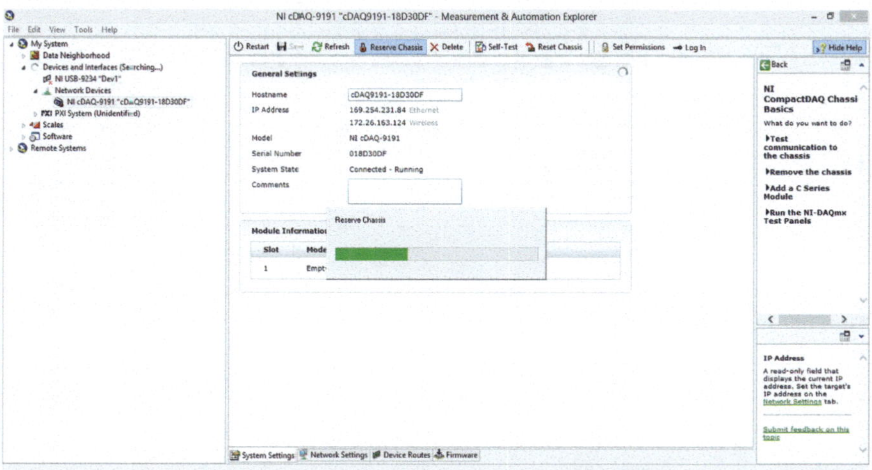

Fig. 2.50 Reserving the chassis for cDAQ 9191

the "obtain IP address automatically", and put the IP address and the subnet noted down in "Use the following IP address". Now disconnect from the network. For our experiment, the addresses were IP address: 169.254.41.171 and Subnet mask: 255.255.0.0 (Fig. 2.51).

Step 2 IP address must be fed in static mode. Any suitable IP address can be used provided it is not the same as the one manually entered in the control panel and has the same subnet (in this example, 169.254.41.172 is used, which is on the same subnet [255.255.0.0] as detected earlier (169.254.41.171). Save the settings (Fig. 2.52).

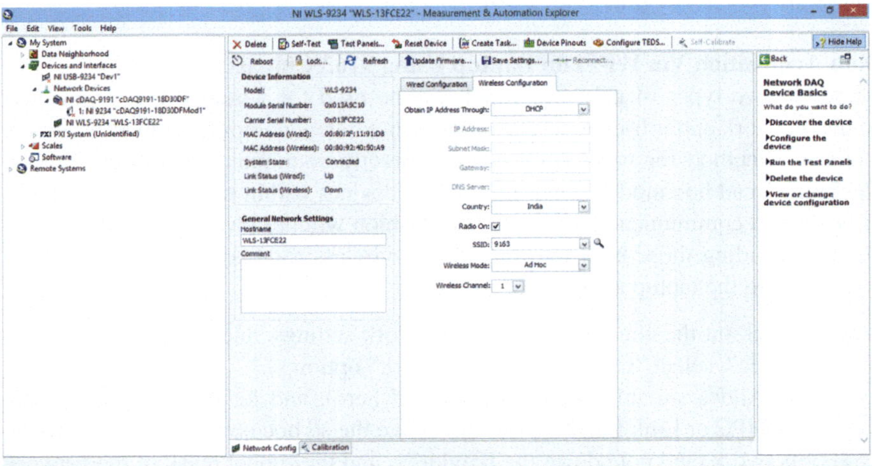

Fig. 2.51 Data acquisition via Wi-Fi using 9163 (Step 1). Image Courtesy: NI

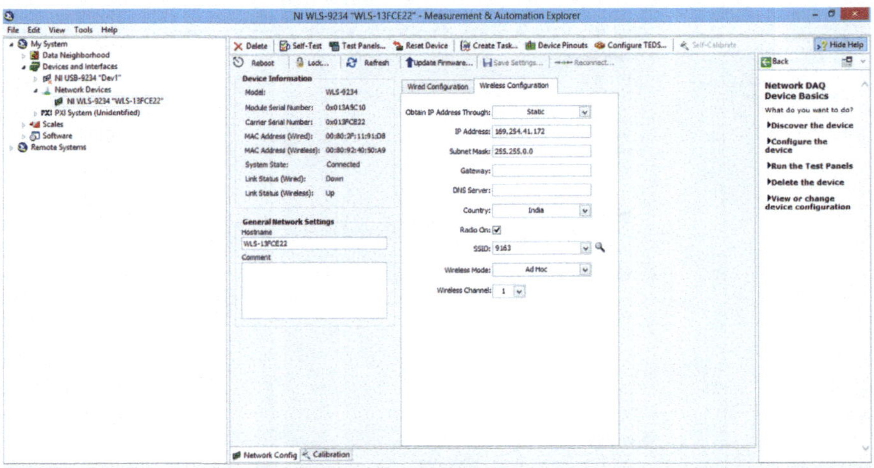

Fig. 2.52 Data acquisition via Wi-Fi using 9163 (Step 2). Image Courtesy: NI

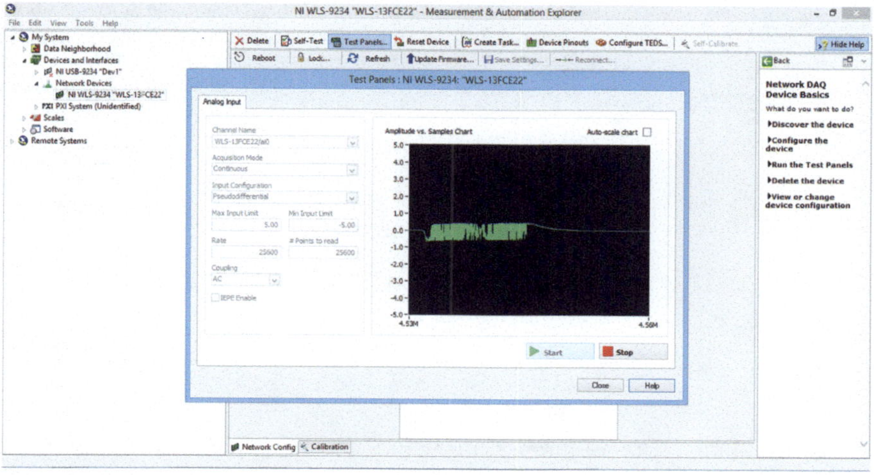

Fig. 2.53 Data acquisition via Wi-Fi using 9163 (Step 3). Image Courtesy: NI

Step 3 Now unplug the Ethernet cable, connect to the newly created network, and click "test panels" to take measurements (Fig. 2.53).

2.2.8.3 Data Acquisition via Wi-Fi on Laptop Using 9191 Chassis

As mentioned earlier for 9163, configurations of the chassis for wireless data acquisition are required given below to complete the configuration.

Step 1 As soon as chassis is connected to the system using Ethernet cable, system detects and it is ready for configuration (Fig. 2.54).

Step 2 Now go to network settings and change the settings of "wireless adapter wlan0" (Fig. 2.55).

Step 3 The IP addresses get updated automatically. As soon as they do, select static from the configure IPv4 Address menu (Fig. 2.56).

Step 4 The chassis has been configured to broadcast the data. Now one can wirelessly connect to the ad hoc network and receive the data (Fig. 2.57).

Step 5 One can see the data received in two ways: one Ethernet and the other wireless whose IP addresses are also different. In case there are two data receiving ways, then it prefers wired over wireless. Now, unplug the Ethernet cable to use the wireless data acquisition (Fig. 2.58).

Step 6 Click on test panel to take the measurements (Fig. 2.59).

Again, over here multiple connections are not possible as the chassis gets reserved to a system.

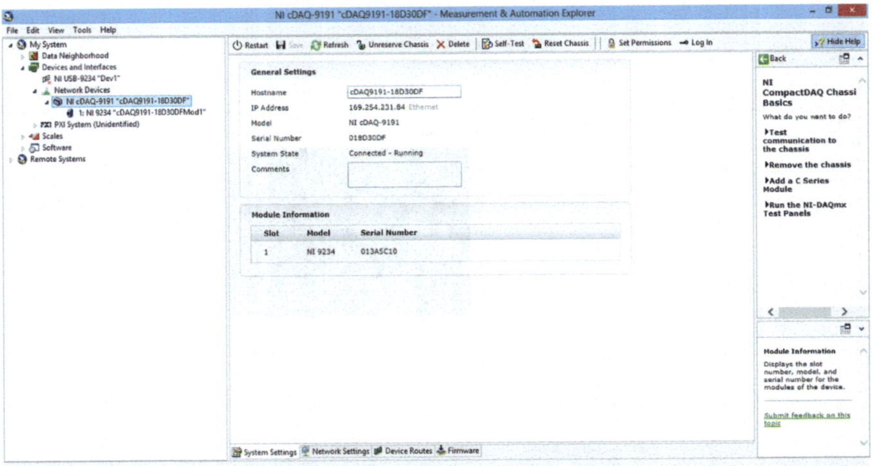

Fig. 2.54 Data acquisition via Wi-Fi using 9191 (Step 1). Image Courtesy: NI

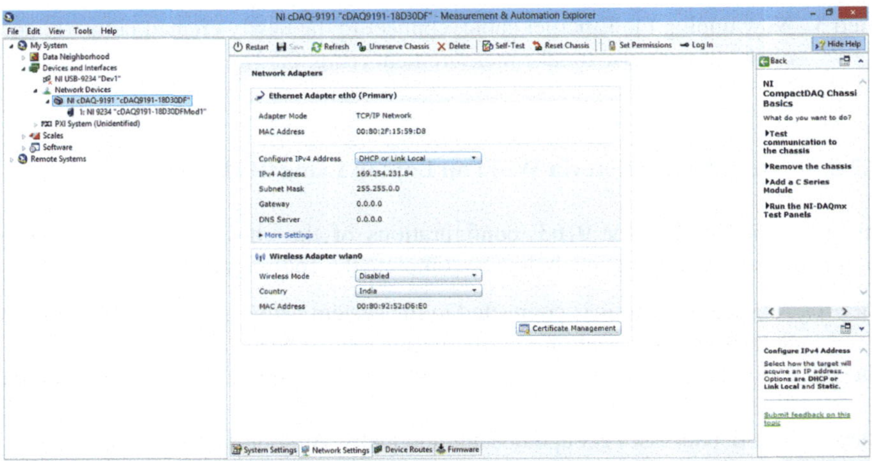

Fig. 2.55 Data acquisition via Wi-Fi using 9191 (Step 2). Image Courtesy: NI

2.2.8.4 Data Acquisition on Android Smartphone

In wireless data acquisition, either of two modes of network can be used ad hoc network or in infrastructure network mode. In ad hoc mode, wireless devices directly communicate with each other. Operating in ad hoc mode allows all wireless devices within range of each other to discover and communicate in peer-to-peer fashion without involving central access points (including those built into broadband wireless routers).

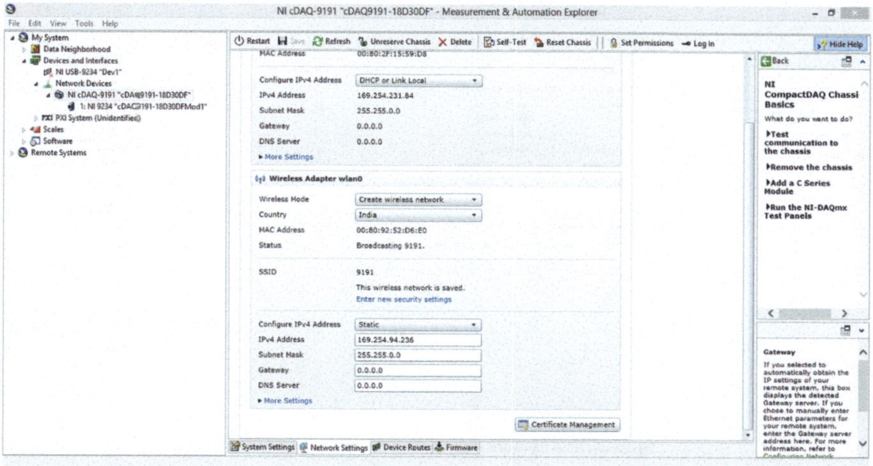

Fig. 2.56 Data acquisition via Wi-Fi using 9191 (Step 3). Image courtesy NI

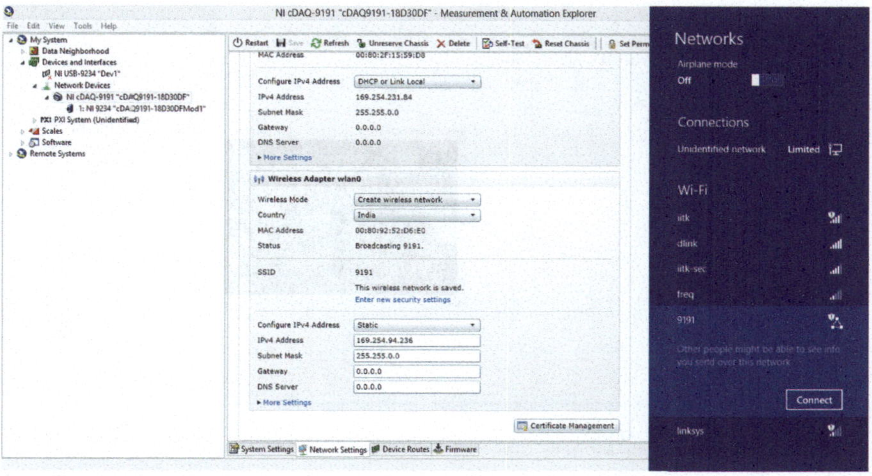

Fig. 2.57 Data acquisition via Wi-Fi using 9191 (Step 4). Image Courtesy: NI

2.2.8.5 Wireless DAQ to Laptop Using an Intermediate Router While Connecting Chassis to the Router via an Ethernet Cable

Step 1 Connect the chassis to the router via a LAN cable. In this case, the router's network is called "dlink". Now, DAQmx shows that the chassis is not wired to the laptop. Connect to the router's network and refresh the program (Fig. 2.60).

Step 2 It should now start showing that the chassis is connected via Ethernet with the address being that of the router. This means that the chassis sends data

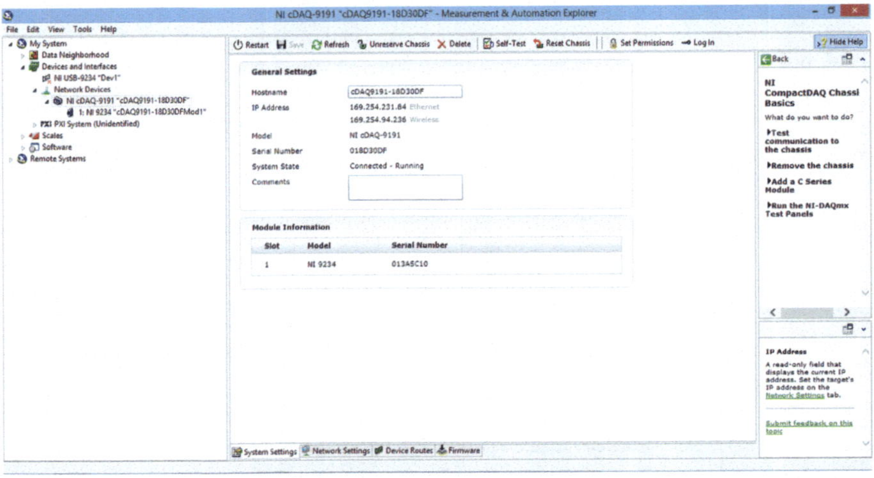

Fig. 2.58 Data acquisition via Wi-Fi using 9191 (Step 5). Image Courtesy: NI

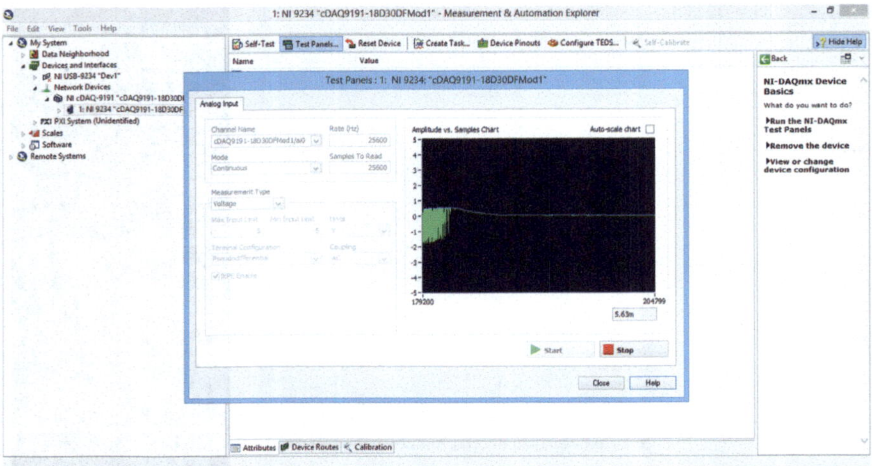

Fig. 2.59 Data acquisition via Wi-Fi using 9191 (Step 6). Image Courtesy: NI

to the router via the cable, which is acquired by laptop via wireless "dlink" (Fig. 2.61).

Step 3 To test the configuration, click the test panels tab and a waveform is obtained indicating successful transmission of data from the chassis to the laptop (Fig. 2.62).

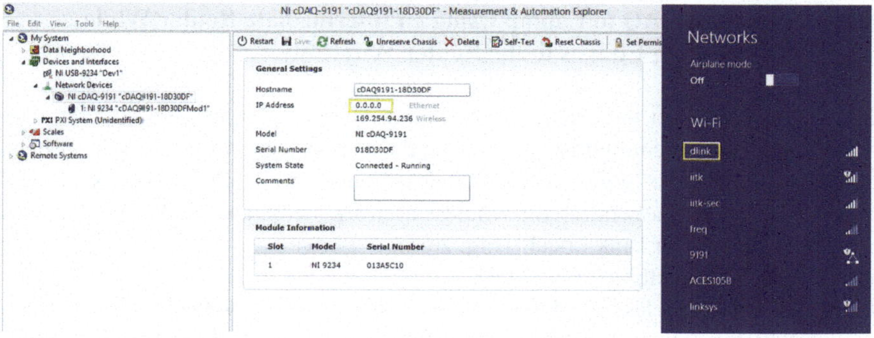

Fig. 2.60 Data acquisition via intermediate router with Ethernet cable using 9191 (Step 1). Image Courtesy: NI

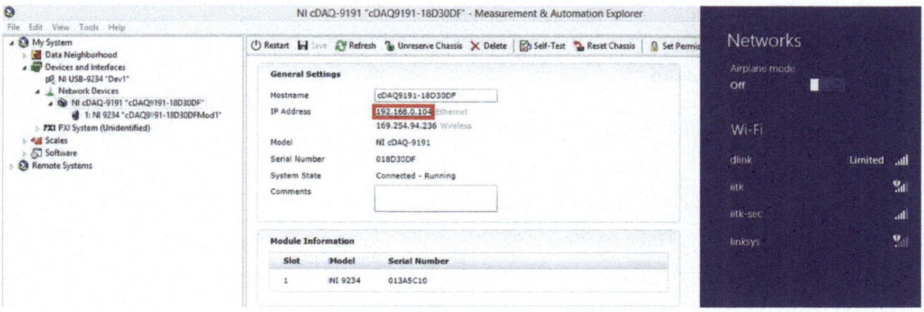

Fig. 2.61 Data acquisition via intermediate router with Ethernet cable using 9191 (Step 2). Image Courtesy: NI

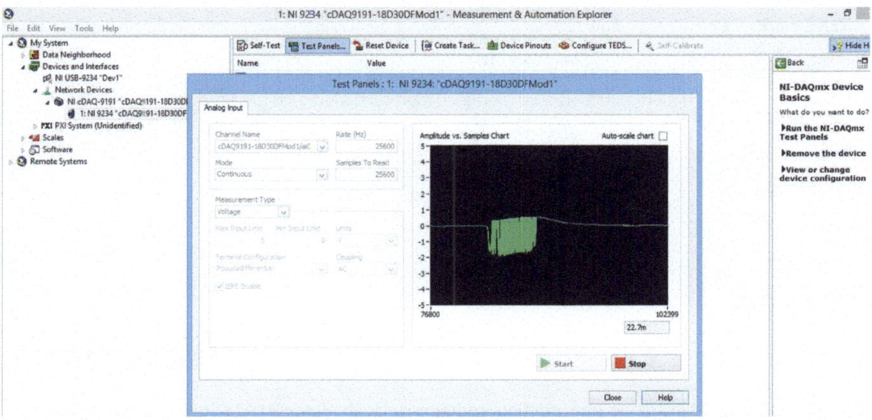

Fig. 2.62 Data acquisition via intermediate router with Ethernet cable using 9191 (Step 3). Image Courtesy: NI

2.2.8.6 Wireless DAQ to Laptop Using an Intermediate Router While Connecting Chassis to the Router Wirelessly

Step 1 First, connect the chassis to the laptop via the Ethernet cable to complete the wireless configuration. Click on the chassis icon, open network settings, and under the "wireless mode", select Connect to wireless network". Select the router's network in the wireless network menu, and let the IPv4 address be "DHPC or Link Local" (Fig. 2.63).

Step 2 Click save settings (Fig. 2.64).

Step 3 Now the wireless IP address of the chassis would show the router's address, which indicates that it was successful in connecting to the router (Fig. 2.65).

Step 4 Now unplug the Ethernet cable, connect to the router's network and click "test panels" to take measurements (Fig. 2.66).

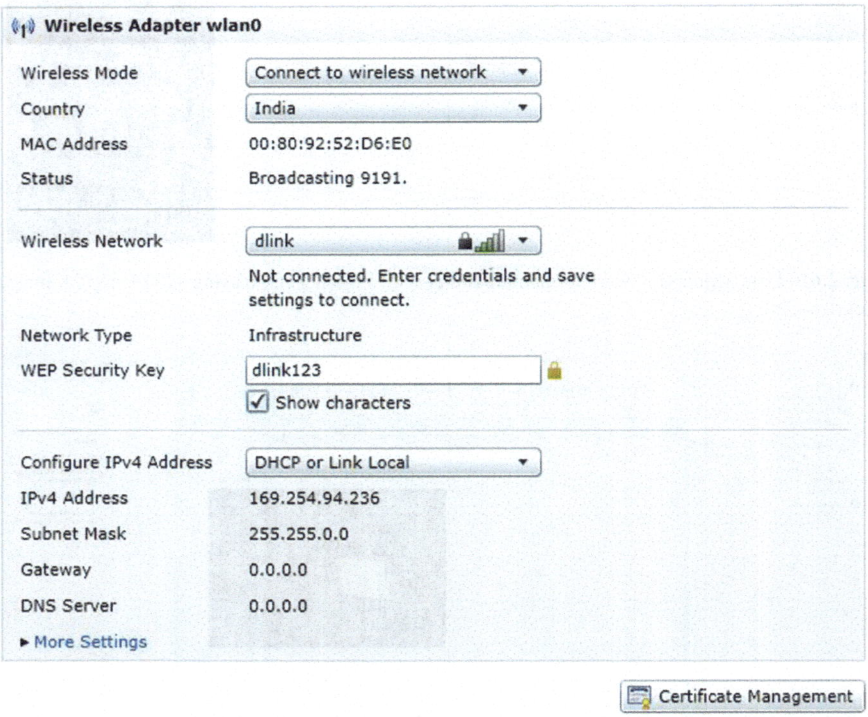

Fig. 2.63 Data acquisition via intermediate router wirelessly using 9191 (Step 1). Image Courtesy: NI

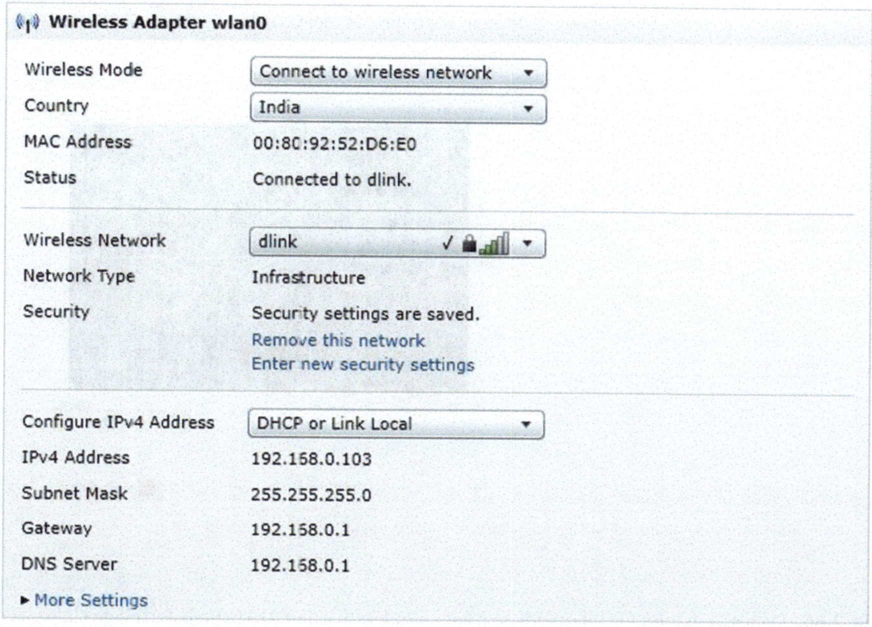

Fig. 2.64 Data acquisition via intermediate router wirelessly using 9191 (Step 2). Image Courtesy: NI

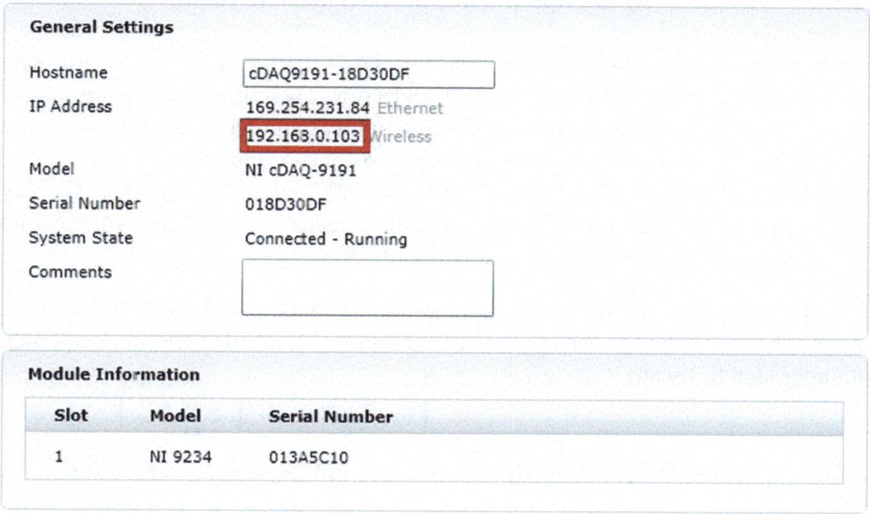

Fig. 2.65 Data acquisition via intermediate router wirelessly using 9191 (Step 3). Image Courtesy: NI

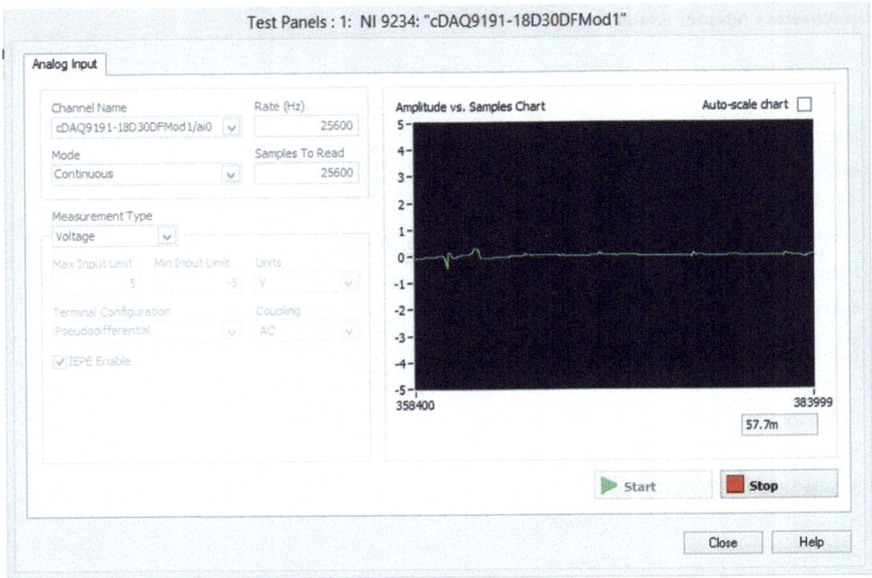

Fig. 2.66 Data acquisition via intermediate router wirelessly using 9191 (Step 4). Image Courtesy: NI

2.2.8.7 Wireless DAQ to Laptop Using an Intermediate Router

Step 1 Since after the firmware upgrade for cDAQ 9191 chassis, the smartphone was not able to detect the network, and a router was used to make the chassis network available to the phone. Its setup was similar as in connecting the chassis wirelessly to the router in the previous example (Fig. 2.67).

Step 2 The IP address of the chassis was manually entered in the Data Display App of the smartphone as shown in the screenshot above. The smartphone returned the following error (Fig. 2.68).

This was because the firmware upgrade was designed for cDAQ 9215 module, whereas the one linked to the chassis was cDAQ 9234, which was not compatible with upgraded firmware.

2.2.8.8 Wireless DAQ to Smartphone Using Laptop as an Intermediate

This requires the Data Dashboard App to be installed on the smartphone. There are two different versions of Data Dashboard App for iPad and for Android smartphone.

Step 1 Connect the chassis to the laptop using an Ethernet cable. In the network settings, select "Connect to wireless network" in the Wireless mode menu. Choose an available network (as in this example "iitk" network). Let the "Configure IPv4 address" be "DHPC or Link Local" (Fig. 2.69).

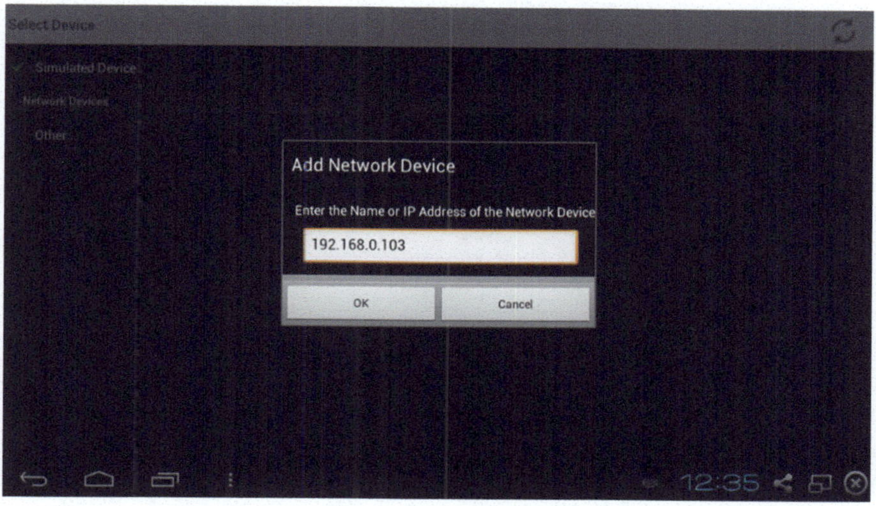

Fig. 2.67 Step 1. Image Courtesy: NI

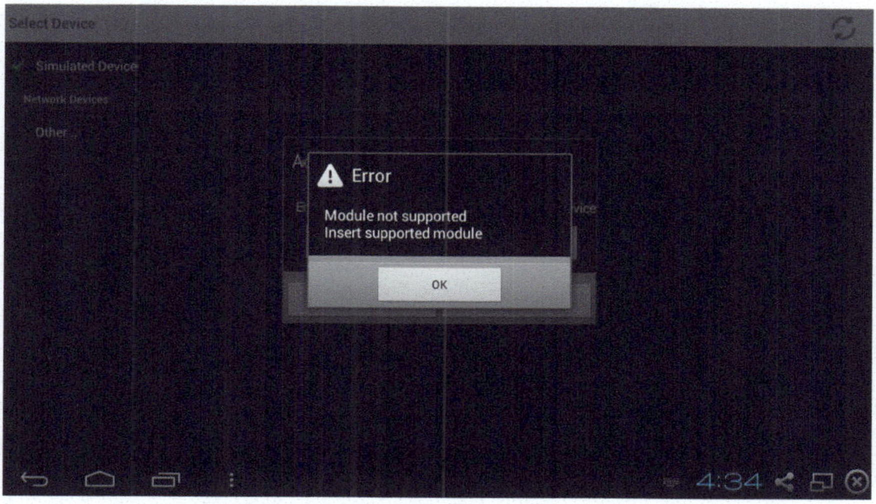

Fig. 2.68 Step 2. Image Courtesy: NI

Step 2 Save the settings to let them take effect, and then select "static" from the "Configure IPv4 address" (Fig. 2.70).

Now, unplug the Ethernet cable and connect the laptop to the same network, i.e., "iitk". Connect the smartphone to "iitk" as well. The chassis sends data to the laptop via "iitk" network. The laptop takes the data in the form of a waveform and with the help of a LabVIEW program, puts the waveform in a variable. This variable is of "Graph" data type depending upon the data dashboard version. As Android version

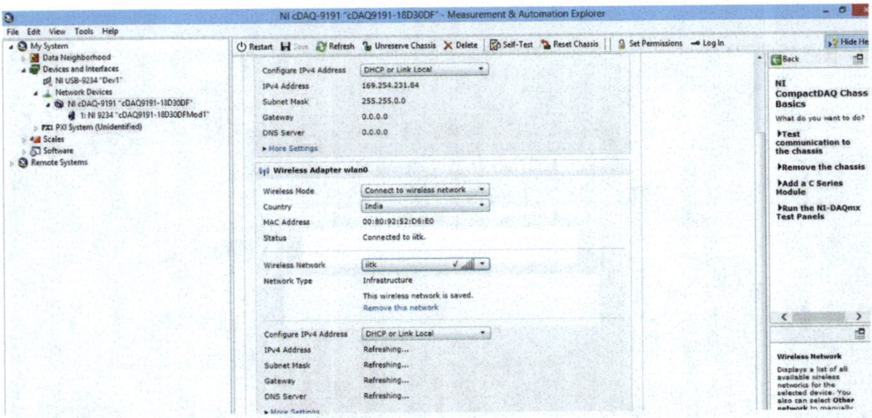

Fig. 2.69 Step 1. Image Courtesy: NI

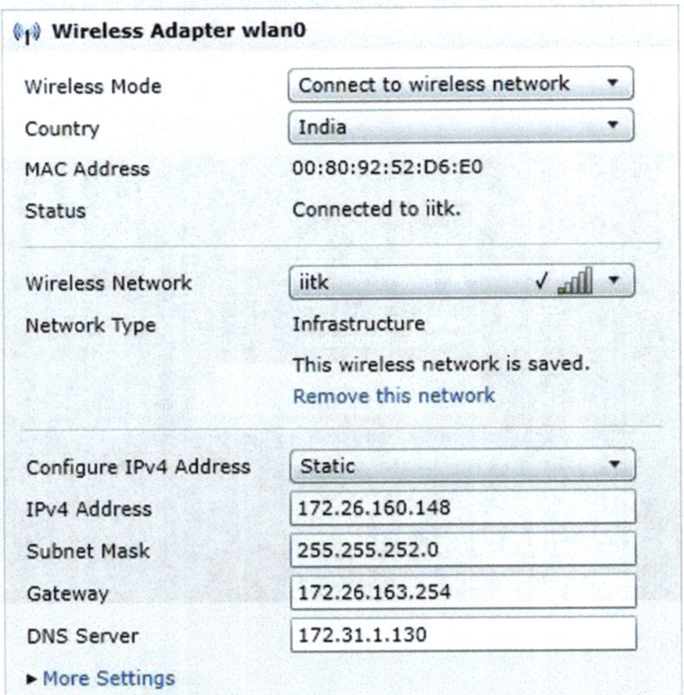

Fig. 2.70 Step 2. Image Courtesy: NI

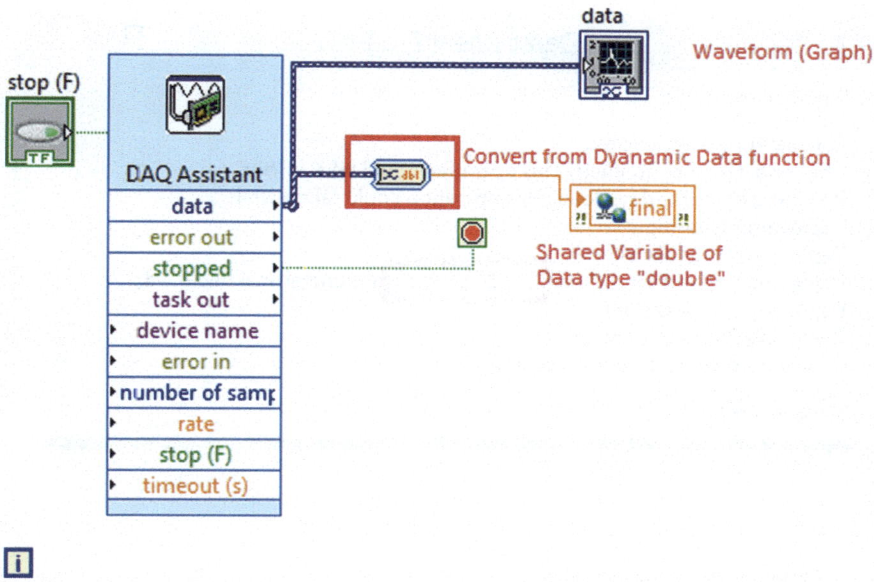

Fig. 2.71 Step 3. Image Courtesy: NI

1.0 does not support graphs so the data (which is of graph data type) is converted to yield a scalar data type, which is put into the shared variable of scalar data type. The shared variable is then deployed on the network, whereas iPad version 2.0 can access the shared variables as it supports graph data type.

Step 3 The waveform is converted to a scalar data type like a "double" using the "Convert from dynamic data" function in LabVIEW (Fig. 2.71).

Step 4 This is put into a shared variable which is then deployed over the network the laptop is connected to (iitk). Note the IP address from the deployment dialogue box shown in Fig. 2.72.

Step 5 Open the Data Dashboard App in the phone and select "Add" (Fig. 2.73).

Step 6 Select "Connect to shared variable" (Fig. 2.74).

Note For the deployment to be successful, the firewall must be turned off on the machine that hosts the shared variable. This is because the firewall blocks all the outgoing connections that are not trusted. In such a sample, the Android app will give the following error (Fig. 2.75).

Step 7 Enter the IP address noted down in Step 4. This would give the list of all the deployed libraries on the server (Fig. 2.76).

Step 8 Select the one which has the shared variable, and run the simulation. The number of samples read, sampling rate, etc., can be configured from the DAQ Assistant in LabVIEW (Fig. 2.77).

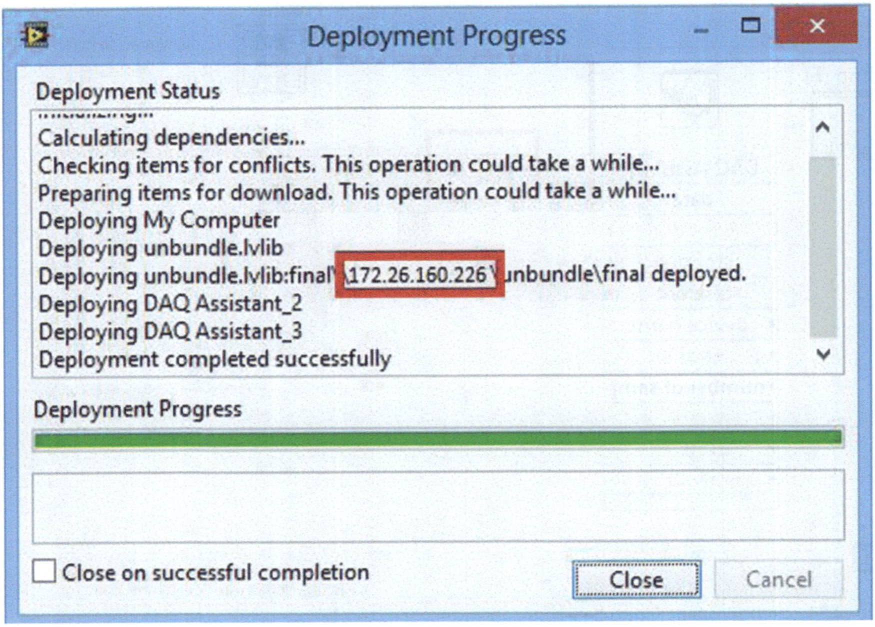

Fig. 2.72 Step 4. Image Courtesy: NI

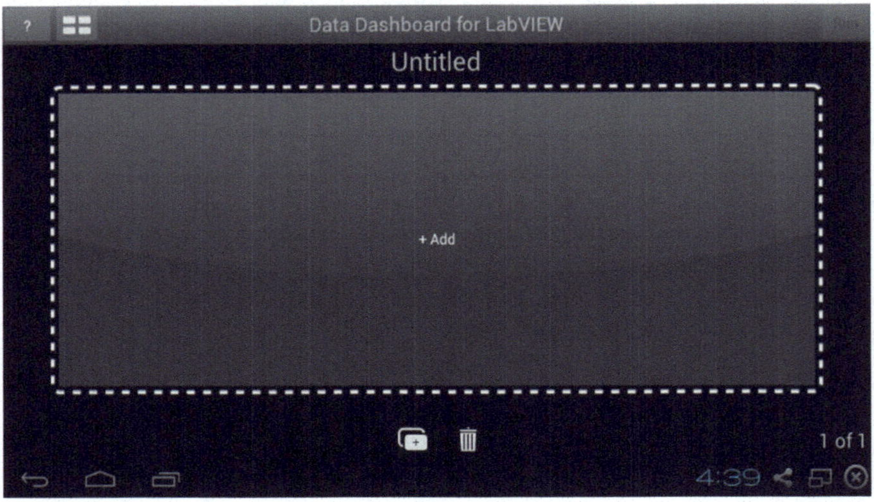

Fig. 2.73 Step 5. Image Courtesy: NI

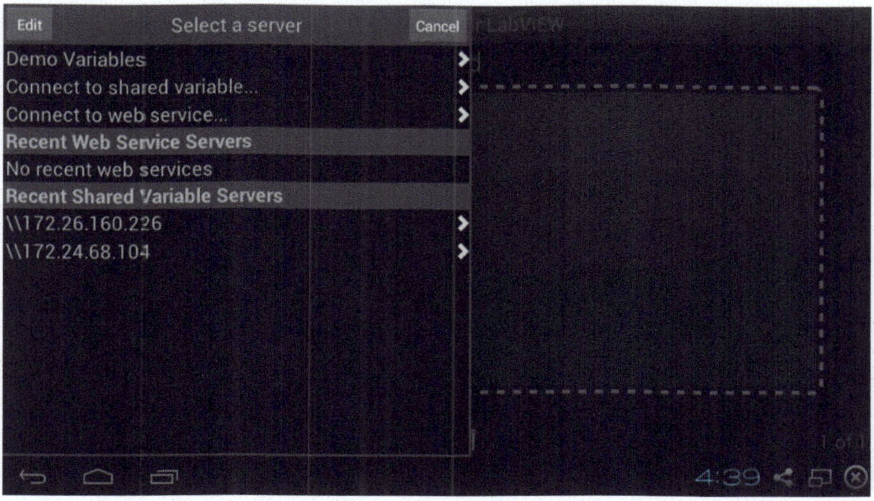

Fig. 2.74 Step 5. Image Courtesy: NI

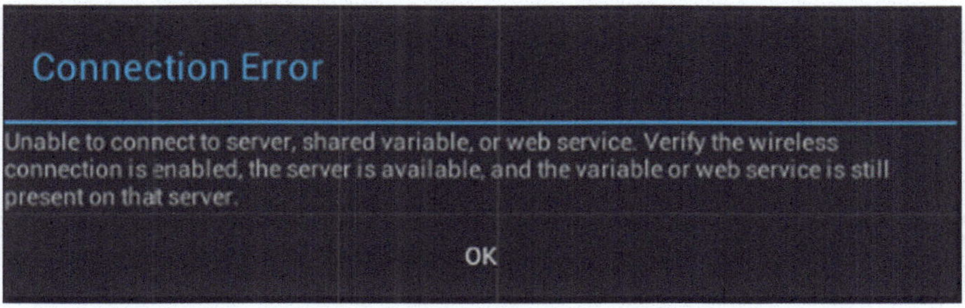

Fig. 2.75 Error in case of unsuccessful deployment. Image Courtesy: NI

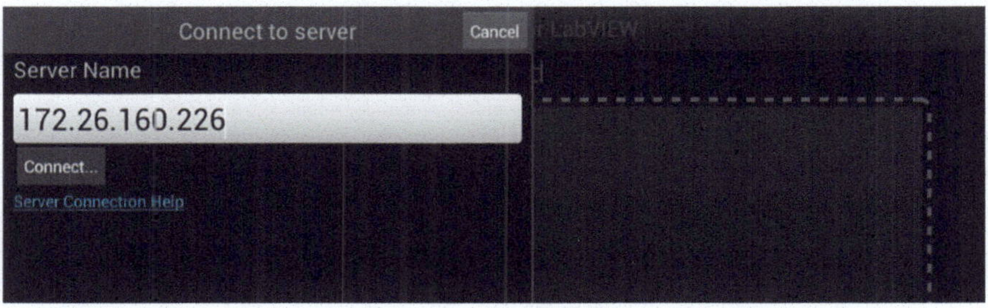

Fig. 2.76 Step 7. Image Courtesy: NI

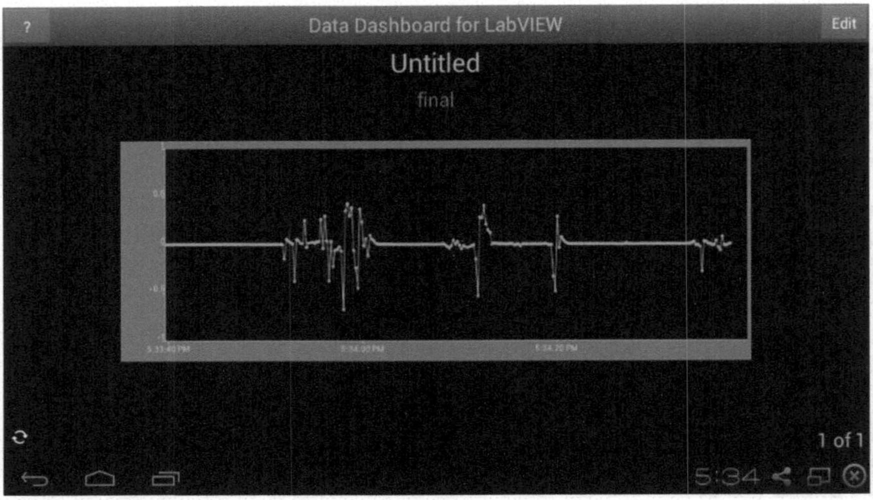

Fig. 2.77 Step 8. Image Courtesy: NI

2.2.8.9 Vibration Data Acquisition using Smartphone

For vibration DAQ on smartphone, initially smartphone's accelerometer was taken. This was a quite tricky problem because smartphone must have an appropriate hardware and software interface to acquire accelerometer data from DAQ module. Then we moved towards other DAQ hardware interfaces as discussed in subsections.

There could be different methods for acquisition of data ranging from ranging from option of wireless sensor to wireless DAQ interface.

- Smartphone
- Wireless sensor
  - Wireless sensor RH503
  - Wireless sensor IMI (model 670A01)
- Wireless data acquisition device NI 9191
- Industrial Solutions
  - YEI Space Sensor
  - BTH 1208LS
- Constant Current Line Drive (CCLD) signal conditioner
- USB accelerometer.

2.2.8.10 DAQ Using Smartphone's Accelerometer

One of the most widely used applications of smartphone is to use the inbuilt sensors like accelerometer gyroscope. Recently, many researches have demonstrated that the usage of dedicated accelerometers can provide good results in the area of activity recognition. Using above fact, we have also explored the possibility of acquiring data from air compressor directly. Inbuilt accelerometer measures acceleration in the units of m/s^2 or g. Main drawback of inbuilt sensor is low resolution. Constant use of smartphone for acquiring training data may damage smartphone itself. As running air compressor vibrates at much high rate, which constantly hits its surface. Second most important concern was the accuracy of this measurement, which may not be with high precision, and hence, this will severely affect the fault detection accuracies. In nutshell, smartphone accelerometer was not an efficient solution.

2.2.8.11 DAQ Using Wireless Sensor

Main idea behind this solution is to acquire vibration data through wireless accelerometer and then to transfer this data via wireless to smartphone. Figure 2.78 depicts the basic flow of DAQ using this method.

Wireless Sensor RH503 It is a vibration and temperature sensor, which can transmit data wirelessly. It is especially used for data acquisition through widespread measurement points. Currently, it is being used for onsite implementation of anti-explosion fields. It can collect data with capacity of 6400 lines spectrum. Its frequency ranges up to 2.4 GHz for communication transmissions. It is designed for lower power consumption. Due to easy maintenance and small size, it is much suitable for industrial workstation. Apart from it also comes up with high-capacity flash memory (Fig. 2.79).

Wireless Sensor IMI (Model 670A01) It is a standalone and battery-powered industrial vibration sensor as shown in Fig. 2.80a. It gives raw vibration data as output. It is software programmable where user can set limits for transmission interval and residual vibration level via RS-232 programming. It also incorporates embedded magnetic switch, whose activation assures transmission of signal. To use IMI accelerometer, a wireless junction box is required as shown in Fig. 2.80b.

(a) Wireless sensors (b) Portable computing device

Fig. 2.78 Vibration data acquisition using accelerometer. Image Courtesy: IDEA Lab, IIT Kanpur

Fig. 2.79 ROZH RH503 wireless accelerometer

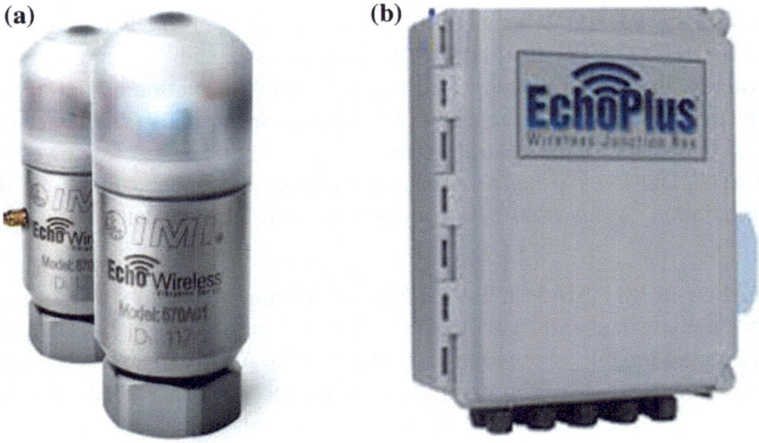

Fig. 2.80 **a** IMI 670A01 wireless accelerometer and **b** IMI echo wireless junction box

Main challenge is to use these sensors with smartphone as currently all of them support only Desktop Windows platform. There are a number of wireless sensors which can transfer data directly from sensor to the device using customized application software.

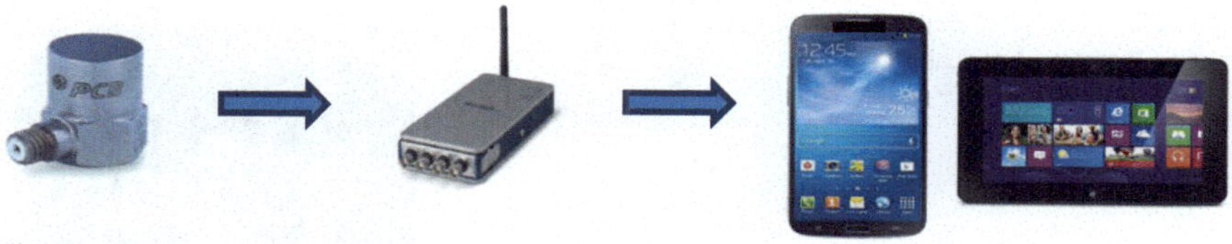

Fig. 2.81 Vibration data acquisition using NI 9191. Image Courtesy: IDEA LAB, IIT Kanpur

2.2.8.12 DAQ Using Wireless Sensor Acquisition Device NI 9191

Instead of using wireless sensor, we can also use wireless data acquisition module which can capture data from dedicated sensor and transfer it to portable computing device. Figure 2.81 shows the example of the same.

2.2.8.13 Industrial Solutions

Apart from these wireless sensors, in market, there are some complete vibration data acquisition devices which can capture data and store or transfer data later using SD card or standard protocol.

Wireless YEI Space Sensor It is a combination of different sensors, namely three-axis accelerometer, three-axis gyroscope, three-axis compass, and high-precision, high-reliability, inertial measurement units (IMU). It contains on-board processing and filtering algorithms to provide exceptionally accurate altitude and heading information based on inbuilt sensors and Micro-SD card as storage devices and a rechargeable lithium-polymer battery. Exceptionally accurate altitude and heading information based on the proprietary multireference vector mode increases accuracy.

Wireless BTH 1208LS DAQ BUNDLE It acquires data over Bluetooth® or USB connection. In this eight 11-bit SE or four 12-bit DIFF, analog inputs can be given. It provides 1 kS/s sampling over Bluetooth, 50 kS/s sampling over the USB connection. Also, it facilitates with two 12-bit analog outputs, eight digital I/O and battery or USB power options. TracerDAQ® software can be used for acquiring and displaying data and generating analog signals, InstaCal software utility for installing, calibrating, and testing. Universal Library includes support for Visual Studio® and Visual Studio® .NET. Universal Library for Android™ includes support and examples for the Android 3.1 platform (API level 12) and later while ULx for NI LabVIEW™ (Figs. 2.82 and 2.83).

Fig. 2.82 YEI Space Sensor. Image Courtesy: YostLabs

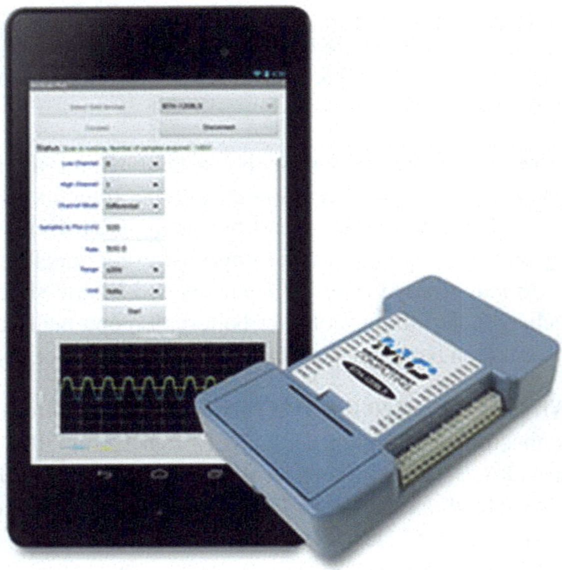

Fig. 2.83 MCC BTH 1208LS DAQ bundle

2.2.8.14 DAQ Using CCLD Signal Conditioner

This is one of the most feasible methods for vibration data acquisition using smartphone and tablets. This method does not support wireless data acquisition. The basic model of this technique is shown in Fig. 2.84. To capture data from the sensor, we have used a signal conditioner CCLD for the acquisition of vibration data. The output

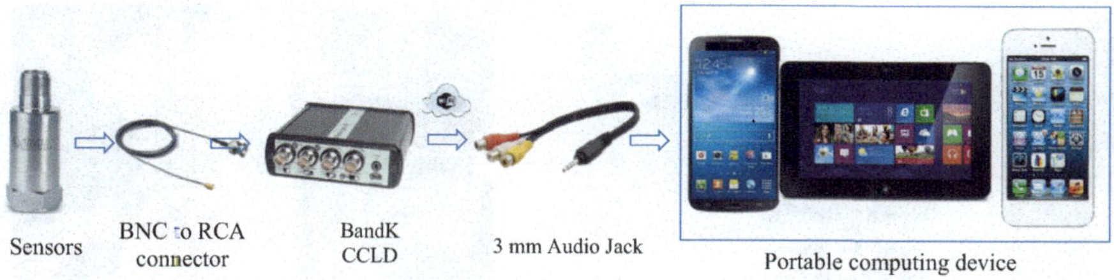

Fig. 2.84 Vibrational data acquisition using CCLD. Image Courtesy: IDEA Lab, IIT Kanpur

of signal conditioner supports audio which can be fed into smartphone using 3bnc to the female jack.

Specification of CCLD Brüel and Kjær CCLD model no: 1704-A-002 provides power for accelerometers, microphones. It comes with LED overload and CCLD cable fault detection. It provides BNC outputs for connection with collecting device. It also provides 3.5-mm stereo socket for soundcard connection. It has a special switch named CCLD on/off which can be used without CCLD power as voltage amplifier and filter. It also provides gain and filters to improve the dynamic range. It has 13-h battery life and USB Micro-B for powering and charging.

2.2.9 DAQ Using USB Digital Accelerometer

It is a single-axis plug and play piezoelectric accelerometer with integrated internal digital data acquisition. Accelerometer has flat response up to 8 kHz. It supports audio sampling rates 48, 44.1, 32, and 16 kHz. Using Android app, we can record audio from an external sound card and store it on the Android file system.

2.3 Experimental Setup

Multiple sensory mechanisms are installed for data collection process to capture fault with higher accuracy. In the following sections, figures have been included which show different positions of microphones to capture the sound signal, accelerometers for vibrations, and the position of the CCD laser for the compressor displacement measurement (Fig. 2.85).

Positions of microphones on compressor are listed below:

- M1: This is considered as the first position. The mic is placed one centimetre away from crankcase on the NRV side.
- M2: This is the second position. The mic is placed one centimetre away from cylinder casing opposite flywheel side (slightly towards NRV side).

Fig. 2.85 Data acquisition setup for microphones. Image Courtesy: ELGI

- M3: In the third position, the mic is placed on the top, one centimetre away from the centre of the cylinder casing.
- M4: This is the fourth position. The mic is placed one centimetre away from crank case opposite NRV side.

Positions of accelerometers on compressor:

- A1: This is considered as the first position. The accelerometer is placed in contact with the crankcase on the NRV side.
- A2: This is the second position. The accelerometer is placed in contact with the crankcase on the piston head.
- A3: This is the third position. The accelerometer is placed in contact with the crankcase opposite to the NRV.
- A4: The fourth position is the opposite side of the pulley. The accelerometer is placed in contact with the crankcase opposite to the pulley (Figs. 2.86, 2.87, and 2.88).

Positions of CCD laser on compressor:

The CCD laser is placed above the cylinder casing, like the third position of the microphone (Refer Figs. 2.89, 2.90, 2.91, and 2.92).

Fig. 2.86 Data acquisition setup for accelerometers. Image Courtesy: ELGI

2.3.1 Sensitive Positions on Machine

2.3.1.1 Introduction

Generally, fault diagnosis of a system involves small data acquisition from various positions of the machines; hence, the process can become too complex, especially for machines which have very large structure. It is practically not possible to study the whole machine; it is more efficient to find best position and then take recordings at those positions. To get best performance w.r.t. accuracy, it is crucial to locate sensitive position on the machine before acquiring data for fault diagnosis. There exists a significant position on and around the surface of machine in respective healthy and faulty conditions, where sensor can pick maximum information. This position is named as sensitive position. These positions are expected to have better

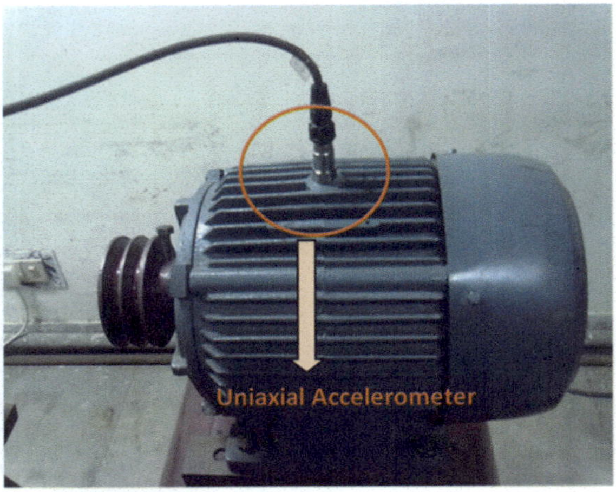

Fig. 2.87 Uniaxial accelerometer placed on top of motor frame. Image Courtesy: ELGI

Fig. 2.88 Triaxial accelerometer placed on motor. Image Courtesy: ELGI

difference in feature values, when the state of machine changes between healthy and faulty state. Following are the methods to find sensitive position on machine:

- Statistical parameter-based method
- Ranking- and correlation-based method
- EMD- and Hilbert transform-based method.

2.3.1.2 Placement of Sensors

To determine sensitive position, at first sensors are placed on different positions on machine. To illustrate the whole process here, air compressor has been taken as

Fig. 2.89 CCD laser setup on ELGI compressor. Image Courtesy: ELGI

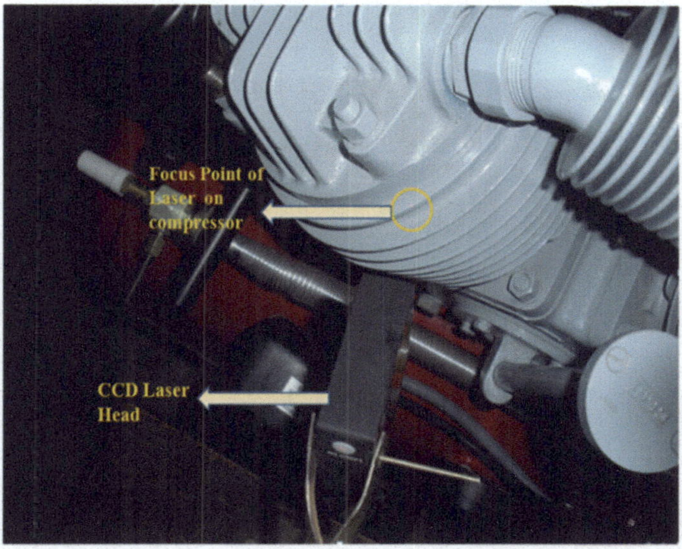

Fig. 2.90 Top view of CCD laser setup. Image Courtesy: ELGI

case study. For convenience, large structure of machine is divided into four sides to represent each position. The four sides chosen were the top of the piston head case, the NRV side crank case, the position opposite the flywheel side crank case, and the one opposite the NRV side crank case as shown in Figs. 2.93 and 2.94 for single-stage air compressor and ELGI air compressor, respectively.

A total of 44 positions were considered based on surface area available on machine and the base area of the sensor. The piston head side consisted of 5 positions, the side

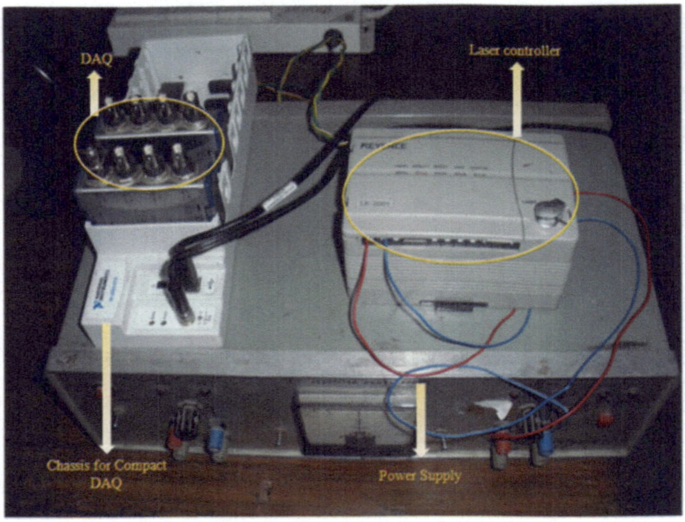

Fig. 2.91 Power source and controller for CCD. Image Courtesy: Keyence

Fig. 2.92 Computer setup for data acquisition. Image Courtesy: Keyence

along the NRV side consisted of 13 positions, the side opposite to the NRV consisted of 12 positions, and the side opposite to the flywheel consisted of 14 positions. Since the extreme corner positions on air compressor are affected by external noise (e.g. the sound emitted from the flywheel while the compressor is running, noise from the motor and exhaust, etc.) and due to this, we modified the number of sensor placements to the total of 24 positions. After processing at each side, six positions were identified where sensor can be placed. As shown in Figs. 2.95, 2.96, 2.97,

Fig. 2.93 Single-stage air compressor. Image Courtesy: ELGI

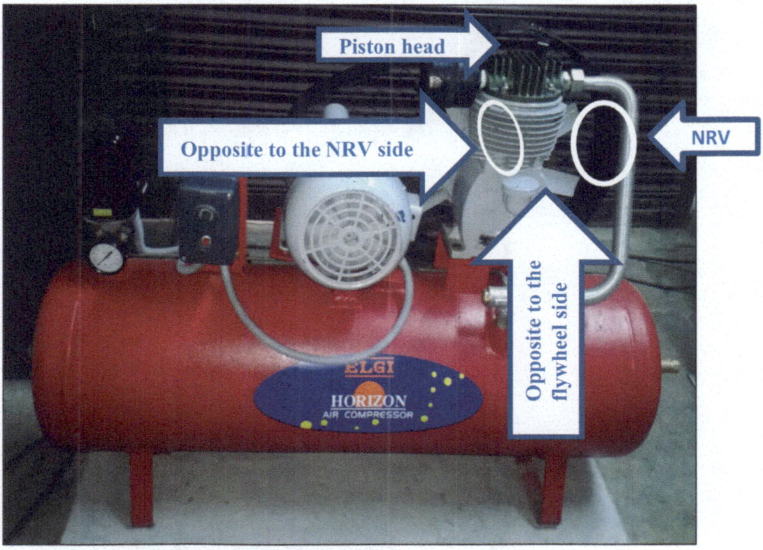

Fig. 2.94 ELGI air compressor. Image Courtesy: ELGI

Fig. 2.95 Sensor positions on piston head. Image Courtesy: ELGI

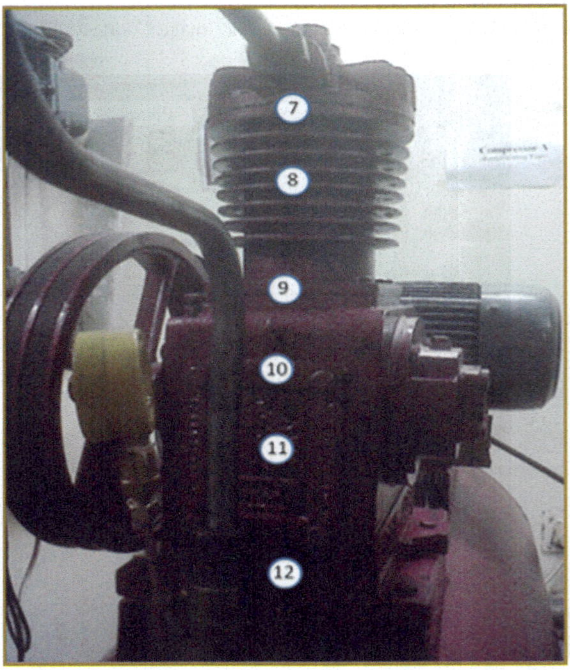

Fig. 2.96 Sensor positions on NRV side crank case. Image Courtesy: ELGI

Fig. 2.97 Positions of sensor opposite to NRV side. Image Courtesy: ELGI

and 2.98, sidewise selected positions for placing accelerometer on single-stage air compressor. Microphones were placed at 1 cm away from their respective positions. The signals were obtained from various positions and processed. Thus, in total we got 24 sensor positions where accelerometers were placed in contact with the compressor to acquire vibration signals.

SIDE I—Top of the piston head—six positions

P1—Microphone is placed at the centre of the right nut piston head (right of the flywheel side).
P2—Microphone is placed at the centre of the left nut piston head (left of flywheel side).
P3—Microphone is placed near the suction end.
P4—Microphone is placed at the outlet pipe at the piston head.
P5—Microphone is placed 1 cm above the heat sink of the piston head.
P6—Microphone is placed at the centre of the two nuts on the piston head casing.

SIDE II NRV side crank case (six positions)

P1—Microphone is placed at the centre of the crank case on the NRV side.
P2—Microphone is placed at the extreme end of the cylinder on the NRV side.
P3—Microphone is placed at the centre of the cylinder on the NRV side.

Fig. 2.98 Positions of sensor opposite to flywheel side. Image Courtesy: ELGI

P4—Microphone is placed at the lower end of the crank case on the NRV side.
P5—Microphone is placed below the extreme end of the cylinder (below position 2) on the NRV side.
P6—Microphone is placed just below the outlet pipe on the NRV side.

SIDE III Opposite NRV side (six positions)

P1—Microphone is placed at the centre of the crank case on the opposite side of NRV.
P2—Microphone is placed at the extreme end of the cylinder opposite to the NRV side.
P3—Microphone is placed at the centre of the cylinder, opposite to the NRV side.
P4—Microphone is placed at the lower end of the crank case opposite to the NRV side.
P5—Microphone is placed below position 2 which is opposite to the NRV side.
P6—Microphone is placed right below the suction end.

SIDE IV Opposite Flywheel side (six positions)

P1—Microphone is placed at the centre of the crank case on the opposite side of the flywheel.

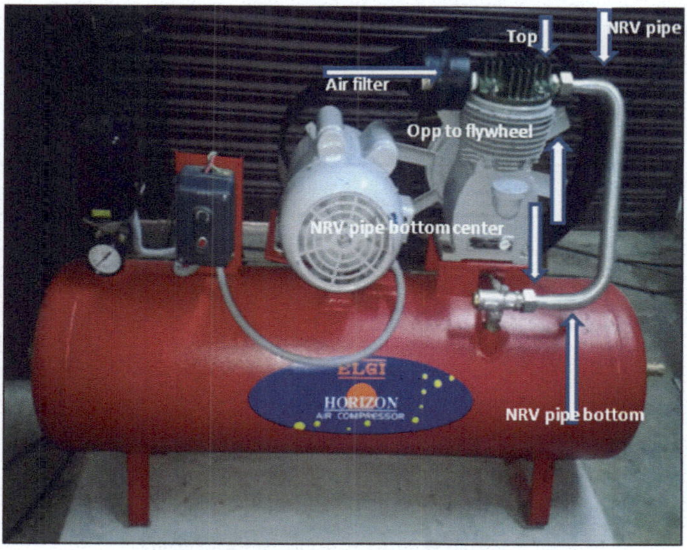

Fig. 2.99 Sensitive positions of ELGI air compressor. Image Courtesy: ELGI

P2—Microphone is placed below the protruding part of the crank case opposite the flywheel side.
P3—Microphone is placed above the protruding part of the crank case opposite to the flywheel side.
P4—Microphone is placed at the extreme end of the cylinder opposite to the flywheel side.
P5—Microphone is placed at the centre of the cylinder opposite to the flywheel.
P6—Microphone is placed at the lower end of the crank case opposite to the flywheel side.

In similar way, for ELGI air compressor, six positions were identified (Fig. 2.99).

P1—Near air filter
P2—Top of head
P3—Opposite to flywheel
P4—NRV pipe top
P5—NRV pipe bottom
P6—NRV pipe bottom centre.

2.3.2 Sensitive Position Using Statistical Parameters

The data samples obtained from various positions are then processed. This method proposes a scheme to determine the most sensitive area using multiple statistical

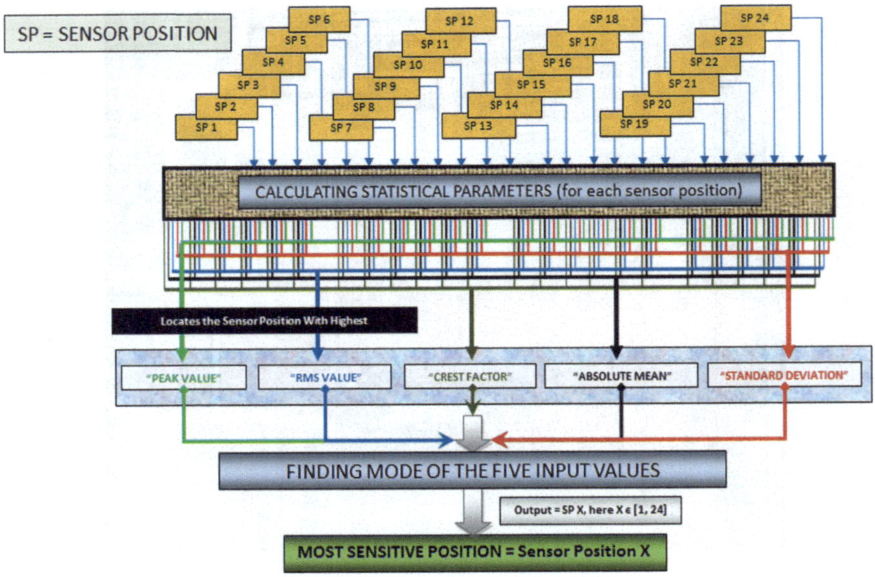

Fig. 2.100 Flow chart of algorithm [6]

parameters, while the system is in different state, i.e., healthy or faulty state. These sensitive positions were determined according to the statistical parameters peak, standard deviation, variance, and the RMS values [6, 7]. Individual positions on each side corresponding to the highest peak are found. The same is done with standard deviation, variance, and RMS values as parameters. The position that ranks higher for most of these parameters is considered as the sensitive position of that respective side. Next step is cross-correlation analysis among sensitive positions of different sides which is additionally performed to avoid redundancy in recordings (Fig. 2.100).

2.3.3 Sensitive Position Using Ranking and Correlational Analysis

In this method, the whole air compressor has been considered as a single unit and has not been divided into four sides [8, 9]. A total of 24 positions have been intuitively selected on the air compressor. All 24 positions were ranked based on peak, absolute mean, standard deviation, and RMS values. The individual ranks for all four parameters were summed up, and then the sensitive positions were ranked based on the final sum. Cross-correlation analysis was performed here, as well to ensure that highly correlated signals are not ranked together at the top of the list. This method is intuitively good except for the fact that these statistical parameters are independent of the actual signal. Hence, this method is highly susceptible to noise. Steps for performing sensitive position analysis were as follows.

Step 1 Raw acoustic data emanating from the machine under running conditions is recorded from several positions of the machine. Acoustic data is recorded for 5 s, at a sampling rate of 50 kHz in ".dat" format.

Step 2 Raw data is then pre-processed with a no. of pre-processing techniques.

- **Clipping**—Clipping of the downsampled version of raw data is done to select the best 1 s data out of 5 s, which has a minimum standard deviation. Minimum standard deviation ensures a relatively more stable signal.
- **Filtering**—Appropriate frequency filters are applied for removal of noise.
- **Smoothing**—A simple averaging filter is used to smooth the data. It is used for reducing the effect of outliers.

Step 3 Next step is to calculate statistical properties, i.e., standard deviation, absolute mean, peak, and root mean square from pre-processed data. The average value of each of these four parameters is calculated over a set of recordings at each position. Positions are then ranked in descending order of the average. The ranking is done independently for each of four parameters, thus giving us four sets of rankings. To consider the effect of all four parameters, four ranks are summed up for each position. The positions are then finally ranked in ascending order of the sum. Cross-correlation analysis has been further done to avoid highly correlated positions to be together at top of the rank list (Fig. 2.101).

2.3.4 Sensitive Position Using Empirical Mode Decomposition (EMD)

This method is more robust w.r.t earlier methods [10]. It is one of the most robust methods for finding sensitive position rankings. The method involves extraction of relevant signal using EMD [11], envelope calculation of relevant signal using Hilbert transform [12], and finally, calculation of sensitive position ranking by performing statistical analysis on the resultant envelope.

Empirical Mode Decomposition EMD was a technique first introduced by Huang et al. which can study non-stationary and nonlinear signals. It decomposes the original signal into its intrinsic mode functions (IMFs), based on local time scale filtering of the original signal. Intrinsic mode functions are locally filtered oscillatory signals that have same number of extrema as the number of zero-crossings. The local time scale here is the scale of one oscillation, which is taken as the time scale between local minima and local maxima. EMD decomposes the IMFs in such a manner that the initial IMFs have a higher frequency of oscillation than the latter ones. It is

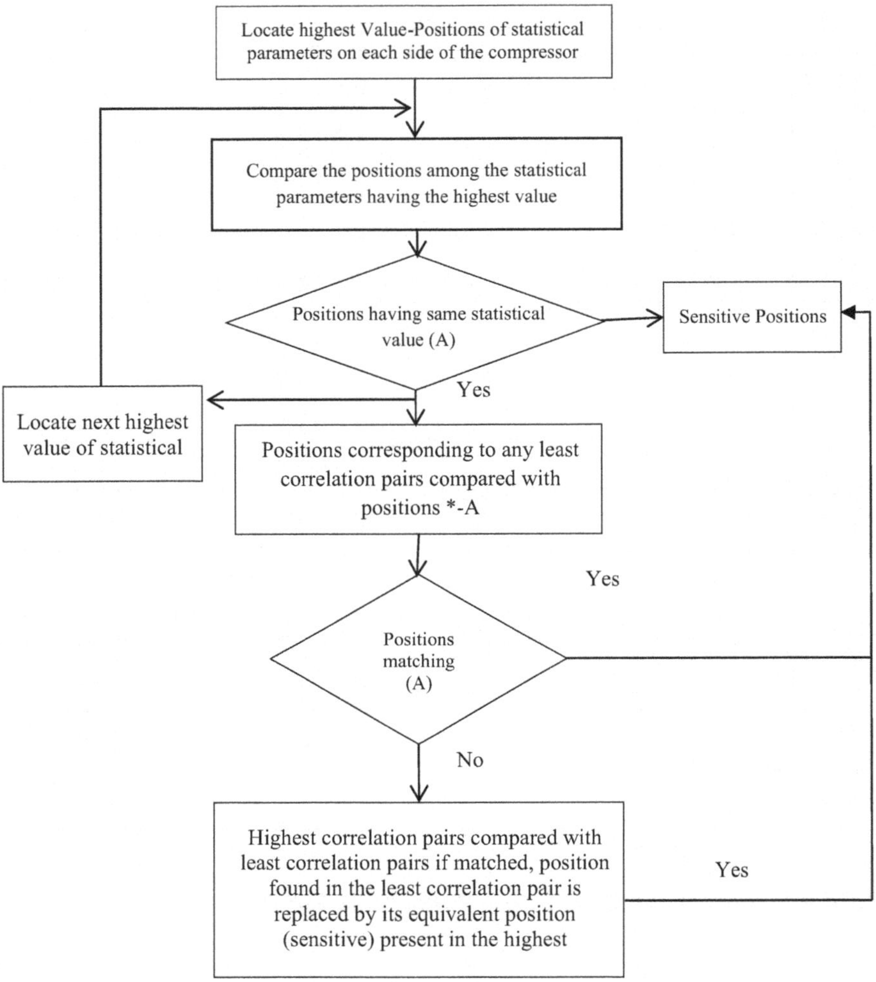

Fig. 2.101 Flow chart [8]

completely data driven. EMD is prone to numerical errors which may occur due to the approximation of spline interpolation for the upper and lower envelopes of the acoustic signal. It appears in extra IMFs which need to exclude from further analysis. For exclusion, correlation of IMFs is computed with original signal higher correlation with original signal would have a richer content of information, and thus, relevant IMFs are found.

Algorithm of finding of sensitive positions using EMD:

Step 1 Take *N* data recordings from different positions, and repeat Steps 2–7 for each data recording.

Step 2 Pre-process the data recording.

Step 3 Use EMD's sifting process to decompose the signal into its IMFs.

$$A = \{imf_1, imf_2, \ldots, imf_n\}$$

$$X(t) = \sum_{i=1}^{n} A + residue$$

Step 4 Select relevant IMFs

for $j = 1 : |A|$

a. find $corr_j = correlation\left(imf_j, x(t)\right)$

b. if $corr_j > threshold$

add $A' \leftarrow \{imf_j\}$

end

Step 5 Reconstruct the signal by summing only the IMFs present in

$$y(t) = \sum A'$$

Step 6 Find the envelope of the reconstructed signal using the Hilbert transform.

Step 7 Calculate two statistical parameters from the reconstructed signal's envelope: the root mean square (RMS) and the absolute statistical mean.

Step 8 In a similar way, find the two statistical parameter values for all recordings of a sensor position and take the statistical average of both parameters across all recordings. Similarly, find the statistical average of the two parameters for all sensor positions individually.

Step 9 Rank all the positions in descending order of individual parameter values.

Step 10 Sum the ranks of individual parameters for each sensor position and rank the positions in ascending order of the sums. The position ranked highest is the most sensitive position.

Graphical User Interface for Sensitive Position

See Figs. 2.102 and 2.103.

2.4 Case Study on Single-Stage Air Compressor

For fault diagnosis purpose, two types of machine data, i.e., acoustic and vibration, were collected. The vibration data was collected with an accelerometer (model # 6030C01), while the acoustic data was sampled with a unidirectional microphone (model # UTP-30). In total, 24 positions were selected to place sensors as mentioned earlier. The total data recorded for analysis = 50 k samples/sec × 5 s (time duration for one reading) × 5(number of readings per pressure slot) × 15 (total number pressure slots). Hence, the total number of data recorded for finding the sensitive positions in every case is 18,750 k. Once all data was collected, different statistical parameters, i.e., peak, standard deviation, root mean square, were calculated, and

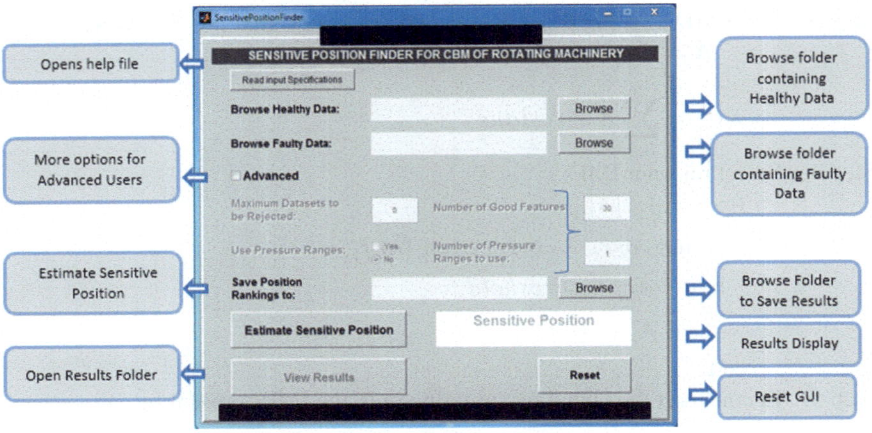

Fig. 2.102 Home page of sensitive position [7]

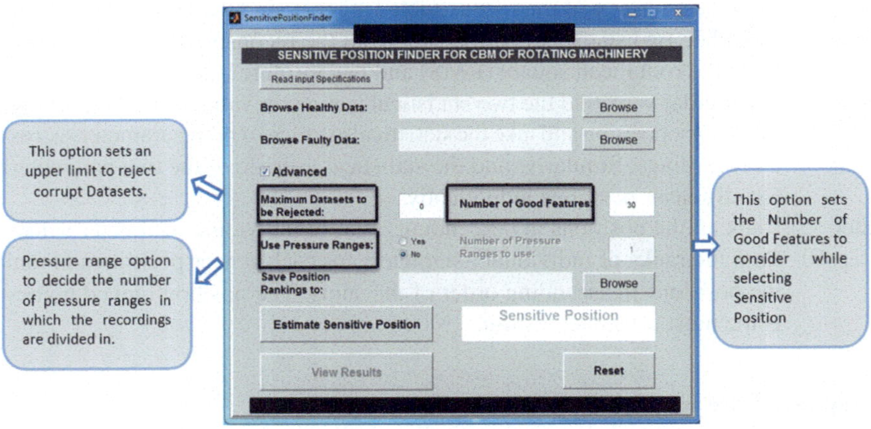

Fig. 2.103 GUI for sensitive position finder [7]

find the positions having the highest values of parameters corresponding to each side was identified.

From experimentation, NRV side is position 4 at which the accelerometer is mounted slightly upper from the centre of the crankcase. Similarly to the piston head side, position 3 is the highest and the accelerometer at this place was mounted on the right nut bolt (right, considering the flywheel side). On the side, opposite to the NRV it was position 4 and the accelerometer was found to be mounted at the centre of the crankcase, and on the side opposite to the flywheel, it was position 3 which was found to be 4 cm below from the centre of the crankcase. So, observing the above highest values their corresponding positions have been taken to be the most sensitive positions. Microphone placed at the centre of the crankcase is position 1 on

the NRV side, microphone placed on the centre of the left nut is position 2 on top of the piston head. On the side opposite to the NRV, position 6 is taken to be just below the suction end, and on the side opposite to the flywheel position 5 is at the centre of the heat sink of the cylinder. It was noticed that position 5, which holds the highest peak value on the side opposite to the flywheel, was affected by the exhaust. Hence, we had to discard this position and choose the best position, which was not affected by the exhaust. The best position according to the crest factor values was chosen to be position 6, after discarding three positions which were affected by exhaust. Thus, according to the above analysis, we obtained the positions on an air compressor using crest factor values as a parameter. Combining the analyses of all the above-mentioned statistical parameters, we observed that the sensitive positions obtained using peak, variance, standard deviation, and RMS were same. Therefore, we chose these parameters as the final parameters of our analysis. Crest factor was not taken into consideration as it did not provide satisfactory results. Hence, we concluded the most sensitive positions to be as follows:
For accelerometer sensors:

- NRV side: position 4
- Top of the piston head: position 3
- Opposite to the NRV side: position 4
- Opposite to the flywheel side: position 3

For microphone sensors:

- NRV side: position 3
- Top of the piston head: position 1
- Opposite to the NRV side: position 4
- Opposite to the flywheel side: position 6.

2.4.1 Sensitive Position Using Ranking and Correlational Analysis

This method was incorporated for finding the measure of similarity between signals from various sensor positions. Our interest was to observe the final sensitive positions that we have chosen using various sensors. Correlation is a statistical measurement of the relationship between two variables. It gives the measure of similarity of two waveforms. It is usually categorized into two types: (a) auto-correlation and (b) cross-correlation. In simple words, auto-correlation is calculated between a signal and a shifted version of itself. Cross-correlation is calculated between two different signals. The correlation will be reported as a number between 0 and 1. A value of 0 implies that there is no correlation. A value of 1 is a perfect positive correlation which does not happen. A value of -1 is a perfect negative correlation which does not happen either. As the values get closer to 1 (either $+1$ or -1), it implies that the measure of similarity is increasing. Correlation coefficients show the strength of the linear relationship between two variables. If we have a series of measurements of X and Y, then the correlation "r" between X and Y can be calculated as follows:

$$r_{xy} = \frac{\sum_{i=1}^{n} (x_i - \overline{x})(y_i - \overline{y})}{(n-1)S_x S_y}$$

where $\overline{x}$ and $\overline{y}$ are sample means of X and Y, respectively; S_x and S_y are the standard deviations of X and Y, respectively.

In our analysis, we have represented the sensor positions according to four different sides of the compressor, as mentioned before. So according to each side, the correlation was determined among various sensor positions. Then consider the number of positions on each side, we calculated the number of all possible permutations accordingly. The results obtained show the signal pairs (or pair of positions) and the correlation between them. This helps us reduce the number of sensor positions, as we can consider only one position in a pair which has high similarity. For example, for a pair of position 3 and position 4, if the measure of similarity is high, we can either consider 3 or 4, instead both. If the similarity is low, it implies that the signals are dissimilar and both the positions in a pair must be considered. So, based on the above explanation, we have analyzed different sets of positions on each side.

The most sensitive positions using accelerometer transducer are as follows:

- Top of the piston head: position—P3
- NRV side: position—P4
- Opposite to flywheel: position—P3
- Opposite to NRV side: position—P4

The most sensitive positions using microphone transducers are as follows:

- Top of the piston head: position—P1
- NRV side crankcase: position—P3
- Opposite NRV side: position—P4
- Opposite flywheel side: position—P6 (Refer Figs. 2.104 and 2.105).

Fig. 2.104 Most sensitive positions on an air compressor using accelerometers under healthy condition [7]

Fig. 2.105 Most sensitive positions on an air compressor using microphones under healthy condition [7]

References

1. Verma, N.K., Sevakula, R.K., Dixit, S., Salour, A.: Intelligent condition based monitoring using acoustic signals for air compressors. IEEE Trans. Rel. **65**(1), 291–309 (2016)
2. Verma, N.K., Sarkar, S., Dixit, S., Sevakula, R.K., Salour, A.: Android app for intelligent CBM. In: 22nd IEEE Symposium on Industrial Electronics, Taipei, Taiwan, pp. 1–6 (2013)
3. Verma, N.K., Singh, J.V., Gupta, M., Sevakula, R.K., Dixit, S.: Windows mobile and tablet app for acoustic signature machine health monitoring. In: 9th International Conference on Industrial and Information Systems, Gwalior, India, pp. 1–6 (2014)
4. Verma, N.K., Dev, R., Dhar, N.K., Singh, D., Salour, A.: Real-time remote monitoring of an air compressor using MTConnect standard protocol. In: IEEE International Conference on Prognostics and Health Management, Texas, USA, pp. 109–116 (2017)
5. Verma, N.K., Sharma, T., Maurya, S., Singh, D., Salour, A.: Real-time monitoring of machines using open platform communication. In: IEEE International Conference on Prognostics and Health Management, Texas, USA, pp. 124–129 (2017)
6. Verma, N.K., Jagannatham, K., Bahirat, A., Shukla, T., Subramaniam, T.S.S.: Statistical approach for finding sensitive positions for condition based monitoring of reciprocating air compressors. In: Proceedings of IEEE Control and System Graduate Research Colloquium Incorporating 2011 IEEE International Conference on System Engineering and Technology, Selangor, Malaysia, pp. 10–14 (2012)
7. Verma, N.K., Jagannatham, K., Bhairat, A., Shukla, T., Salour, A.: Finding sensitive sensor positions under faulty condition of reciprocating air compressors. In: IEEE Recent Advances in Intelligent Computational Systems, Trivandrum, India, pp. 242–246 (2011)
8. Verma, N.K., Kumar, P., Sevakula, R.K., Dixit, S., Salour, A.: Ranking of sensitive positions based on statistical parameters and cross correlation analysis. In: 6th International Conference on Sensing Technology (ICST), Kolkata, India, pp. 815–821 (2012)
9. Verma, N.K., Sevakula, R.K., Dixit, S., Salour, A.: Ranking of sensitive positions based on statistical parameters and cross correlation analysis. Int. J. Smart Sens. Intell. Syst. **6**(4), 1745–1762 (2013)
10. Verma, N.K. Singh, N.K., Sevakula, R.K., Salour, A.: Ranking of sensitive positions using empirical mode decomposition and Hilbert transform. In: Proceedings of the 7th IEEE Conference on Industrial Electronics and Applications, Hangzhou, China, pp. 1926–1931 (2014)
11. Li, R., He, D.: Rotational health machine monitoring and fault detection using EMD-based acoustic emission feature quantification. IEEE Trans. Instrum. Meas. **61**(4), 990–1001 (2012)

12. Hai, H., Pan, J.: Speech pitch determination based on Hilbert-Huang transform. Int. J. Signal. Process. **86**(4), 792–803 (2006)

Chapter 3
Pre-processing

Abstract As mentioned in Chap. 1, the real-time machine data is collected, which may be corrupt or inconsistent due to the presence of environmental noise. Therefore, the cleaning of data is required to remove unwanted frequencies as well to reduce the size of data for further analysis. This chapter details the second most important step of fault diagnosis framework, i.e., pre-processing of data. Low-quality data leads to misleading results, therefore, to make a better, robust, and more accurate fault classification model, pre-processing is required. The pre-processing involves filtering, clipping, smoothing, and normalization methods. Further, a graphical representation of the acoustic signal has been introduced. The chapter ends by a briefing of the development of pre-processing tool.

3.1 Introduction

Pre-processing describes the processing performed on raw data to retain useful information in the input data. It transforms the data into a format which will be more easily and effectively processed. Data collected in real-time environment is frequently found to be noisy containing inconsistent information. The data needs to be pre-processed to make a better, robust, and more accurate fault classification model. If the collected data is not of good quality, misleading results may be obtained. The process of improving quality of input data with the removal of unwanted noise is known as pre-processing. It transforms data into the format which can be more easily and effectively processed. Figure 3.1 shows the block diagram of pre-processing approach.

Raw data is normally referred to as the data collected from machines using various mechanisms. The samples of acoustic and vibration raw data are shown in Figs. 3.2 and 3.3, respectively. For example: for a single-stage reciprocating air compressor, machine's acoustic and vibration noise data were collected with the help of DAQ hardware, smartphone, and LabVIEW software.

N. K. Verma and A. Salour, *Intelligent Condition Based Monitoring*, Studies in Systems, Decision and Control 256, https://doi.org/10.1007/978-981-15-0512-6_3

Fig. 3.1 Block diagram of pre-processing approach [1]

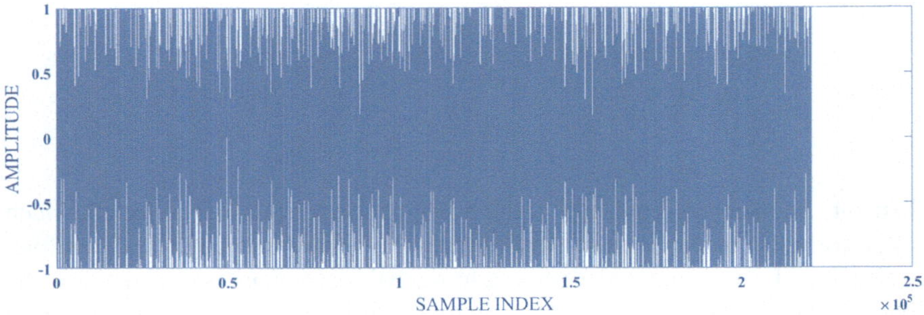

Fig. 3.2 Time domain plot of raw data (acoustic) [1]

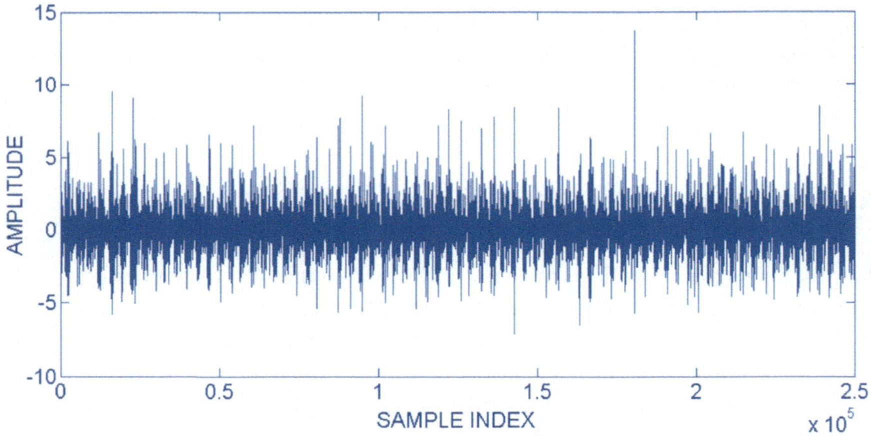

Fig. 3.3 Time domain plot of raw data (vibration). Image Courtesy: IDEA LAB, IIT Kanpur

The specifications of acquired acoustic data are as follows:

- Sampling frequency = 44.1 kHz.
- Total time of recordings = 5 s.
- Total number of samples = 220,500.

The specifications of acquired vibration data are as follows:

- Sampling frequency = 50 kHz.
- Total time of recordings = 5 s.
- Total number of samples = 250,000.

The four major pre-processing steps used, shown in Fig. 3.1, are as follows.

i. **Filtering**: It is a process of removing unwanted frequencies from the signal.
ii. **Clipping**: It is a process of reducing data size by "cutting" certain amount of data from given input dataset.
iii. **Smoothing**: It is done to remove the effect of high-frequency noise and outliers.
iv. **Normalization**: It is a process which scales the values from different recordings to same level.

3.2 Filtering

Filtering is the process of selecting a frequency range for which different types of filters are used. Herein, it is used to remove unwanted signals from the original signal. Noise can be attributed to various environmental causes along with the disturbances added due to imperfections in the equipment. Frequency-based filtering is done by first identifying the region in frequency domain containing noise or regions including the desired signal only. Later, a band-pass filter is applied such that it attenuates unwanted regions of the frequency spectrum. It must be noted that noise can only be removed by this method if it exists in a frequency region other than the desired signal. The two-stage filtering has been applied while performing the experiment on single-stage reciprocating air compressor. The two stages of filtering have been explained below.

Stage 1: External Fan Filtering
To remove compressor's cooling fan noise (which primarily constitutes low-frequency components) from the acoustic data, a high pass filter is applied. To determine its cut-off frequency, the acoustic data is first filtered using low pass filter at various frequencies lower than 500 Hz. The cut-off frequency for desired high pass filter is set to that frequency where the sound of compressor becomes just audible. This means that the compressor's sound cannot be neglected onwards the cut-off frequency. It is found that the noise due to cooling fan, which is prominently made of low-frequency component, can be significantly reduced by this method.

i. **Finite Impulse Response (FIR) Filter**

Finite impulse response (FIR) filter is a linear time-invariant (LTI) system, which generates the filtered signal by convolving the impulse response with the input signal. It is a band-pass filter that attenuates unwanted regions of frequency spectrum. FIR refers to the length of impulse response.

Let the impulse response of system is $h(n)$ and input to the system is $x(n)$, then the system said to be FIR if it satisfies the following conditions:

$$
\begin{aligned}
h(n) &\neq 0 \quad N_1 \leq n \leq N_2 \\
&= 0 \quad \text{everywhere else}
\end{aligned}
$$

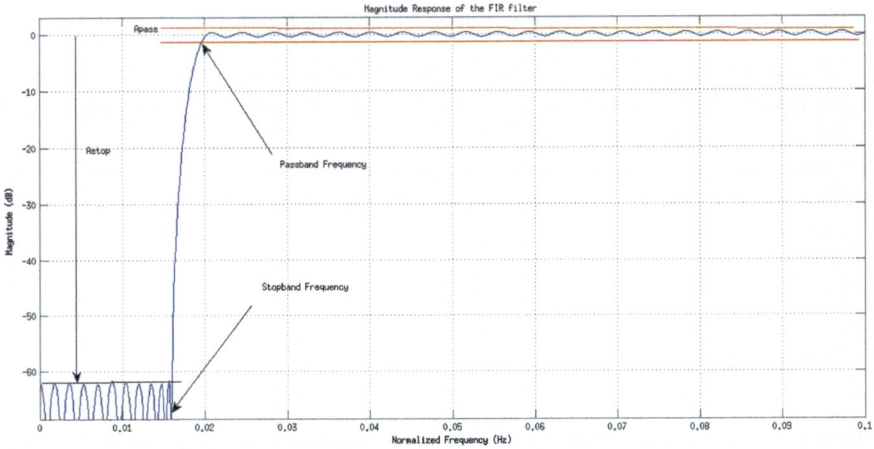

Fig. 3.4 Magnitude response of FIR high pass filter. Image Courtesy: IDEA LAB, IIT Kanpur

It indicates that the range of vision is finite on n axis. If the system output is $y(n)$, then it can be expressed by convolution theorem as follows:

$$y(n) = \sum_{i=N_1}^{N_2} h(n)\ x(n-i) \tag{3.1}$$

In case of infinite impulse response (IIR) filter, either N_1 or N_2 or both are infinite (Fig. 3.4).

The frequency spectra of implemented high pass FIR filter are of order $N = 1065$. The parameter values are: $F_{\text{stop}} = 300$ Hz, $F_{\text{pass}} = 400$ Hz, $A_{\text{stop}} = 60$ dB, and $A_{\text{pass}} = 1$ dB.

Stage 2: Butterworth Filter

The Butterworth low-pass filters are defined by the property that the magnitude response is maximally flat in band-pass. For nth-order low pass filter, this means that the first $(2n-1)$ derivatives of the magnitude-squared function are zero at $w = 0$. Another property is that the magnitude response is monotonic in band-pass as well as for stopband. The magnitude-squared function for a continuous-time Butterworth low pass filter can be expressed as follows:

$$|H(j\omega)|^2 = \frac{G_0^2}{1 + \left(\frac{\omega}{\omega_c}\right)^{2n}} \tag{3.2}$$

where

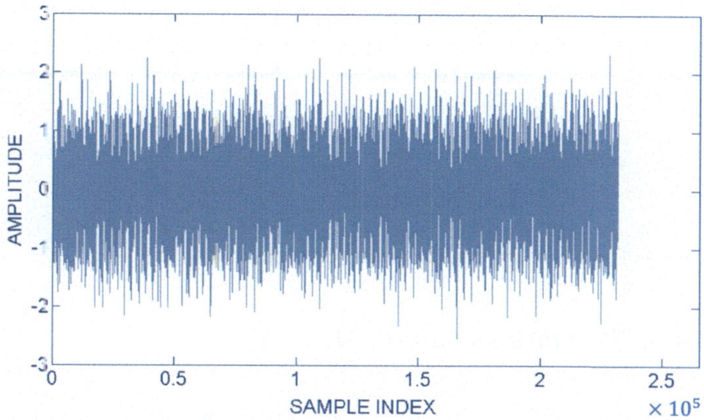

Fig. 3.5 Time domain plot of filtered data (acoustic) [1]

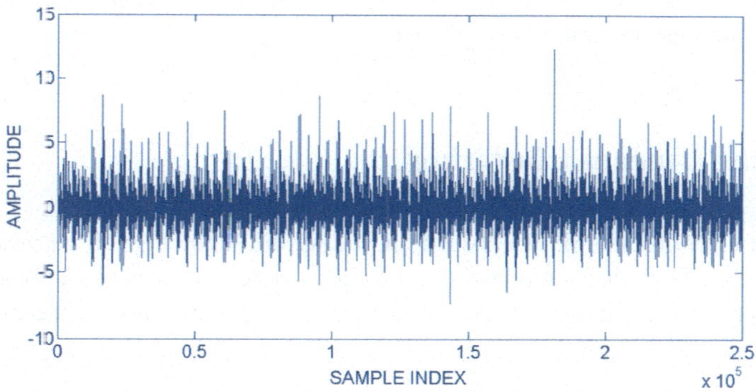

Fig. 3.6 Time domain plot of filtered data (vibration). Image Courtesy: IDEA LAB, IIT Kanpur

n order of the filter,
ω_c cut-off frequency (approximately -3 dB frequency), and
G_0 is the DC gain (gain at zero frequency) (Figs. 3.5 and 3.6).

3.3 Clipping

The amount of data in an input signal is generally too large for efficient computation and contains many regions of no importance. Clipping is the process of identifying a good segment from the signal which is enough to replace the complete signal for further purposes. In the areas of speech processing, segmentation techniques based on the edge detection and the dynamic programming have been used for detecting speech areas and locating the segments based on their audio types. Here, we have experimented with two different methods of clipping.

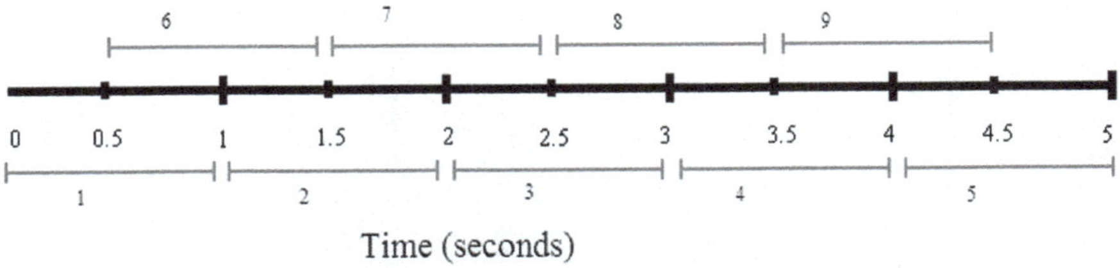

Fig. 3.7 Division of total time interval into multiple regions for clipping [1]

3.3.1 Using Standard Deviation Method

A simple clipping method is to first divide the signal into multiple regions and then select the region which has minimum standard deviation to avoid strong variations, i.e., obtain a steady signal. Standard deviation is calculated for every region of the signal as indicated in Fig. 3.7 (0–1, 0.5–1.5, 1–2, and so on). Then, part of the signal with least standard deviation was selected.

3.4 Using *k*-means Clustering

The earlier clipping method was based on finding the segment with minimum standard deviation. The other parameters of the signal were not taken into consideration. In signal processing, it is very common that for applying various transforms on

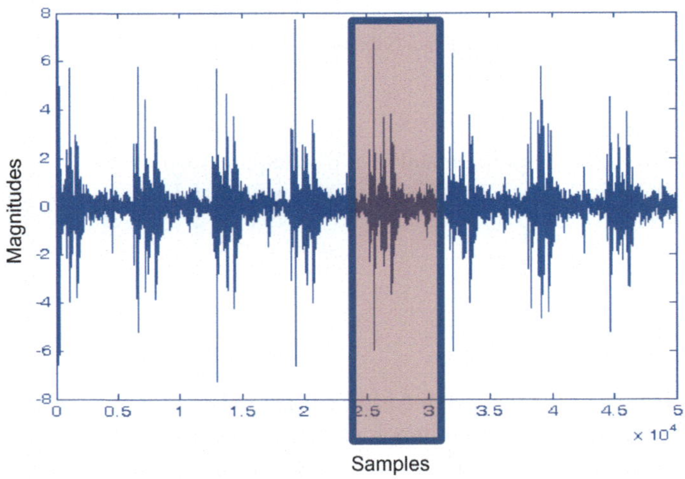

Fig. 3.8 Shaded portion depicts the window [2]

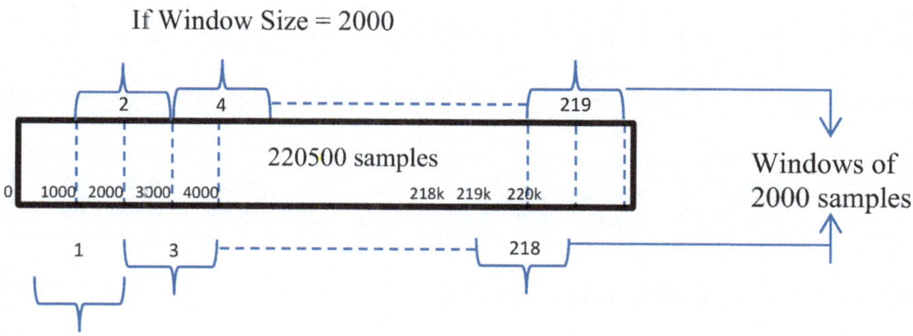

Fig. 3.9 Window selections based on 50% overlapping [2]

non-stationary signal, windowing operation is performed. At any given time, the transform is applied on the samples, which lie in that window as shown in Fig. 3.8.

Figure 3.9 shows the methodology of creating windows from a 5 s audio sample with a sampling rate of 44.1 kHz, i.e., 220 and 500 samples in total. Here, we have applied 50% overlapping for window selections. For example, 219 windows of 2000 samples in each window is shown in Fig. 3.9. Now, to identify which window size is the best for fault recognition, features extracted from each windowed signal are obtained, and classification as explained later is used. If a window size is given, an option from the following can be considered:

i. Extract features for all such windows in the entire duration of signal and then take the mean/median of these features.
ii. Find the best window among all and then extract features from only that window.

After observation, it has been found that the size of window has strong connection with the consistency of feature values. The plot of absolute mean feature for different windows of the same signal is shown in Fig. 3.10.

Smaller windows generally give us inconsistent feature values. This is because the small window gives a very local picture of the signal; hence, feature values vary a lot. Spikes/outliers are commonly found in such feature value plots. Similarly, having a very large window is also not appropriate. This is because the transforms tend to

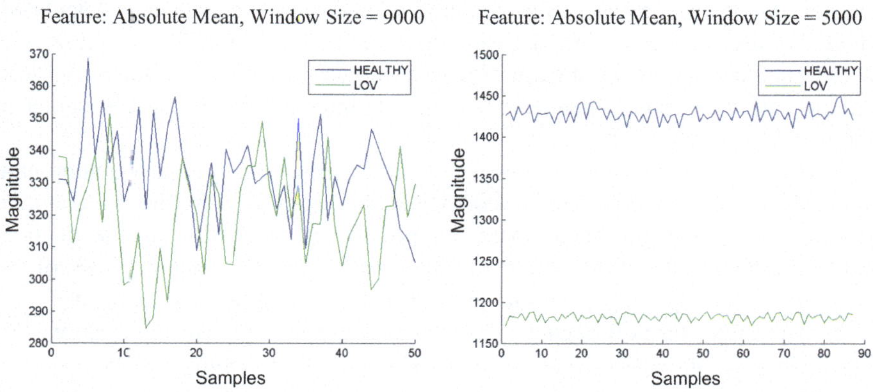

Fig. 3.10 Feature plots for different window size of the same signal [2]

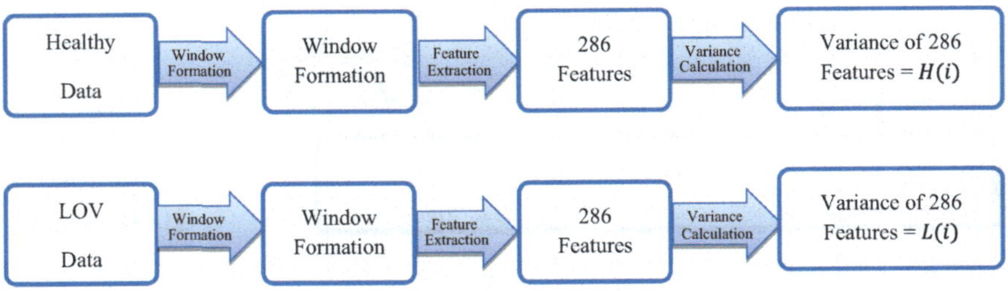

Fig. 3.11 The flow diagram for finding the best window size [2]

average out the signal characteristics over the large range. Thus, some important signal variations which are unique to a certain fault/state of air compressor may be lost.

After observations, it is also found that the feature-level distinction among different states of an air compressor is improved by selecting appropriate window size. The features are extracted for all windows of signal with different window size ranging from 1000 to 10,000. This has been done for two states of the air compressor—healthy and leakage outlet valve (LOV). For selecting the best window size, the variance of all features is calculated independently over entire time period after normalization. As shown in Fig. 3.11, the flow chart shows the methodology involved in finding the best window size.

To find the best window size, variance has been chosen because it is a measure of how far a set of data points are spread out. A smaller variance indicates that the data points tend to be closer to the mean and to each other. A higher variance indicates that the data points are more spread out from the mean and from each other. Ideally, the value of each feature should be almost constant during the entire duration of the signal. Hence, the best window size is one whose variances of both: feature values of a healthy state and that of LOV fault state is minimum. So, "Sum of Variance" is chosen as a deciding factor for finding the optimum window size. The window size whose maximum number of features have minimum variance is chosen as the best window size. For example, in this case for the two states, i.e., sum of variance is denoted as $H(i) + L(i)$, where $H(i)$ denotes the variance of feature i for healthy data, and $L(i)$ denotes the variance of ith feature for LOV data. The variance is calculated for all 286 features on window size ranging from 1000 to 10,000. The window size where a maximum number of features have minimum variance is chosen as the best window size. The result is tabulated in Table 3.1, and graph for the same is depicted in Fig. 3.12.

Now, out of all window sizes considered, the maximum features have a minimum variance when the window size = 5000. On further analysis, it has been found that window size of 5000 for feature extraction provides better feature level distinction among different states of the compressor as compared to earlier window sizes of 1 s of sampled data. Fig. 3.13 shows the flow chart for finding the best segment.

To determine the best segment for clipping a new signal in the feature, 18-dimensional feature space has been used. Though, more features can be used for this purpose, 18 are enough. They are chosen specifically to represent the shape of the curve. They are as follows:

Table 3.1 Comparison to study the effect of noise on both methodolcgies

Window size	Number of features having minimum sum of variance
1000	21
2000	28
3000	76
4000	29
5000	86
6000	12
7000	10
8000	11
9000	8
10,000	5

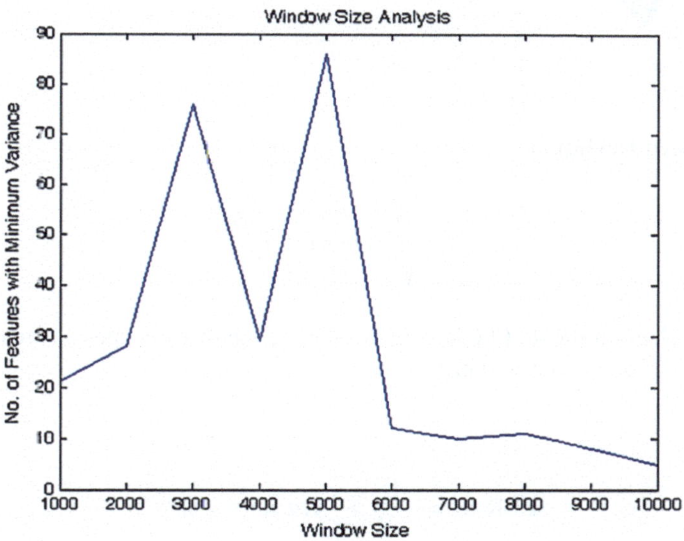

Fig. 3.12 Plot of features having minimum variance [2]

i. **Absolute Mean**:

$$\overline{x} = \frac{\sum |x[i]|}{N}. \tag{3.3}$$

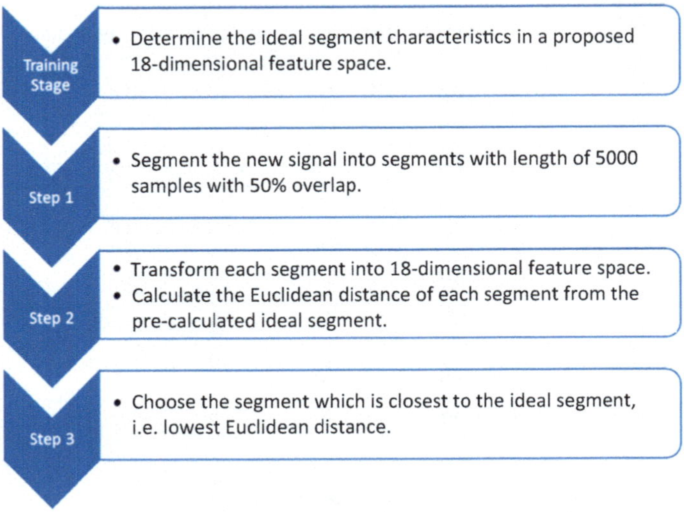

Fig. 3.13 Flow chart to find the best segment [2]

ii. **Standard Deviation**:

$$\sigma = \sqrt{\frac{\sum (x[i] - \overline{x})^2}{N - 1}}. \tag{3.4}$$

iii. **Zero-Crossing Rate (ZCR)** : A measure of number of times a signal crosses zero axis, and it is calculated as

$$\mathrm{ZCR}(n) = \frac{1}{2N} \sum_{m=0}^{N-1} x(m) w(n - m)$$

where $x(m)$ is the signal and $w(n)$ is window size.

iv. **Time Centroids**:

$$C_{\mathrm{T}} = \frac{\sum_{n=0}^{N-1} n * x(n)}{\sum_{n=0}^{N-1} x(n)} \tag{3.5}$$

where "n" is the sample number, and $x(n)$ is the time domain value of nth sample. The time domain signal is divided into five equal, non-overlapping bins. The weighted mean of each of the bins with their amplitudes as weights is found to give five features. This is shown in Fig. 3.14.

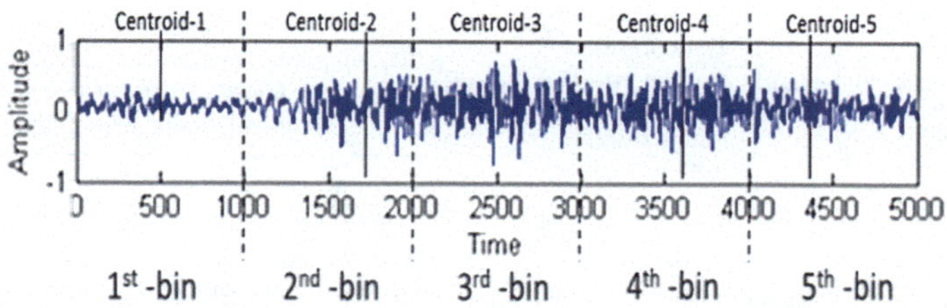

Fig. 3.14 Five bins of signal to calculate time centroid features [2]

v. **Spectral Centroid**:

$$C_S = \frac{\sum_{n=0}^{N-1} f * X(f)}{\sum_{n=0}^{N-1} X(n)} \tag{3.6}$$

where $X(f)$ represents the magnitude of frequency component f in the frequency spectrum which is found using fast Fourier transform. The frequency spectrum of the signal is divided into five equal non-overlapping bins. The weighted mean of the frequencies with their spectral coefficients as weights is found for each of the bins to give five features as shown in Fig. 3.15.

vi. **Absolute Mean of Bins**: The time domain signal is divided into five equal, non-overlapping bins. The absolute mean of each of these bins gives us five features. Now, the training of clipping model is done to find the best segment in the 18-dimensional feature space as shown in Fig. 3.16. As this is an unsupervised step, training dataset includes recordings of all the states of air compressor. To find out the characteristics of the best segment in 18-dimensional feature space, *k*-means clustering is first performed on the training dataset while varying $k =$ 2 to 10. Global Silhouette index (GS), a heuristic used for judging the cluster quality, is calculated for different values of "*k*".

The *k*-means clustering is performed with the value of "*k*" for which GS is the highest. The good features are those whose values over multiple segments for different conditions/states of an air compressor, show significant difference.

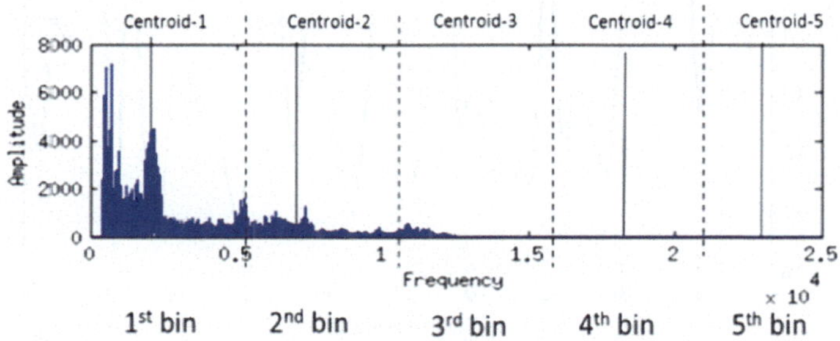

Fig. 3.15 Bins of signal to calculate spectral centroid features [2]

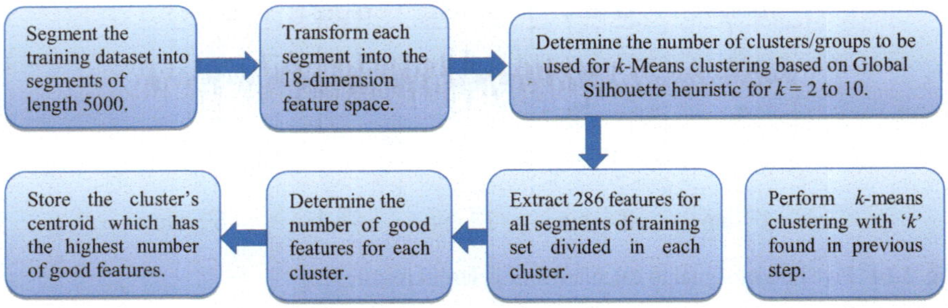

Fig. 3.16 Flow chart of the training of clipping model [2]

1. **Separation**: The difference of feature values should be above some threshold over multiple segments. This is given by the ratio of absolute mean of feature value(s) for different conditions.

$$m_1/m_2 = 1.25$$

where m_1 and m_2 are the mean values of the feature for the two classes. This upper limit is decided after observing the plots for various features as shown in Fig. 3.17.

2. **Consistency of Separation**: The difference should be consistent over multiple readings. This can be checked by calculating ZCR of the difference of the feature values for two conditions for multiple readings. These thresholds were determined by observing the behaviour of multiple features. In Fig. 3.18a, even

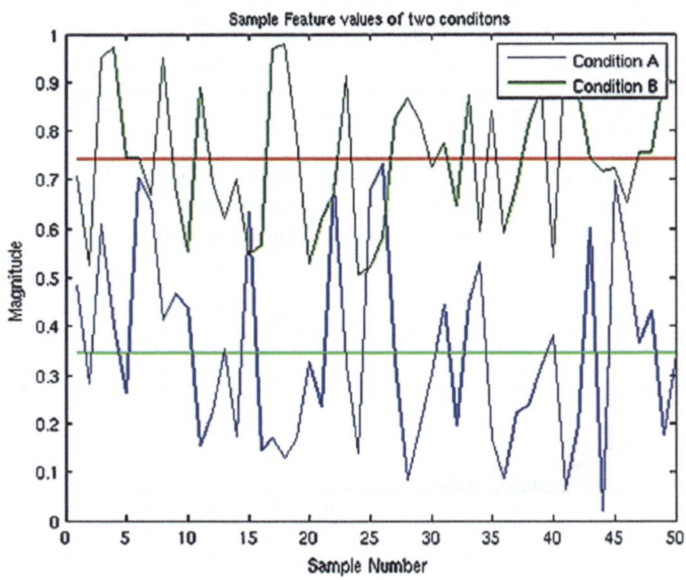

Fig. 3.17 Plot of features for two different conditions of the machine [3]

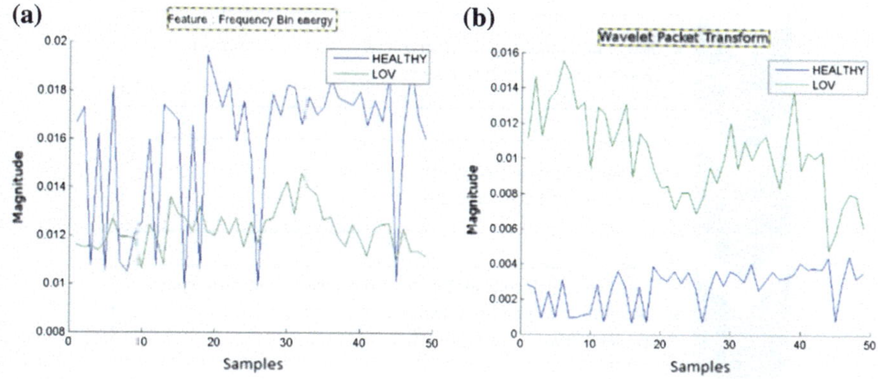

Fig. 3.18 **a** Bad feature, **b** good feature [3]

the difference of absolute mean is greater than the threshold but ZCR of the difference is too high; hence, it will be considered as a bad feature. In Fig. 3.18b, it shows good value for the ratio of absolute mean and low ZCR.

After determining "*k*", following has been done: a test dataset is considered (as explained in the section of Experimentation Procedure), and clipping is performed based on each cluster centroid. A total of 286 feature values are obtained from clipped segment for 50 readings of two conditions, namely healthy and LOV of an air compressor. The next step is to find the number of good features from 286. This is done based on the measures described earlier for clipping. This is done for each cluster centroid. The cluster providing the highest number of good features is chosen, and its centroid value is stored as reference for the best segment in the 18-dimensional feature space. The overall process used for clipping is shown in Figs. 3.19, 3.20, and 3.21.

3.5 Smoothing

High-frequency and small amplitude variations in the signals cause distortion in them. Smoothing reduces the effect of such high-frequency noise superimposed on data. Thus, the smoothing process can be considered like applying a low pass filter. Some popular linear smoothing techniques are moving average and weighted moving average, while median smoothing and polynomial fitting are well-known nonlinear techniques. Here, moving average, geometric mean, median, and Savitzky–Golay smoothing techniques are chosen for comparison because of their deterministic nature and lower time complexity.

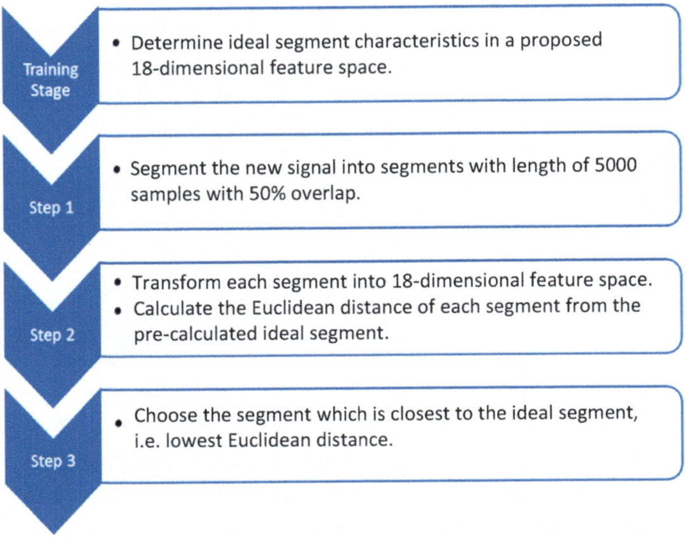

Fig. 3.19 Flow chart of the clipping process [2]

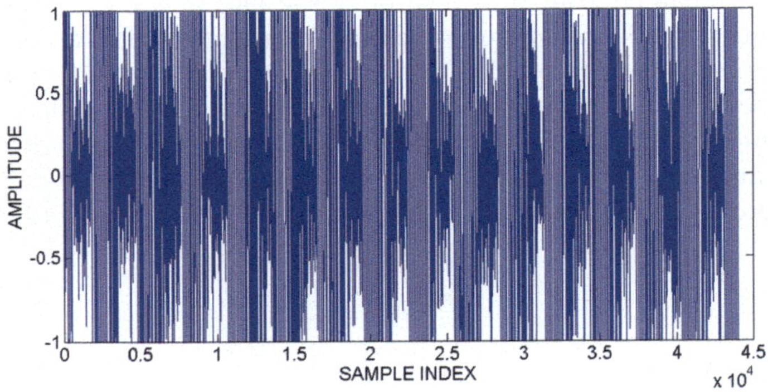

Fig. 3.20 Time domain plot after clipping of acoustic data [2]

3.5.1 Moving Average

It is done by calculating series of mean values of fixed "n" consecutive points in the given dataset. For example, if $n = 5$, kth point of the resultant series will be

$$y_k = \frac{(X_{k-2} + X_{k-1} + X_k + X_{k+1} + X_{k+2})}{5}. \tag{3.7}$$

Here, "X" is the original series of data points, while "y_k" is the resultant series after moving average with $n = 5$. Over smoothing is one of the disadvantages of moving

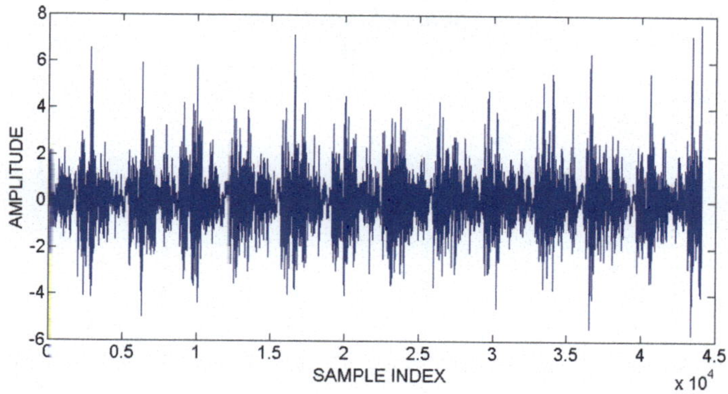

Fig. 3.21 Time domain plot after clipping of vibration data. Image Courtesy: IDEA LAB, IIT Kanpur

average. It is not possible to find an optimal value of "*n*" by this algorithm. Hence, one needs to take an empirical way to decide the best value of "*n*".

3.5.2 Locally Weighted Scatter-Plot Smoothing with First and Second Degree of Polynomials

It is also abbreviated as "Lowess" and as the name suggests it uses locally weighted linear regression to smoothen the data. The smoothing process is considered local because each smoothed value is determined by the neighbouring data points defined within a span. The process is weighted because a regression weight function is defined for the data points contained within a span. The polynomial for regression can be a first degree or second degree (abbreviated as "Loess"). It gives better results than the moving average but its computational time is more. Also, by using the mean absolute deviation in weight function, it can be made more robust for outliers but with a considerable increase in computational time. Fig. 3.22 shows the Lowess smoothing. Randomly generated signals, as shown in Fig. 3.23, have three outliers which are shown as data points. After applying the smoothening process, outliers are smoothed as indicated by a red line.

3.5.3 Savitzky–Golay Filter-Based Smoothing

It is a polynomial fitting method which performs local polynomial regression on a set of equally spaced points. The polynomial coefficients are independent of the value of signal at that point and thus can be determined beforehand. A flow chart for this method has been depicted in Fig. 3.24.

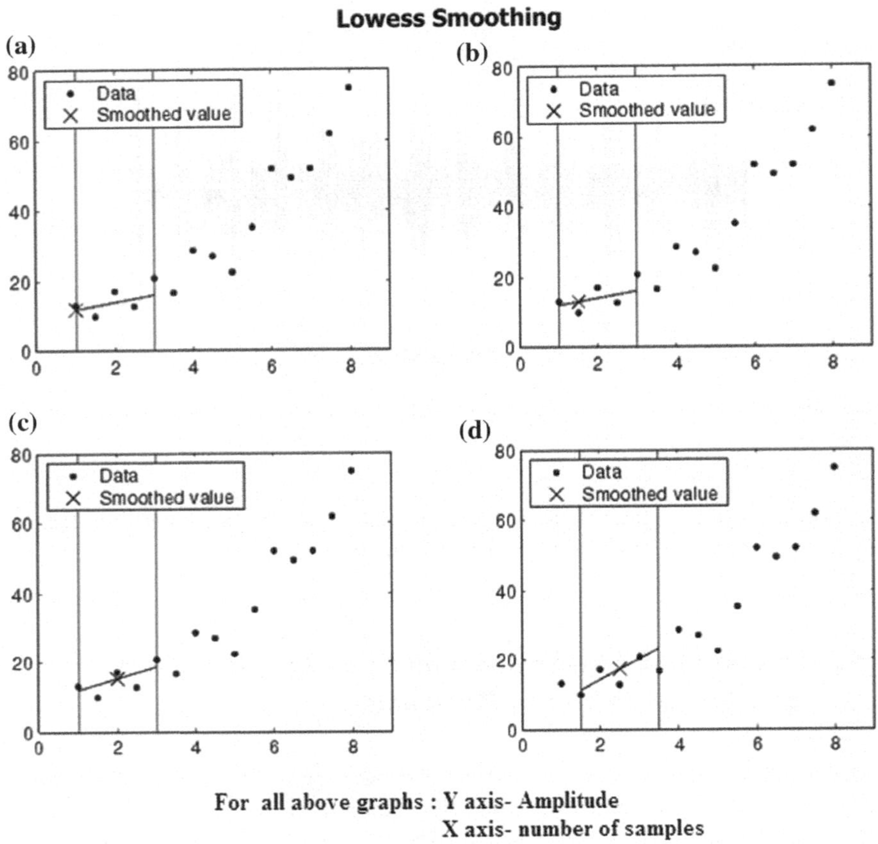

Fig. 3.22 Determination of value for next sample by neighbouring data points within the span. Image Courtesy: IDEA LAB, IIT Kanpur

3.5.4 Geometric Mean

This smoothing technique replaces each point by the geometric mean of points in a window of size 3 centred about that point (Figs. 3.25, 3.26, and 3.27).

$$s[i] = (x[i-1] * x[i] * x[i+1])^{(1/3)}. \tag{3.8}$$

3.6 Normalization

Multiple sensors generally provide output in different scales; therefore, it requires normalizing the signal for proper comparison. Normalization is also essential for calculating distances in multidimensional feature spaces. Commonly used normalization techniques are max-min normalization and z-score normalization. The presence

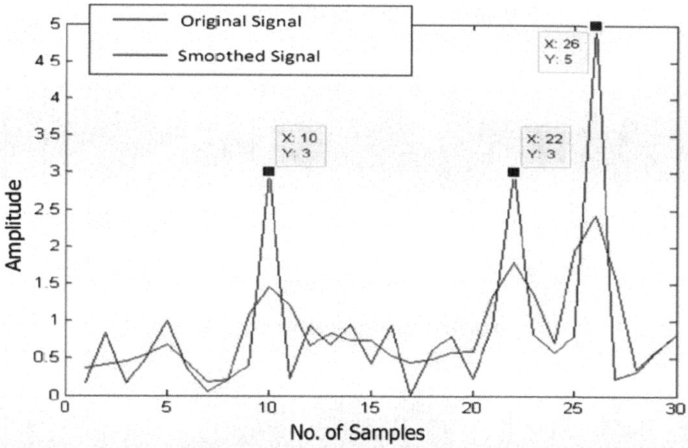

Fig. 3.23 Effect of smoothing on random signal (random sequence on random scale). Image Courtesy: IDEA LAB, IIT Kanpur

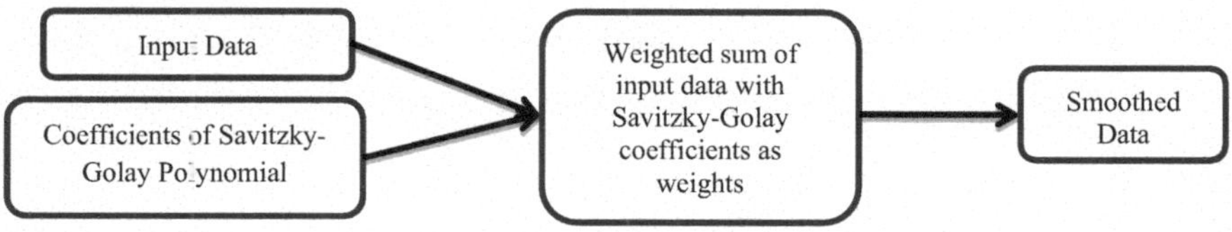

Fig. 3.24 Flow chart of Savitzky–Golay filter-based smoothing. Image Courtesy: IDEA LAB, IIT Kanpur

of outliers give false maximum and minimum values in the case of max-min normalization, while z-normalization though more robust than the max-min normalization still does not take explicit steps towards nullifying outlier's effect. Normalization also improves processing time for some algorithms, especially those involving optimization problems like support vector machine (SVM).

3.6.1 (0–1) Normalization

The (0–1) normalization (also called linear normalization) is nothing but shifting of data range from any bound to lower bound (0) and upper bound (+1). In generalized form, it can be calculated as:

$$Y_i = \frac{(X_i - X_{\text{min}})}{(X_{\text{max}} - X_{\text{min}})} \times \text{UB} + \text{LB} \tag{3.9}$$

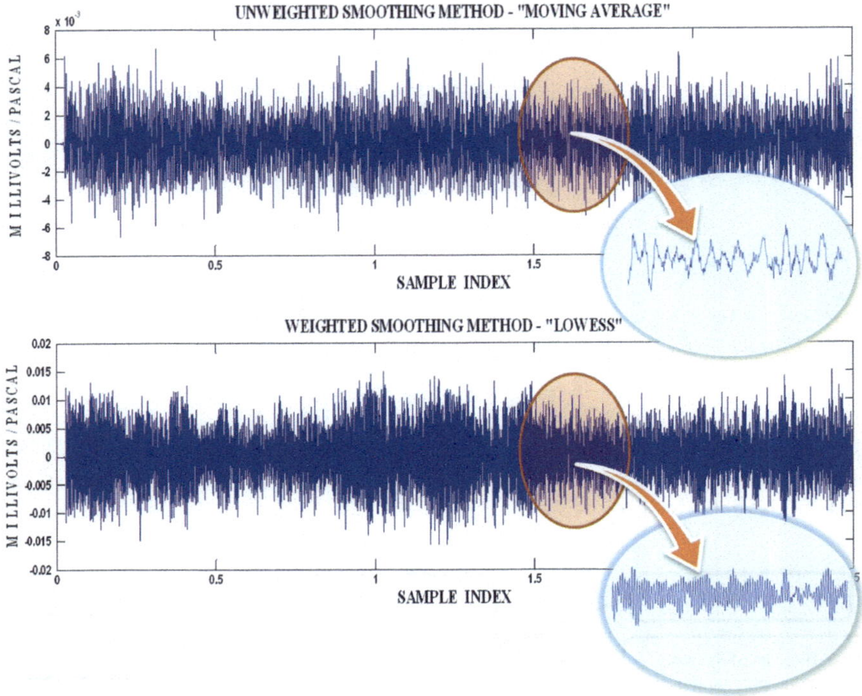

Fig. 3.25 Unweighted versus weighted smoothing methods. Image Courtesy: IDEA LAB, IIT Kanpur

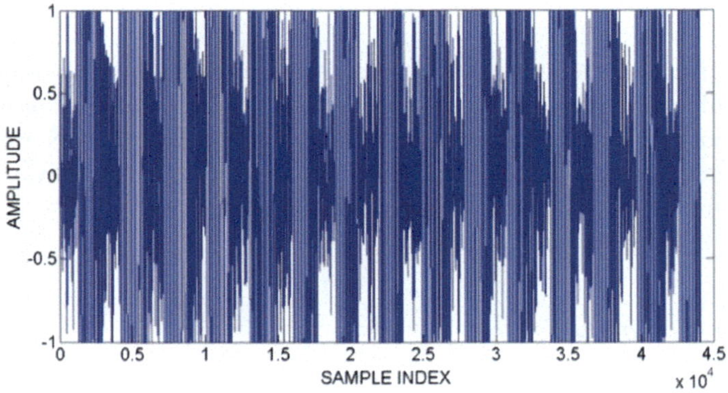

Fig. 3.26 Time domain plot after smoothing of the acoustic data [1]

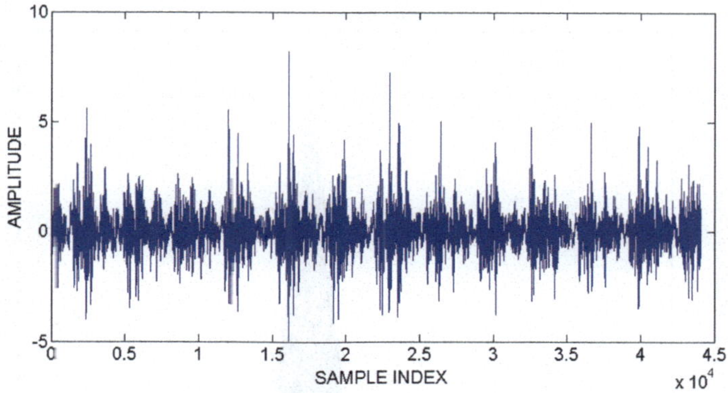

Fig. 3.27 Time domain plot after smoothing of the vibration data. Image Courtesy: IDEA LAB, IIT Kanpur

where Y_i is the data point in normalized series, and X_i is the original data point with $X_{\min}$ as minimum and $X_{\max}$ as maximum values. UB and LB denote the desired upper bound and lower bound between 1 and 0.

3.6.2 Mean and Variance Normalization

The mean and variance normalization are achieved by subtracting the mean from each data point and then dividing it with its variance.

$$Y_i = \frac{(X_i - \mu_x)}{(\sigma_x)} \tag{3.10}$$

where μ_x is mean, and σ_x is the variance of signal. The modified signal ensures zero mean and unity variance distribution. This normalization is found to be more useful in the developed algorithm.

3.6.3 Max-Min Normalization Using Outliers

The max-min normalization described before uses the maximum and minimum elements of the dataset. The presence of few outliers can distort the normalization process. This distortion may cause similar signals to vary greatly because of the outliers present in one of them. In the normalization method used in this work, this fact is taken into consideration while still using max-min normalization.

For analysing machine signals, the histogram of sampled signal values is plotted. Interestingly, the histogram plot is found to be like that of the Gaussian distribution

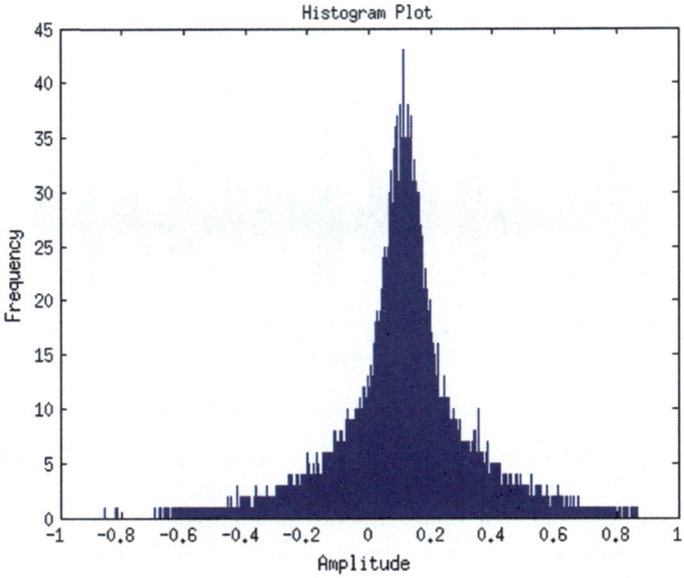

Fig. 3.28 Histogram plot of sampled signal values [1]

as shown in Fig. 3.28. In a Gaussian distribution, 99.95% of points lie within the range of 6σ. Using this fact, an assumption is made that 0.05% of points in such a distribution will be outliers. Eliminating 0.025% points from top and bottom of the sorted values of signal, the signal which is presumably free of outliers is obtained. Maximum and minimum values of the remaining dataset serve as inputs for the max-min normalization applied on the original signal. The 0.025% points are not removed as they cannot be surely decreed as outliers but their effects are cancelled somewhat. The proposed scheme is different from the z-score normalization on grounds that z-score centres the data on its mean. However, in the proposed scheme, majority of the data will lie in the bounds of [0, 1] and is not necessarily centred on the mean (shown in Figs. 3.29 and 3.30).

3.7 Graphical Representation of Acoustic Signal

After applying all the pre-processing stages, the resultant signal is shown in Figs. 3.31 and 3.32 along with the original signal. The details of pre-processing stages are given in the table.

The complete procedure has been tabulated in Table 3.2.

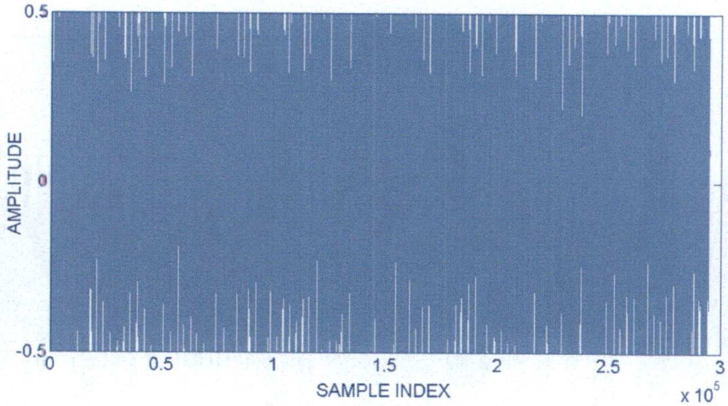

Fig. 3.29 Time domain plot after normalization of the acoustic data [1]

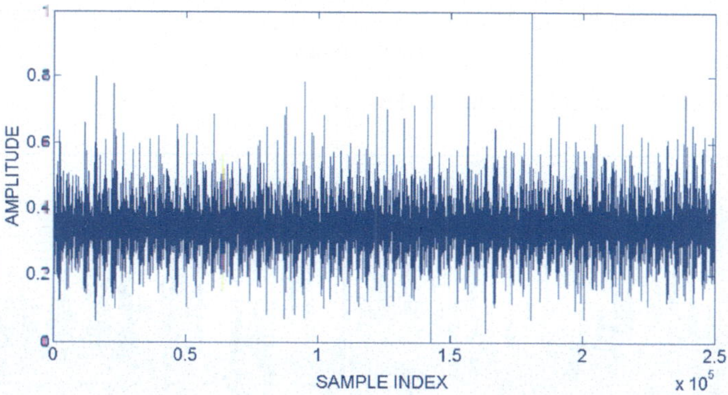

Fig. 3.30 Time domain plot after normalization of the vibration data. Image Courtesy: IDEA LAB, IIT Kanpur

3.7.1 Spectrogram of the Pre-processed Signal in Different Conditions

See Figs. 3.33, 3.34, 3.35, and 3.36.

3.8 Development of Pre-Processing Tool

3.8.1 Development Pre-Requisites and Memory Access

3.8.1.1 Android Platform

To develop the Android application, the following software packages were required:

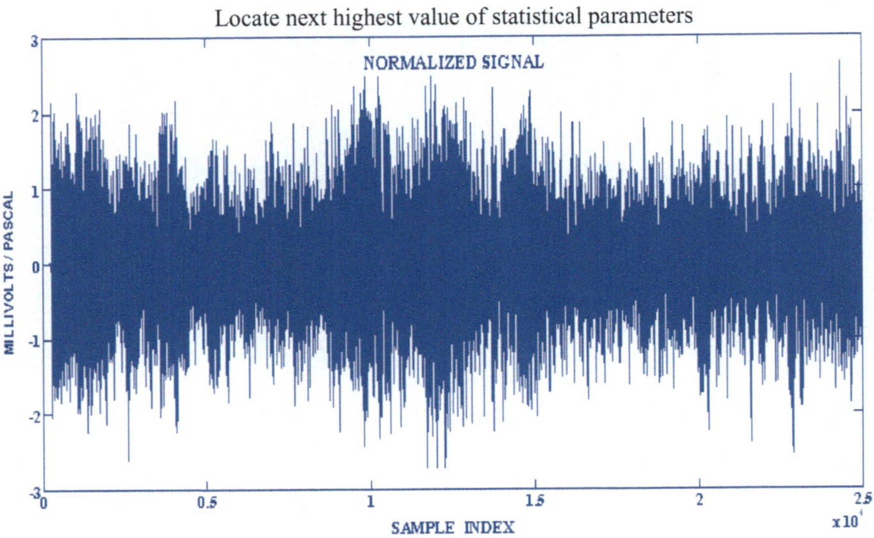

Fig. 3.31 Normalized signal for healthy condition [1]

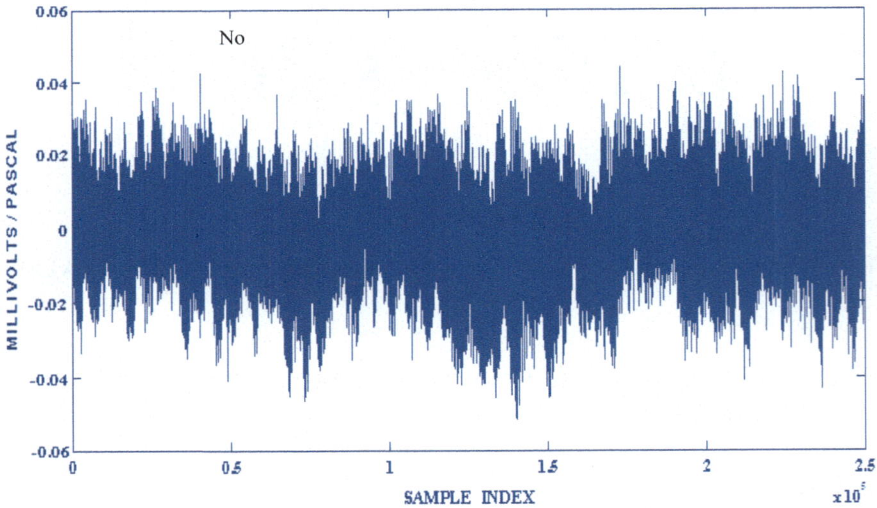

Fig. 3.32 Raw data signal [1]

Table 3.2 Pre-processing steps

S. No.	Steps	Parameter setting
1.	Downsampling	Sample rate: 25–50 kHz
2.	Clipping	Length of signal: 1 s
3.	Filtering	Band pass: pass band as 400–12 kHz
4.	Smoothing	Lowess (1st degree polynomial)
5.	Normalization	Mean and variance normalization

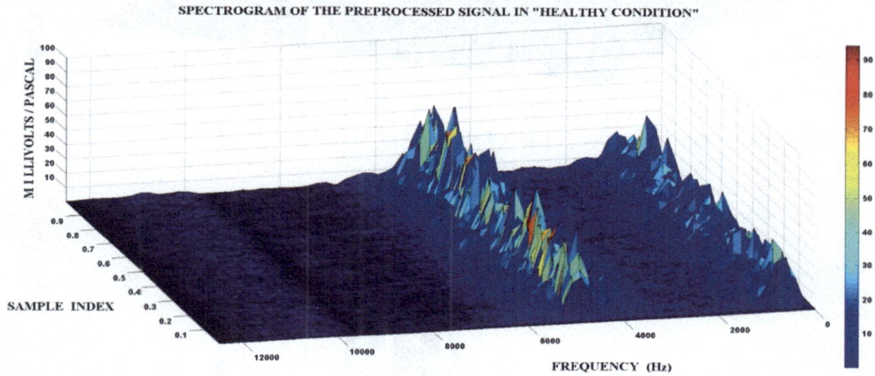

Fig. 3.33 Spectrogram of the pre-processed signal in healthy condition [1]

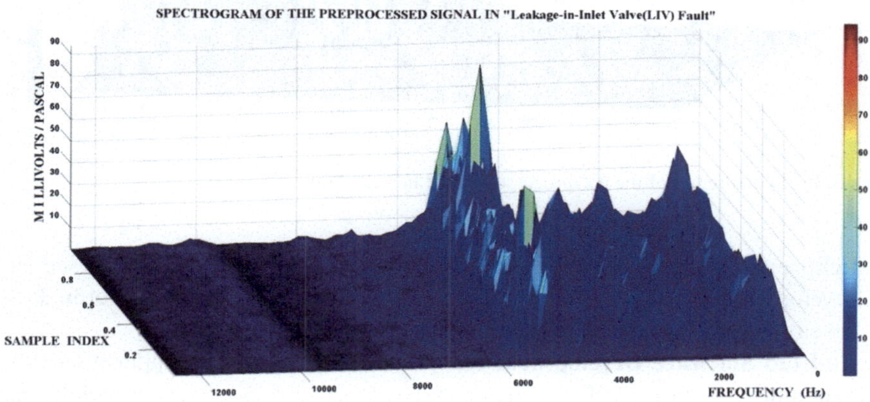

Fig. 3.34 Spectrogram of the pre-processed signal in LIV fault [1]

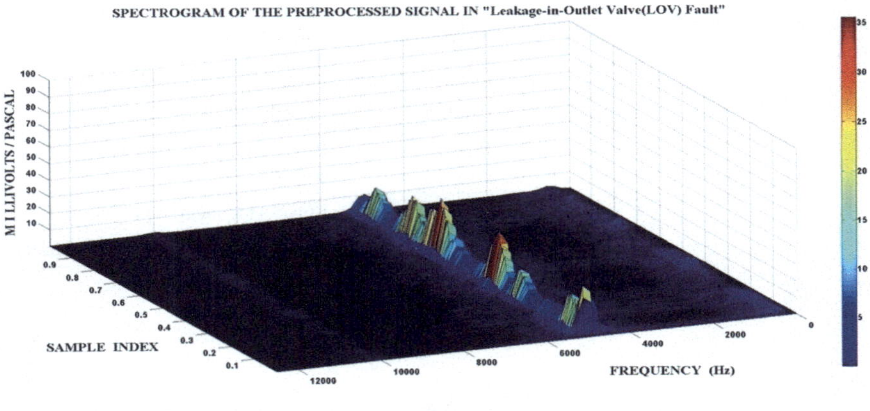

Fig. 3.35 Spectrogram of the pre-processed signal in LOV fault [1]

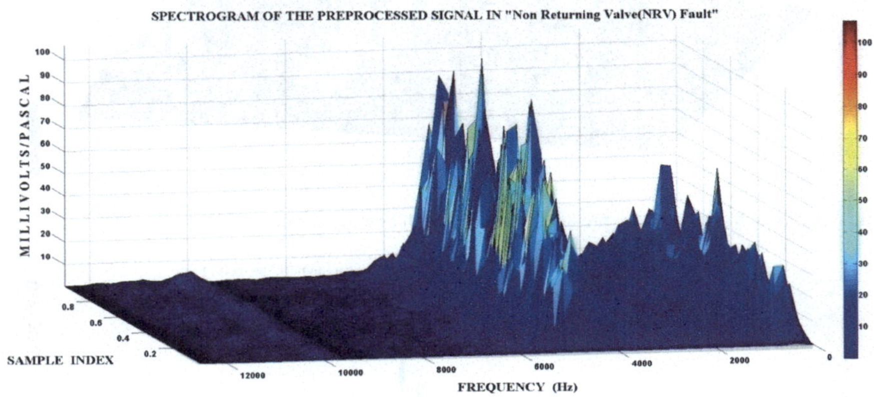

Fig. 3.36 Spectrogram of the pre-processed signal in NRV fault [1]

- **Eclipse Classic 3.7.2**: An Integrated Development Environment (IDE) used for development of variety of applications. It comprises of Eclipse platform, Java development tools, and plug-in development environment.
- **Android Software Development Kit (SDK)**: It provides the Application Programming Interface (API) libraries and developer tools necessary to build, test, and debug apps for Android. The main components of SDK are Android Virtual Device, Android Debug Bridge, Dalvik Virtual Machine, Android Emulator, and Resource Editors.
- **Android Developer Tools (ADT)**: It is a combined set of components (plug-ins) which extend the Eclipse IDE with Android development capabilities, SQLite Database Browser 2.0, an open source database which is embedded into Android, Programming Language, and Java.

In the memory access, Android is flexible enough and it allows storing application data in databases, files, or preferences in internal or removable storage. Android allows files to be read/write directly on a device known as internal storage. These files are private to the application which means that it cannot be accessed by the other applications unless special permission is given. In the case of external storage, Android allows files to be read and write on a removable storage device, i.e., SD card.

3.8.1.2 Windows Mobile App Platform

To develop a Windows Mobile application, the following software packages were required:

- **Visual Studio Express 2013 for Windows Professional**: An IDE consists of tools to create Windows apps. The tools include a full-featured code editor, a powerful debugger, a focused profiler, and rich language support for HTML5/JavaScript, C++, C#, or Visual Basic.
- **Windows 8.1**: To develop and store the apps, this operating system is required.
- **Windows Phone (WP) 8.1 Phone/Emulators**: It comes as version Express 2013 package. It provides a facility to test Windows Store apps on multiple kinds of device including WP. To test apps on a phone, one must register the phone for development and get a developer licence.
- **Programming Language**: Visual C#.
- **WP Power Tools**: It allows access of isolated storage from Desktop PC.

Windows platform has strict rules for developers, especially with respect to reading and writing files in the device. In the case of internal storage, Android app allows files to be read/write only from specific internal storage space. Each app is allotted some internal space which can be used to store app data. This internal storage is also knows as isolated memory and when it comes to external Storage, WP App provides read-only access to the SD card, and it is limited to file types for which your app has registered with a file association. Since data analytics tools require several operations where the ability of reading and writing files is needed, a user may wish to look at the content of the memory for checking the intermediate and final results. A separate module "Access Isolated Storage" has been developed for the same. It gives flexibility to the user to create and delete folder in the current directory. The custom bar shown in the screenshot is designed for this.

3.8.1.3 Windows Tablet Platform

To develop Windows Mobile application, the following software packages were required:

- **Visual Studio Express 2013 for Windows Professional**: It is an IDE using which one can design, code, test, and debug apps.

- **Windows 8.0 Pro**: A Windows version compatible with Windows 7, designed for touch apart from mouse and keyboard.
- **Windows SDK**: A package of compilers, headers, libraries, code samples, and tools that helps to create applications.
- **Programming Language Used**: Visual C#.

Like WP, Tablet apps do not allow full memory access. Tablet permits user to access, i.e., read and write in only Media libraries namely Pictures, Music, Videos, and Documents libraries.

3.8.2 Pre-Processing Tool for Android Platform

Step 1 As the user runs pre-processing App, its Welcome Page appears as shown in Fig. 3.37. Here, the user is asked to tap anywhere on the screen to move onto the main activity page.

Step 2 As the user taps on the Home Page, first activity page opens as shown in Fig. 3.38. From here, user can select the file format and sampling rate based on which the input file is chosen, and the pre-processing starts.

Step 3 On clicking the "General Instructions" button, following screen (Fig. 3.39) appears showing the guidelines of pre-processing App.

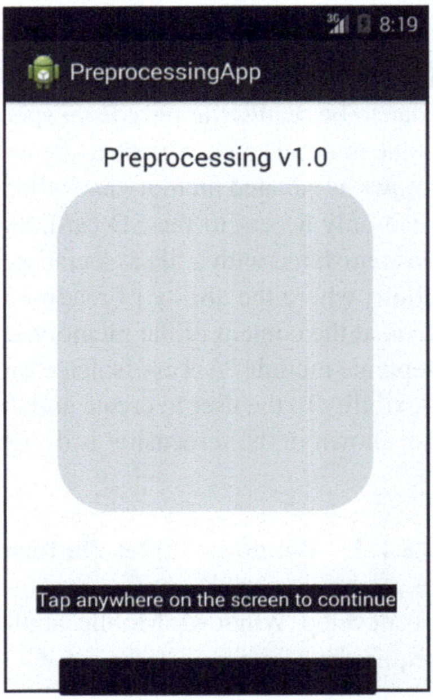

Fig. 3.37 Home page of the app [4]

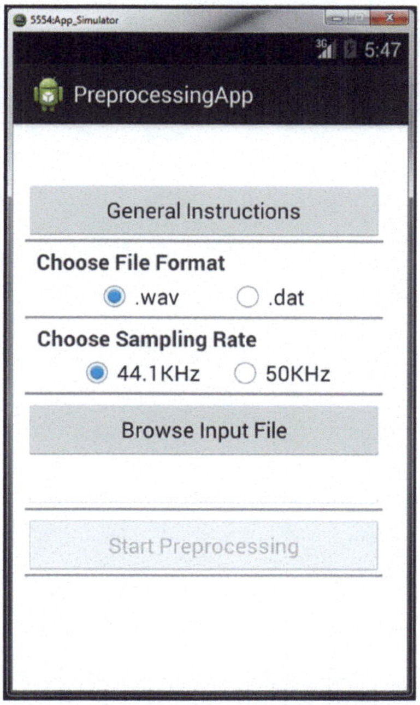

Fig. 3.38 Pre-processing activity on page 1 [4]

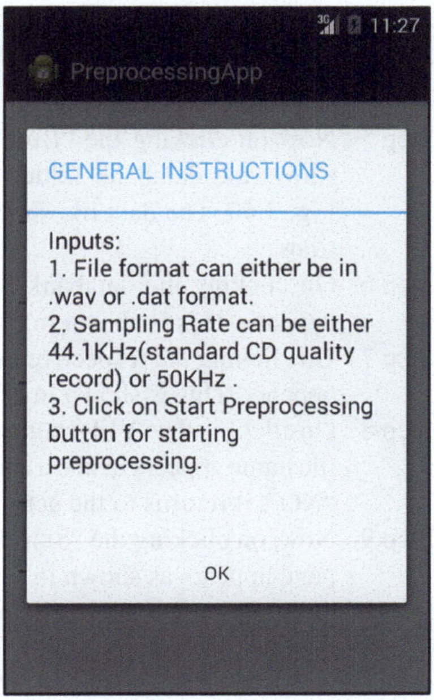

Fig. 3.39 General instruction page of the app [4]

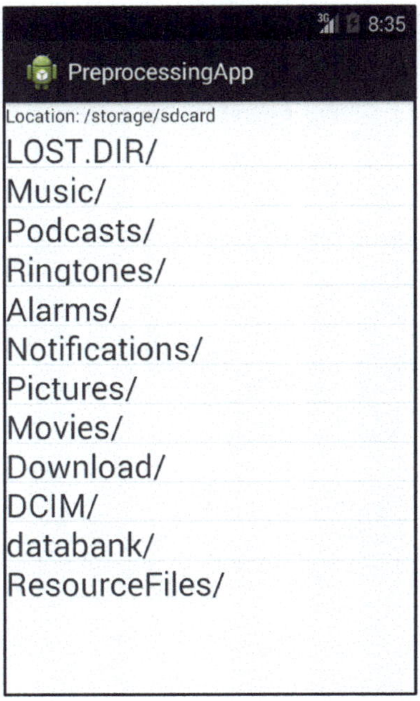

Fig. 3.40 Browse activity on page 1 of the app [4]

Step 4 The user can select the file format for which pre-processing is to be done. It may be either ".dat" or ".wav" file. Then, the user can also select the sampling rate. It may be either 44.1 kHz or 50 kHz.

Step 5 Now on clicking the "Browse Input File" button, a screen appears which shows the databank folder containing the required file. This is shown in Fig. 3.40. The data file should be chosen according to the above specifications.

Step 6 On clicking the databank folder, a screen showing the required input file appears. This is shown in Fig. 3.41.

Step 7 On clicking the respective input file ("Reading.wav"), a confirmation screen appears. This is shown in Fig. 3.42.

Step 8 On clicking the "YES" button, page moves to the activity page where selected file name appears inside a textbox as shown in Fig. 3.43, whereas on selecting "NO", it returns to the activity page screen as shown in Fig. 3.44.

Step 9 Now, on clicking the "Start Pre-processing" button, next processing activity page appears as shown in Fig. 3.44. This page does not take any user input. Only a progress spinner appears on the screen showing that pre-processing is going on.

Now as soon as the pre-processing is done, the current activity page moves to the result activity page as shown in Fig. 3.45. This page displays a message that result has been saved. Here, user is given an option whether he wants to perform the pre-processing again or not.

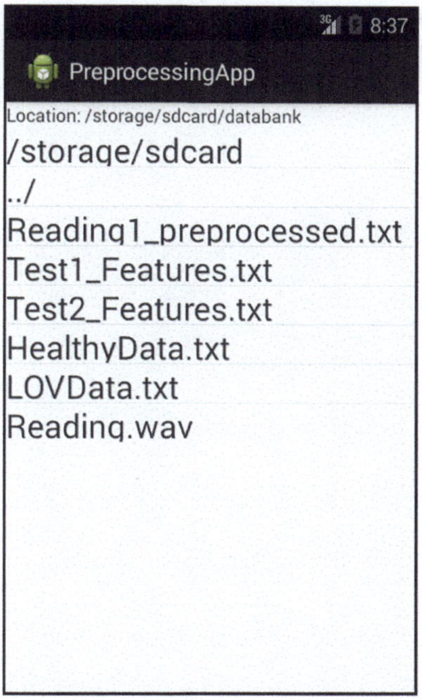

Fig. 3.41 Browse activity on page 2 of the app [4]

Fig. 3.42 File confirmation page [4]

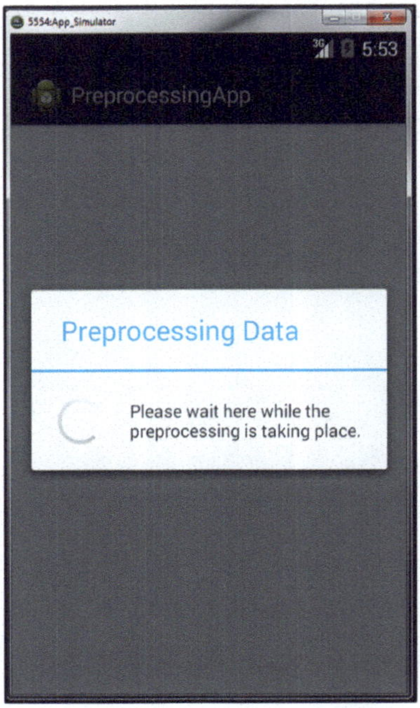

Fig. 3.43 Activity page 1 with selected parameters [4]

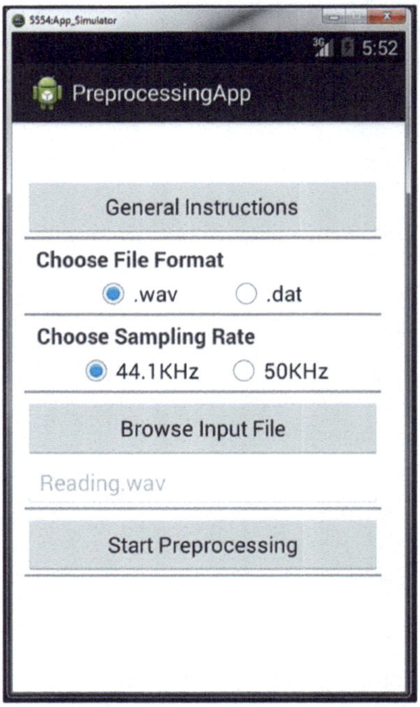

Fig. 3.44 Data pre-processing activity page [4]

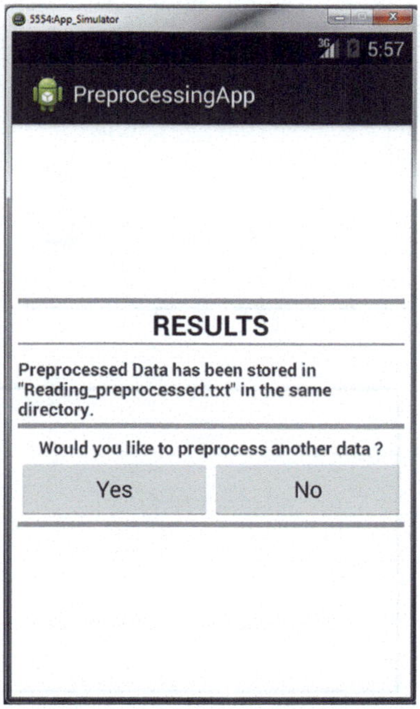

Fig. 3.45 Result activity page of the app [4]

3.8.3 Pre-processing Tool for Windows Mobile and Windows Tablet Platform

Similar application has been developed on Windows Mobile and Windows Tablet Platform as shown in Figs. 3.46 and 3.47, respectively. It takes raw data in the form of ".wav" and ".dat" file and pre-processes the whole data as per the user choice. Once the data is pre-processed, the application writes result into a text file, named as <filename>_Preprocessed.txt". This file is saved in the same directory where the input data file exists.

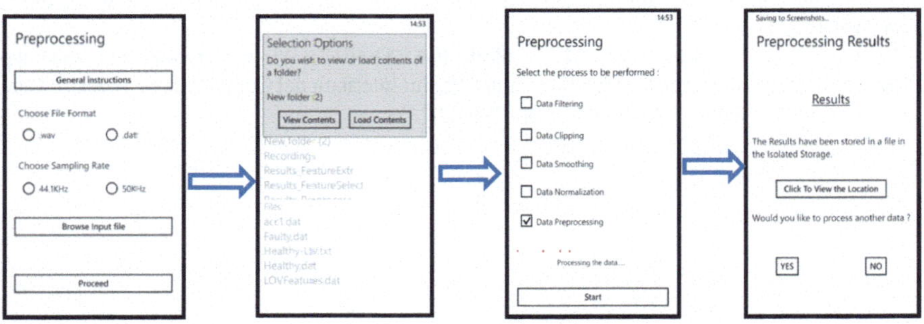

Fig. 3.46 Snapshots of pre-processing tool on Windows mobile [5]

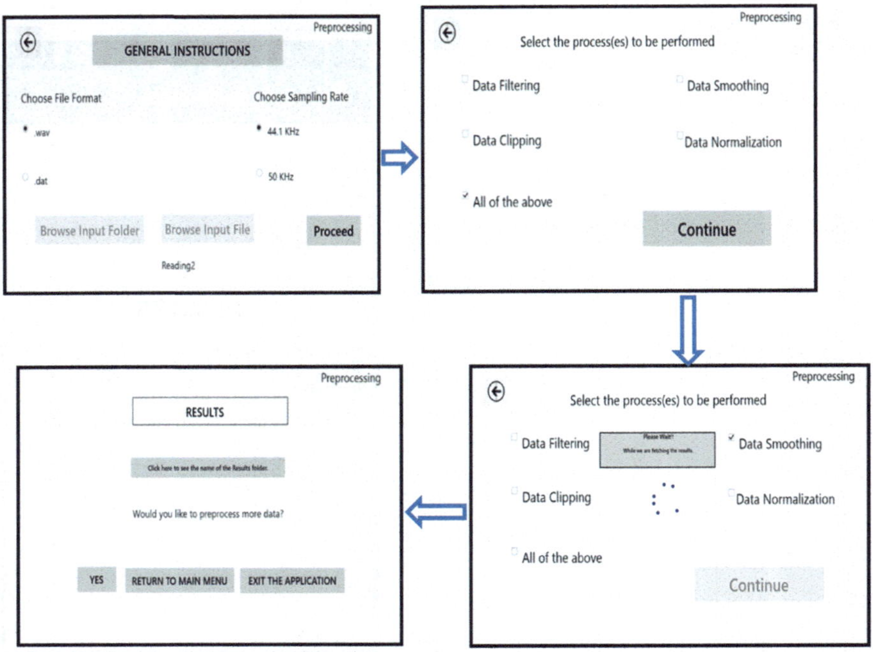

Fig. 3.47 Snapshots of pre-processing tool on Windows tablet platform [5]

References

1. Verma, N.K., Sevakula, R.K., Dixit, S., Salour, A.: Intelligent condition based monitoring using acoustic signals for air compressors. IEEE Trans. Rel. **65**(1), 291–309 (2016)
2. Verma, N.K., Agarwal, A., Sevakula, R.K., Prakash, D., Salour, A.: Improvements in preprocessing for machine fault diagnosis. In: IEEE 8th International Conference on Industrial and Information Systems, Kandy, Sri Lanka, pp. 403–408 (2013)
3. Verma, N.K., Sevakula, R.K., Thirukovalluru, R.: Pattern analysis framework with graphical indices for condition-based monitoring. IEEE Trans. Rel. **66**(4), 1085–1100 (2017)
4. Verma, N.K., Singh, S., Gupta, J.K., Sevakula, R.K., Dixit, S., Salour, A.: Smartphone application for fault recognition. In: 6th International Conference on Sensing Technology, Kolkata, India, pp. 1–6 (2012)
5. Verma, N.K., Singh, J.V., Gupta, M., Sevakula, R.K., Dixit, S.: Windows mobile and tablet app for acoustic signature machine health monitoring. In: International Conference on Industrial and Information Systems, Gwalior, India, pp. 1–6 (2014)

Chapter 4
Feature Extraction

Abstract As detailed in Chap. 3, pre-processing of data improves the quality of data. Now, this chapter proceeds towards the next step of fault diagnosis framework where key characteristics of the data are found. For this purpose, data is analyzed in different domains and thus we obtain a new set of data which we call that good features are obtained. This chapter details how features can be extracted out from the pre-processed data. When the input data to an algorithm is too large to be processed and it is suspected to be redundant (much data but not much information), then the input data will be transformed into a reduced set of features (also named features vector). Transforming the input data into the set of features is called feature extraction. In order to extract useful information from the captured data, it is necessary to represent the data in a suitable form. Generally, for feature extraction purpose, three forms of signal representation as below are utilized.

- Time domain
- Frequency domain
- Time–frequency/wavelet domain

4.1 Time Domain Representation

Generally, the real-world data in raw form exists in time domain. The time domain analysis of a signal gives its variation over time [1, 2] as shown in Fig. 4.1. The importance of time domain analysis becomes more prominent while dealing with stochastic signals and processes [3]. Here, eight statistical parameters of the signal have been extracted from time domain. They are absolute mean, maximum peak, root mean square, variance, kurtosis factor, crest factor, shape factor, and skewness of data as explained below [4].

i. **Absolute mean**: It is the mean of absolute magnitude of a signal defined as

$$\bar{x} = \frac{1}{N}\sum_{i=1}^{N}|x_i| \tag{4.1}$$

N. K. Verma and A. Salour, *Intelligent Condition Based Monitoring*, Studies in Systems, Decision and Control 256, https://doi.org/10.1007/978-981-15-0512-6_4

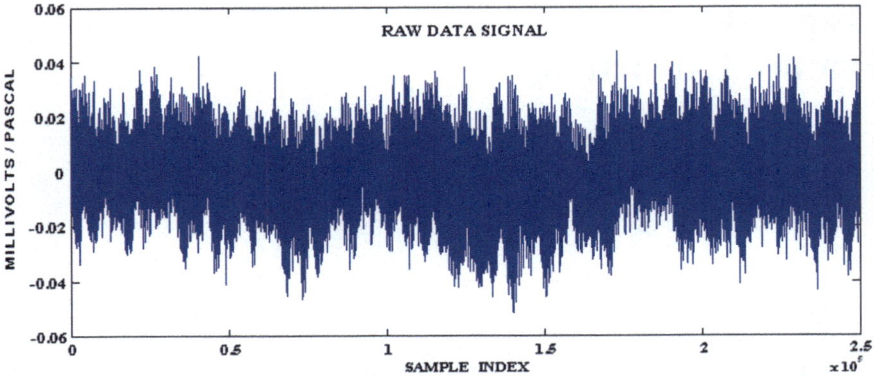

Fig. 4.1 Raw data signal in time domain [5]

where

N is the number of samples,
x_i is amplitude for given "i", and
$\bar{x}$ is absolute mean square of given signal.

ii. **Maximum peak value**: It is the peak value of a signal given by

$$x_{\mathrm{p}} = \max|x_i| \tag{4.2}$$

x_i is amplitude for given"i", and
x_{p} is the peak value.

iii. **Root mean square**: It is defined as square root of the arithmetic means of the squared signal given by

$$x_{\mathrm{rms}} = \sqrt{\frac{1}{N}\sum_{i=1}^{N} x_i^2} \tag{4.3}$$

N is the number of samples,
x_i is amplitude for given "i", and
x_{rms} is root mean square value.

iv. **Variance**: It is defined as the measure of average distance between each data points of the dataset and their mean value given by

$$x_{\mathrm{var}} = \frac{1}{N-1}\sum_{i=1}^{N} (x_i - \bar{x})^2 \tag{4.4}$$

where

N is the number of samples,
x_i is amplitude for given "i",
$\bar{x}$ is the mean value, and
x_{var} is the variance.

v. **Kurtosis**: It is the measure of the "Peakedness" of the probability distribution of a real-valued signal. It is like fourth moment of the signal defined as

$$\beta = \frac{E(X - \bar{x})^4}{\sigma^4} \tag{4.5}$$

where

σ is the standard deviation,
β is kurtosis, and
$E(t)$ is the expected value function given by

$$E(X - \bar{x}) = \frac{1}{N} \sum_{i=1}^{N} (x_i - \mu)^4$$

vi. **Crest factor**: It is the ratio of peak value to RMS value of the signal as

$$x_{\text{cf}} = \frac{x_{\text{p}}}{x_{\text{rms}}} \tag{4.6}$$

where

x_{p} is the peak value,
x_{rms} is the root mean square value, and
x_{cf} is the crest factor value.

vii. **Shape factor**: It is the ratio of RMS value of the signal to the absolute mean as

$$S_{\text{f}} = \frac{x_{\text{rms}}}{\bar{x}} \tag{4.7}$$

where

$\bar{x}$ is the mean value,
x_{rms} is the root mean square value, and
S_{f} is the shape factor.

viii. **Skewness**: It is defined as a measure of asymmetry in probability distribution given by

$$S = \frac{E(X - \bar{x})^3}{\sigma^3} \tag{4.8}$$

where

σ	is the standard deviation,
$\bar{x}$	is the mean value, and
$E(X - \bar{x})$	is the expected value function.

4.2 Frequency Domain Representation

In general, using time domain signal, it is difficult to detect the system condition. The energy distribution in frequency domain can reflect different system conditions to a certain extent. Therefore, a signal can be better analysed in frequency domain. Figure 4.2 shows the frequency response of a pre-processed signal. For studying the properties of a signal in frequency domain, the time domain signal needs to be converted to frequency domain by using Fast Fourier transform (FFT). FFT converts the signal information to a magnitude and phase component for each frequency. It shows how much of the signal's energy is present as a function of frequency. Here, the discrete form of Fourier Transform (FT) is used. Discrete Fourier transform (DFT) of signal $x(n)$ is defined as:

$$X(k) = \sum_{n=0}^{N-1} x(n)e^{-\frac{2\pi kn}{N}} \tag{4.9}$$

where $k = 0, 1, \ldots, N - 1$.

Similarly, the inverse DFT can be expressed as follows:

$$x(n) = \frac{1}{N}\sum_{k=0}^{N-1} X(k)e^{-\frac{2\pi kn}{N}} \tag{4.10}$$

where $n = 0, 1, \ldots, N - 1$.

Thus, a spectrum of frequency components are obtained. Then, the entire frequency band of transformed signal is divided into certain number of bins (here eight bins). Each frequency bin collects the energy or amplitude for a small frequency range. The total spectral energy and individual bin spectral energy can be formulated as

$$E_s = \sum_{F_s/2} |X(k)|^2 \text{ and} \tag{4.11}$$

$$\{E_B\}_i = \sum_{f_B/2} |X(k)|^2 \tag{4.12}$$

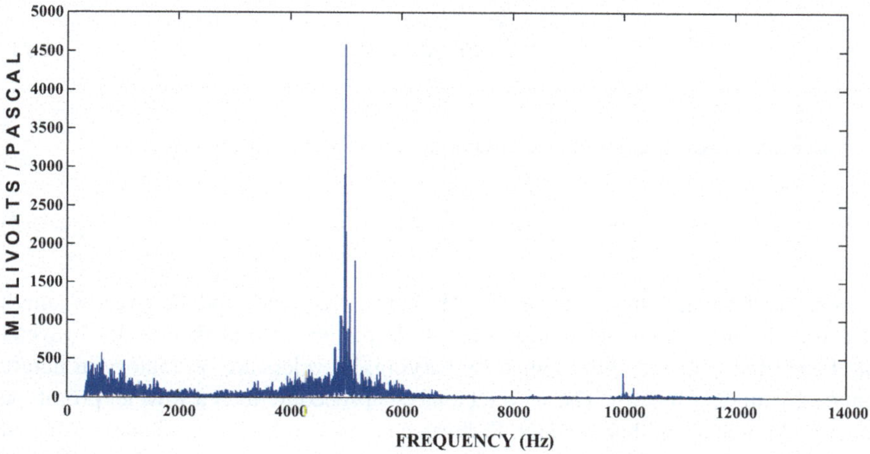

Fig. 4.2 Frequency response of the pre-processed signal [5]

The spectral energy in each bin is calculated and divided by total energy along the entire spectrum. These energy ratios form eight features of frequency domain. The ratio of spectral energy of each bin to the total spectral energy provides desired number of features in frequency domain as

$$F_i = \frac{\{E_{\mathrm{B}}\}_i}{E_{\mathrm{s}}} \tag{4.13}$$

where

$X(k)$ is magnitude of samples,
E_{s} is total spectral energy,
$\{E_{\mathrm{B}}\}_i$ is spectral energy of ith bin,
F_{s} is sampling frequency,
f_{B} is range of bin frequency, its value is always less than $F_{\mathrm{s}}/2$, and
i represents the bin number, i.e., 1, 2, … up to the number of bins.

4.3 Time–Frequency Domain Representation

As machine generated acoustic signals are non-stationary in nature, their spectral components change with time. To observe spectral content of a signal with time, time–frequency representation of a signal is needed. It represents data in terms of a set of basic functions known as wavelets. Basically, wavelets are mathematical functions that divide data into different frequency components and then analyse it at different scales and resolution. The wavelet function of zero average ψ is defined as

$$\int_{-\infty}^{\infty} \psi(t)\,\mathrm{d}t = 0 \tag{4.14}$$

The wavelet is dilated with a scale parameter s, and translated by u as

$$\psi_{u,s}(t) = \frac{1}{\sqrt{s}}\psi\left(\frac{t-u}{s}\right) \tag{4.15}$$

A wavelet function has finite energy. The centre frequency and the window length of a wavelet are varied by changing the scale parameter and thereby the wavelet can be shifted over the entire signal by varying. Wavelets are oscillatory in nature and decay quickly to zero. The wavelet analysis procedure uses a wavelet prototype function known as mother wavelet. Performance of wavelet is significantly affected by mother wavelet. The features in wavelet domain are calculated using following three wavelet transforms [6]. Each of them is explained in detail as follows.

4.3.1 Continuous Wavelet Transform (CWT)—Morlet Wavelet

Morlet function is used as "Mother wavelet" (also called as prototype wavelet) for calculating CWT [7]. A Morlet wavelet is a cosine signal decaying exponentially on both sides. Mathematically, it is defined as

$$y_{a,b}(t) = \mathrm{e}^{\frac{-b^2(t-b)^2}{a^2}} \cos\left(\frac{p(t-b)}{a}\right) \tag{4.16}$$

where a and b are mother wavelet. The signal is first transformed from time domain to time–frequency domain by convolving it with the Morlet wavelet. Standard deviation, wavelet entropy, kurtosis factor, skewness, variance, zero-crossing rate, and sum of peaks are calculated from the resultant wavelet coefficients. This generates seven more features [8] defined as follows:

i. **Standard deviation**: Standard deviation is a measure of the variability of a dataset. A low standard deviation indicates that coefficient tends to be very close to the mean value while high standard deviation indicates that coefficients are spread out over a large range of values.
ii. **Entropy**: The diversity of a possibility series can be measured with Shannon entropy. Thus, sparsity of wavelet coefficients may be measured with the entropy of the wavelet coefficients. The entropy here is termed as "wavelet entropy". The expression obtained on dividing each coefficient m_i by $\sum_{i=1}^{N} x_i$ gives

$$\{d_i\} = \frac{m_i}{\sum_{i=1}^{N} x_i} \tag{4.17}$$

where $i = 1, 2, 3, \ldots, N$.
The wavelet entropy En is calculated using the following expression

$$\mathrm{En} = -\sum_{i=1}^{N} d_i \log d_i \quad (4.18)$$

where

m_i is the amplitude for a given "i" of vector "M" and
N is the total number of samples.

iii. **Kurtosis**: Kurtosis is a measure of the peakedness. Higher kurtosis indicates more variance due to infrequent extreme deviations. It is used for the detection of faults because it is sensitive to sharp variant structure, such as impulses.
iv. **Skewness**: Skewness is a measure of asymmetry. Lower the value of skewness more is the symmetric signal.
v. **Variance**: Variance is a measure of dispersion of coefficients of Morlet wavelet about their mean value.
vi. **Sum of peaks**: The feature components of the mechanical signals are impulse in nature. Finding the dominant peaks and their sum becomes important feature for fault diagnosis.
vii. **Zero-crossing rate**: It is the rate of change of sign along the coefficient vector, i.e., the rate at which the value of coefficients changes from positive to negative and back.

4.3.2 Discrete Wavelet Transform

In Discrete Wavelet Transform (DWT), the signal is analysed at different scales using digital filtering techniques. Variation in the filtering operation changes the resolution of signal. The variation in scale is caused by upsampling and downsampling. DWT uses filter banks [9] for the computation of wavelet transform [4]. The filter banks consist of filters which separate a signal into various frequency bands. Filters are interpreted as wavelet functions at different scales in DWT, which represent the frequency content of wavelet function. There are several families of these filters like Haar, Daubechies, Coiflets, etc. The signal is convoluted with one of these filters to obtain two sets of coefficients. The convolution expression is given by:

$$h[n] = \sum_{m} f[n - m] * g[m] \quad (4.19)$$

where $h[n]$ is convolved output of functions f and g. The coefficients obtained from low pass filter are known as approximation coefficients and those obtained from

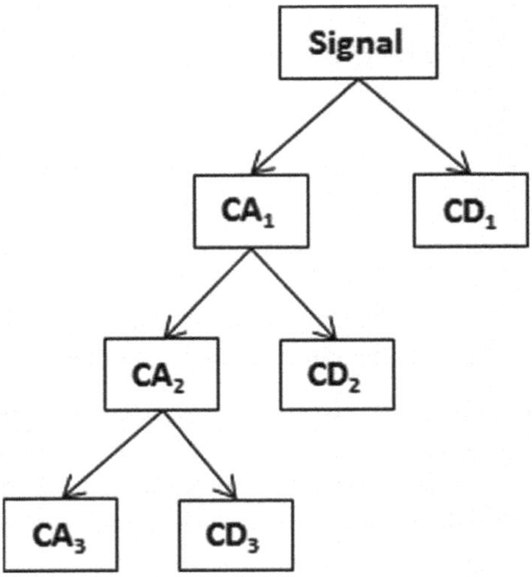

Fig. 4.3 DWT composition of a signal [10]

high pass filter are known as detail coefficients. The transform follows a one-sided tree-like structure as shown in Fig. 4.3.

In this figure, the signal is decomposed in CA_i and CD_i (where $i = 1, 2, \ldots$). These are approximation and detail coefficients, respectively, of ith level. The approximation coefficients obtained are further decomposed to get new set of approximation and detail coefficients. On the other hand, detail coefficients are not decomposed further. Daubechies-4 (db4) decomposition filter has been used for the model used here. Both, db4 low pass and db4 high pass filter consists of eight coefficients.

The level of decomposition depends on the signal and task to be performed. Herein, the signal is decomposed till sixth level which results in six detail coefficients and one approximation coefficient. The detail coefficients at level 1, 2, and 3 have higher frequency content as compared to detail coefficients at level 4, 5, and 6. The initial decomposition levels have more abrupt variations as compared to higher decomposition levels. These observations are useful in extraction of features from DWT as out of the nine features extracted, three are variances of the detail coefficients at level 1, 2, and 3. For finding next three features, autocorrelation of detail coefficients at level 4, 5, and 6 is performed. Variance of autocorrelation coefficients at three levels forms the next three features [11]. Autocorrelation is a mathematical tool for finding repeating patterns, thus capturing similarity within the observations.

$$R_{xx}(l) = \sum_{n=i}^{N-|K|-1} x[n] \times x[n-1] \tag{4.20}$$

where

$i = l, k = 0$ for $l > 0$ and

$i = 0, k = l$ for $l < 0$

$l =$ time shift (or lag) parameter.

Three more features can be found by taking the absolute mean of smoothened versions of detail coefficients at level 1, 2, and 3. The averaging filter is used for smoothing of rapid fluctuations of detail coefficients defined as:

$$y(k) = \sum_{l=k-m}^{k+m} |X(l)| \tag{4.21}$$

where m decides the degree of smoothness which is 5 in our case considered here. Again, the decomposition of approximation coefficients is done using both filters and the process is repeated till nth level. These coefficients are further used for calculation of features. Figure 4.4 shows the decomposition at different levels using DWT. Figure 4.5 shows the pre-processed signal in time domain.

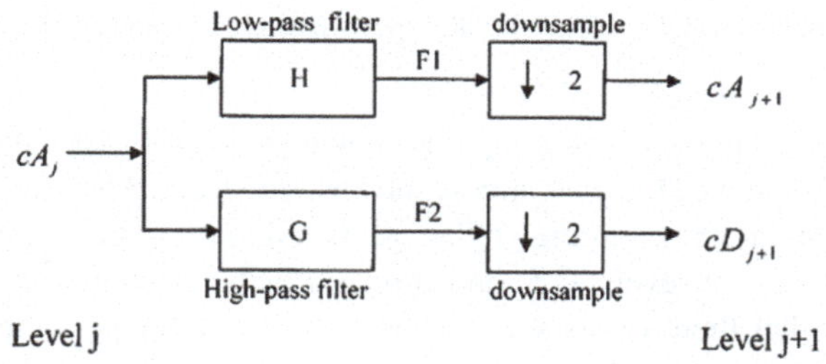

Fig. 4.4 DWT decomposition at different levels [12]

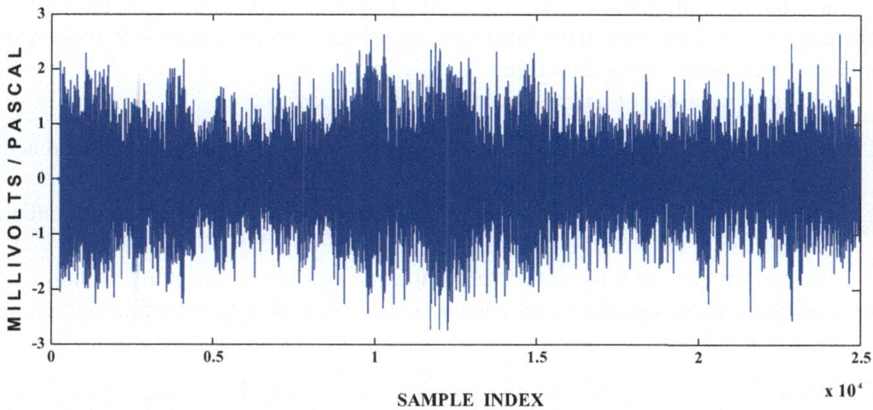

Fig. 4.5 Pre-processed signal in time domain [5]

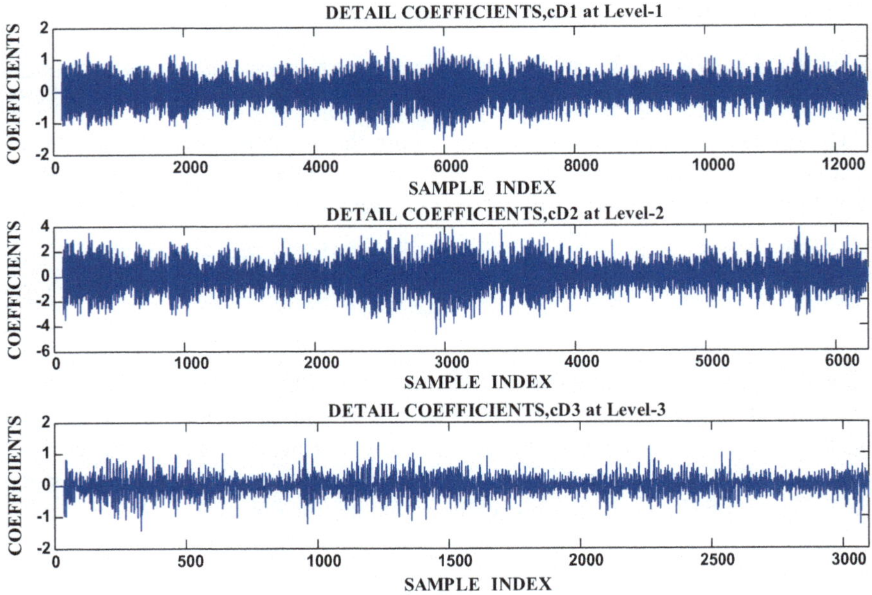

Fig. 4.6 Detail coefficients CD_1, CD_2, CD_3 at respective levels 1, 2, and 3 [5]

The decomposed pre-processed input is shown in Fig. 4.6. The detail coefficient vectors CD_1, CD_2, and CD_3 at decomposition levels 1, 2, and 3 have relatively higher frequency content, as shown in Fig. 4.7 as compared to the detail coefficient vectors CD_4, CD_5, and CD_6 at levels 4, 5, and 6 respectively, as shown in Fig. 4.8. It is readily verified that there exists some overall similarity (or periodicity) at higher decomposition levels while abnormal rapid effects are observed as abrupt variations in the initial decomposition levels. These observations are explored in order to extract useful features from the set of coefficients.

The Autocorrelation Function (ACF) has been used to express the signal similarity, whereas a form of maximum deviation on smoothed signals is employed to express rapid changes in the signal structure.

The autocorrelation of a signal has following properties:

i. Autocorrelation function is defined as a measure of similarity or coherence between a signal and its shifted version.
ii. The two versions of a signal "match" yielding maximum autocorrelation under no shift.
iii. Increase in time shift decreases the similarity, and hence, the autocorrelation.
iv. As shift approaches infinity, all traces of similarity vanish and the autocorrelation decays to zero.

The signal-averaging filter has been used for smoothing the rapid fluctuations in coefficient vectors CD_1, CD_2, and CD_3. A signal-averaging filter is also called as smoothing or moving average filter. The moving average filter is implemented as

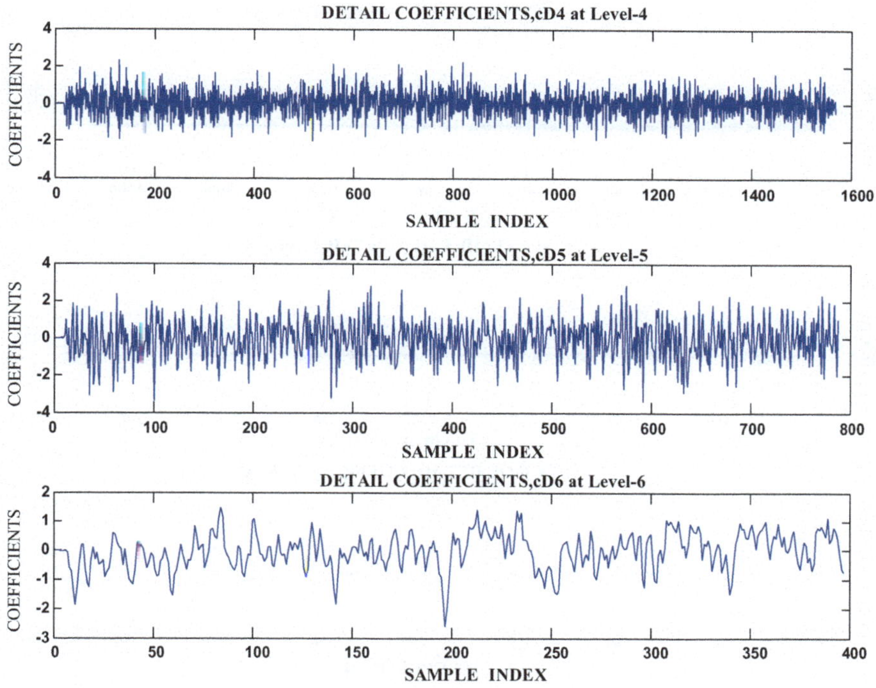

Fig. 4.7 Detail coefficients CD_4, CD_5, CD_6 at respective levels 4, 5, and 6 [5]

follows. Let $x(n)$, $y(n)$ be the input and output signals, respectively, then for each data point, the moving average can be given as

$$y(k) = \frac{\sum_{l=k-m}^{k+m} |x(l)|}{2m} \tag{4.22}$$

The smoothed signals are referred to as S_1, S_2, and S_3 of detail coefficients CD_1, CD_2, and CD_3 at levels 1, 2, and 3, respectively. Figure 4.9 shows the smoothed signals S_1, S_2, and S_3.

4.3.3 *Wavelet Packet Transform (WPT)*

One possible drawback of wavelet transform is the poor frequency resolution in the high-frequency region. Therefore, it is difficult to discriminate between signals having close high-frequency components. Wavelet packets (a generalization of wavelet bases) are alternative bases that are formed by taking linear combinations of the usual wavelet functions [10, 13]. A wavelet packet function (W) is a function with three indices as

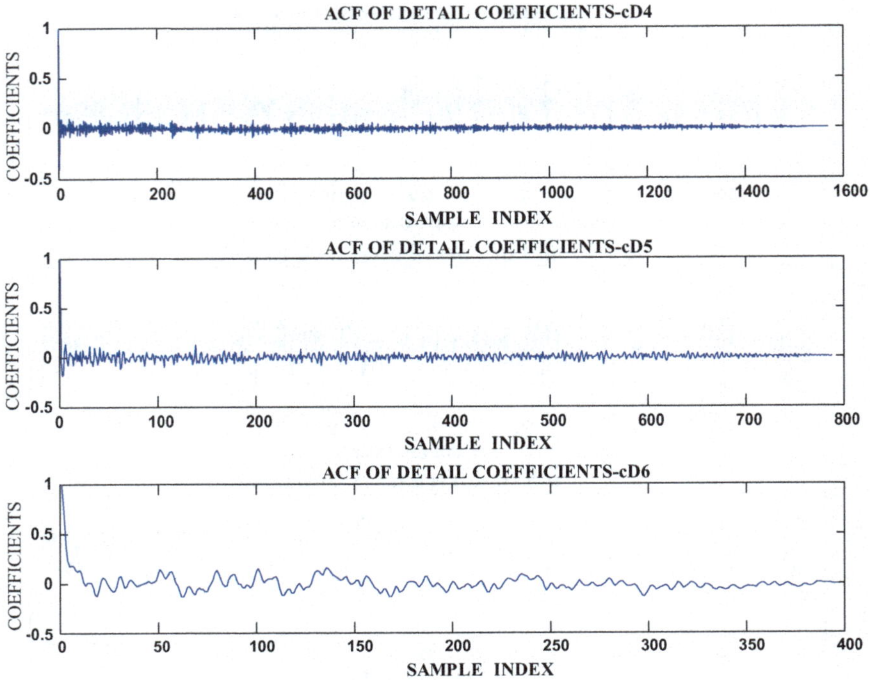

Fig. 4.8 ACF signal of detail coefficients CD_4, CD_5, CD_6 [5]

$$W_{j,k}^{n}(t) = 2^{j/2} W^{n}\left(2^{j} t - k\right) \tag{4.23}$$

where integers j and k are the index scale and translation operator, respectively, which are similar to wavelets. The index n is called the modulation parameter or the oscillation parameter.

The first two wavelet packet functions are the usual scaling function and mother wavelet function, respectively defined as

$$\begin{aligned} W_{0,0}^{0}(t) &= \varphi(t), \text{ and} \\ W_{0,0}^{1}(t) &= \psi(t) \end{aligned} \tag{4.24}$$

The wavelet packet functions for $n = 2, 3, \ldots$ are then defined by the following recursive relationships.

$$W_{0,0}^{2n}(t) = \sqrt{2} \sum_{k} h(k) W_{1,k}^{n}(2t - k) \tag{4.25}$$

$$W_{0,0}^{2n+1}(t) = \sqrt{2} \sum_{k} g(k) W_{1,k}^{n}(2t - k) \tag{4.26}$$

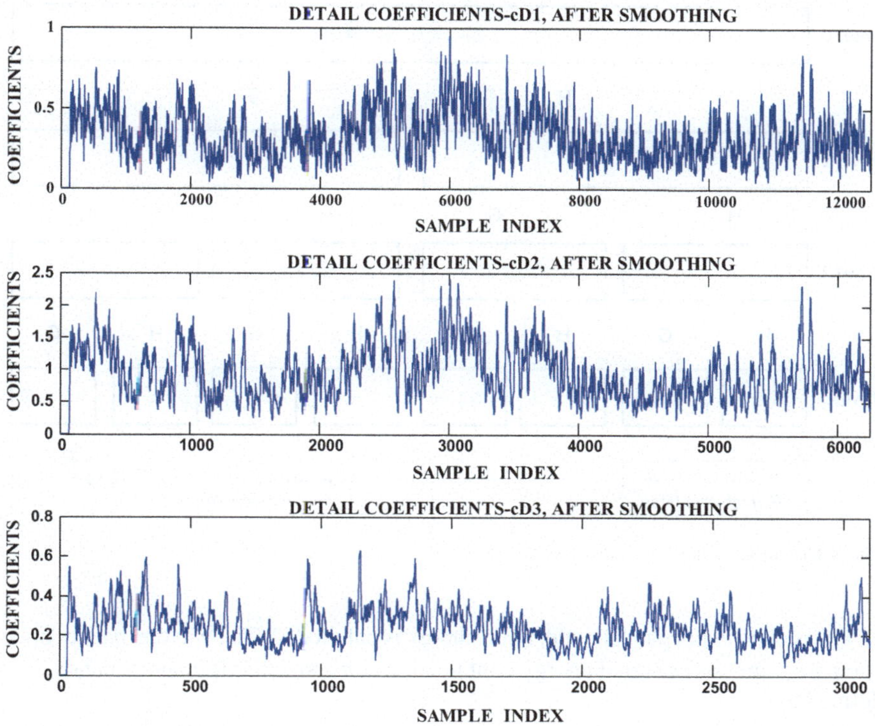

Fig. 4.9 Smoothed signals S_1, S_2 and S_3 [5]

where $h(k)$ and $g(k)$ are the Quadrature Mirror Filter (QMF) associated with the predefined scaling function and mother wavelet function. To measure specific time–frequency information in a signal, we simply take the inner product of the signal along with that particular basis function. The wavelet packet coefficients of a function f can be computed as

$$w_{j,n,k} = \langle f, W^n_{j,k} \rangle = \int f(t) \cdot W^n_{j,k}(t)\, \mathrm{d}t \tag{4.27}$$

The idea of the usual wavelet decomposition is generalized to describe the calculation of wavelet packet coefficients $w_{j,n,k}$ of a discrete-time signal. Computing the full Wavelet Packet Decomposition (WPD) of a discrete-time signal involves the application of both filters to the discrete-time signal $[x_1, x_2, \ldots, x_n]$ and then recursively to each intermediate signal. The procedure is illustrated in Fig. 4.10.

Each wavelet packet coefficient is also referred as wavelet packet node. It is given as the inner product of the signal function and wavelet bases defined by

$$w_{j,n,k} = \langle f, W^n_{j,k} \rangle = \langle f, 2^{j/2} W^n (2^j t - k) \rangle \tag{4.28}$$

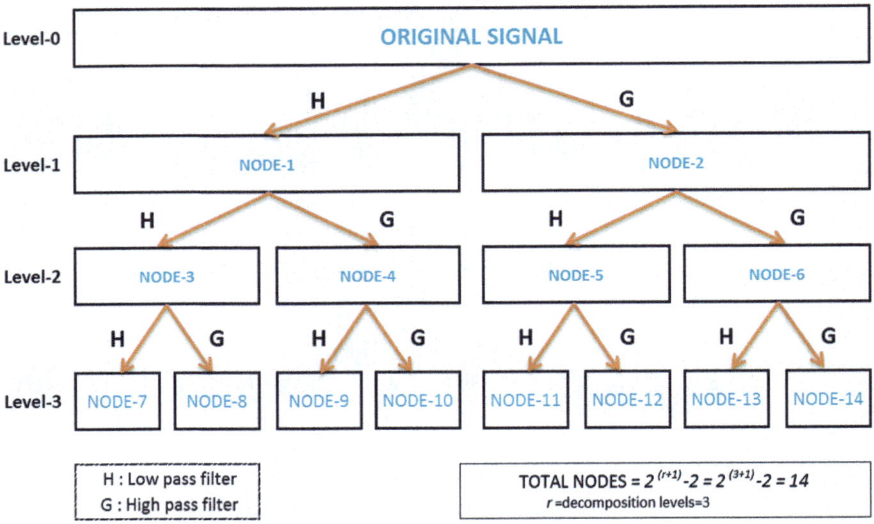

Fig. 4.10 Implementation of discrete WPD [5]

For each wavelet packet node, node energy is calculated. The wavelet packet node energy $e_{j,n}$ measures signal energy contained in some specific frequency bands. It is defined as

$$e_{j,n} = \sum_{k} w_{j,n,k}^{2} \tag{4.29}$$

These node energies characterize a robust signal feature from classification point of view rather than using the wavelet coefficients alone directly [10]. Hence, it is used as a feature component. WPT allows a signal decomposition to examine different time–frequency resolution components in a signal. By computing the full WPD on a signal segment with $n = 2^J$ points for r resolution levels (where J and r are integers), it results in a group of $2^1 + 2^2 + \cdots + 2^r = 2^{r+1} - 2$ sets of coefficients where each set corresponds to wavelet packet node. Here, we are computing wavelet node energy which is computed as a feature component which results in $2^{r+1} - 2$ sets of features [4].

4.4 Expanding New Set of Features

Recorded signals are real-time signals which are mainly non-stationary in nature, which means statistical properties of signals changes with respect to time [14–16]. To analyse these kinds of signals, numbers of algorithms have been developed where both time and frequency information of signal can be determined simultaneously [17].

Here, following transforms [18] have been studied and applied to extract some important properties from signals based on which different states of machines can be recognized [19, 20].

i. Short-Time Fourier Transform
ii. Wigner–Ville Distribution
iii. Pseudo-Wigner–Ville Distribution
iv. Choi–William Distribution
v. Born–Jordan Distribution
vi. S-Transform
vii. Discrete Cosine Transform
viii. Autocorrelation
ix. Updated Morlet Transform
x. Convolution with Sine Wave

The sample plots of transform and feature extracted using them for random vibration recording from healthy, LOV, and LIV datasets have been studied and compared here. Also, for feature level study, we did a comparison using smartphone acoustic readings dataset having 150 recordings each for healthy, LOV, and LIV.

4.4.1 Short-Time Fourier Transform (STFT)

As signal is non-stationary, frequency domain information is defined over infinite period of time. STFT performs FT on small sections of signal as shown in Fig. 4.11. Hence, it is termed as windowed FT or STFT. In STFT, it is assumed that each section/window of signal is stationary. Thus, first signal is segmented into narrow time intervals (which are narrow enough to be considered stationary) and then FT is performed on each section.

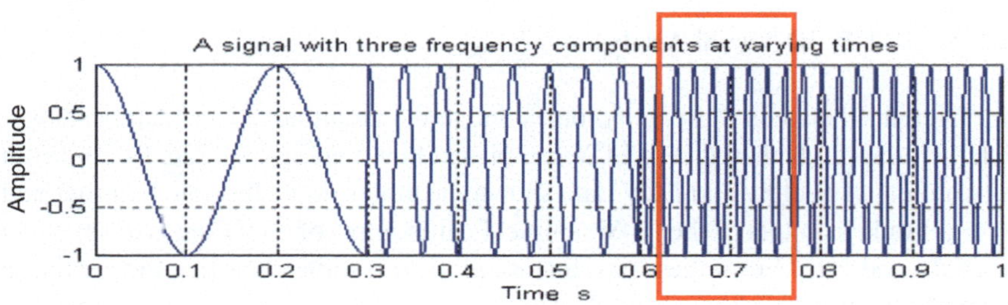

Fig. 4.11 Window in signal chosen for STFT [21]

STFT can be calculated as follows:

$$\mathrm{STFT}(\tau, f) = \int_{-\infty}^{\infty} [x(t) \cdot w(t-\tau)] \cdot \mathrm{e}^{-\mathrm{j}2\pi f t}\, \mathrm{d}t \tag{4.30}$$

where

$x(t)$ signal to be analysed,
$w(t)$ window function centred on time,
τ time parameter, and
f frequency parameter.

4.4.1.1 Steps to Extract Feature Using STFT

The whole procedure of calculating STFT in four steps are

Step 1 In this step, we set the window parameter, i.e., size of analysis window, amount of overlap between windows (e.g. 30%), and windowing function are set.
Step 2 Once all these parameters were set, signal is multiplied by the windowing function which gives windowed segment.
Step 3 Apply FFT to each windowed segment which maps signal into two-dimensional space of time and frequency.
Step 4 For extracting features from this two-dimensional form, first find maximum values of matrix along the column (i.e., maximum value for given frequency in the entire time scale of the signal) to get signal spectrum in one-dimensional format. The resultant signal spectrum is then divided into 72 equal segments for calculating energy. The ratio of energy of each segment to the total energy of resultant spectrum is considered as a feature. In this manner, 72 features from STFT are obtained.

4.4.1.2 STFT: 3D Visual Analysis

The difference between these two plots is due to the amplitude. The healthy sample has larger amplitude than that of LOV sample as shown in Figs. 4.12 and 4.13. From the two plots of healthy and LOV data shown, healthy sample has much outlier and less amplitude than that of the LOV sample. Feature plots of STFT are also shown in Figs. 4.14 and 4.15. From these two feature plots, it is quite clear that there exists a separation between healthy and LOV class.

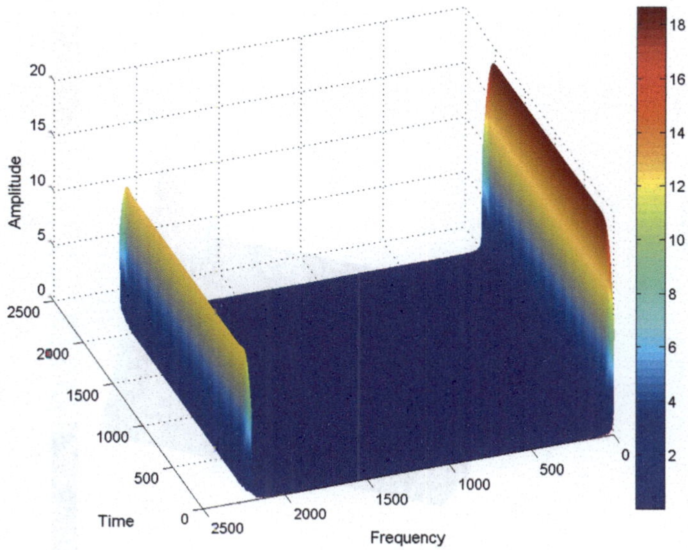

Fig. 4.12 3D plot of healthy sample [21]

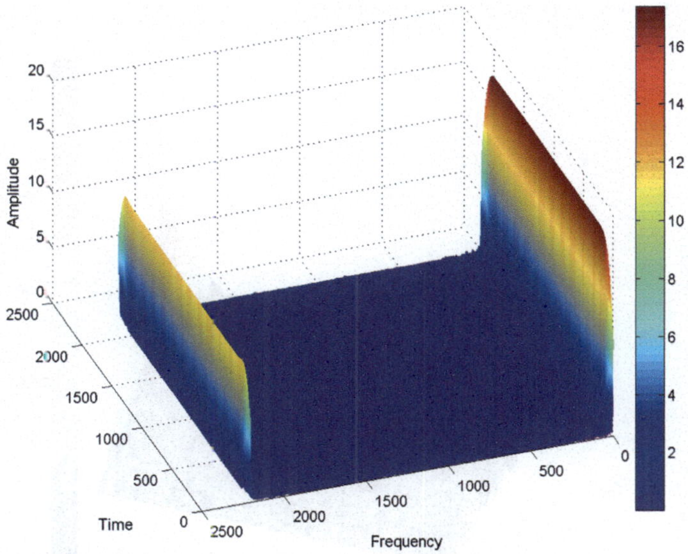

Fig. 4.13 3D plot of LOV sample [21]

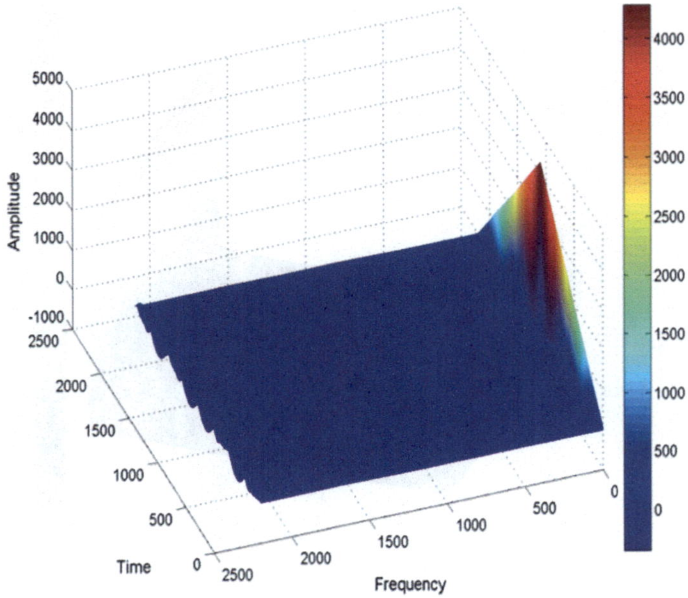

Fig. 4.14 STFT feature 3D plot of healthy sample [21]

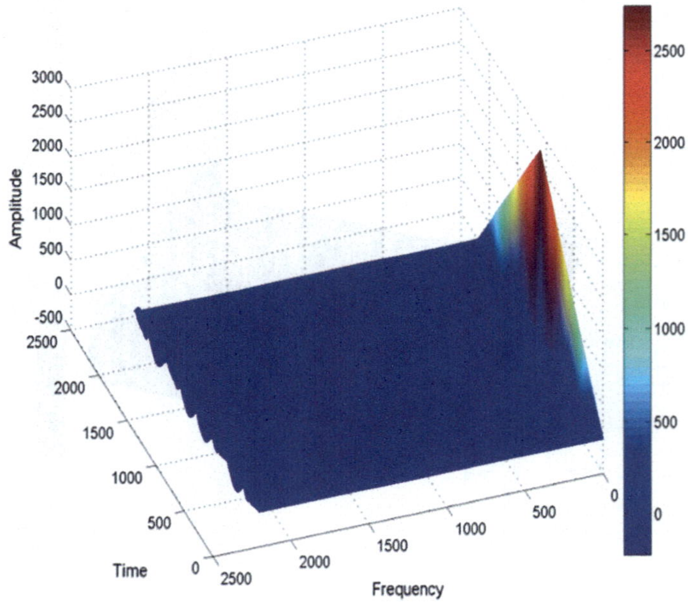

Fig. 4.15 STFT feature 3D plot of LOV sample [21]

4.4.1.3 Drawbacks of STFT

One of the major drawbacks of using STFT is that it has fixed window size. Narrow window in STFT provides poor frequency resolution and use of wide window makes poor time resolution in time–frequency analysis. The second issue is of resolution which also occurs using STFT in time–frequency distribution. To overcome the problem of resolution, other STFT transforms used.

4.4.2 *Wigner–Ville Distribution (WVD)*

WVD is based on the autocorrelation function (quadratic approach) for a time–frequency representation. It basically compares the signal with itself at different times and frequencies.

For a signal, $x(t)$ with its analytic associate $s(t)$, the WVD $(W(t, f))$ is defined as

$$W(t, f) = \int_{-\infty}^{\infty} s\left(t + \frac{\tau}{2}\right) * s\left(t - \frac{\tau}{2}\right) \mathrm{e}^{-\mathrm{j}2\pi f} \, \mathrm{d}\tau \tag{4.31}$$

where $s(t) = x(t) + \mathrm{i}H[x(t)]$ and $H[x(t)]$ is the Hilbert transform of input signal.

The Hilbert transform is a kind of quadrature filter which simply shifts phase of all frequency components of its input by $\frac{-\pi}{2}$ rad.

The analysis of WVD spectra of continuous and discrete signals with time-limited windows demonstrates better frequency concentration and less phase dependence than real or analytic signal Fourier spectra. WVD maps the signal into two-dimensional functions of time and frequency. For extracting the key features from this two-dimensional form, like STFT, find maximum values of the matrix along column. Thus, one finds the maximum amplitude over time corresponding to every frequency. The one-dimensional spectrum is then divided into 72 equal bins. The energy of each bin is then calculated. The ratio of each bin's energy to the total spectral energy provides 72 features.

4.4.2.1 WVD: 3D Visual Analysis

From the plots of healthy versus LOV and healthy versus LIV samples shown in Figs. 4.16 and 4.17, respectively, it can be observed that both samples are separable in terms of amplitude as well as outliers. Figures 4.18 and 4.19 show the WVD feature plots of healthy versus LOV and healthy versus LIV samples, respectively.

The WVD features of the dataset of different class (healthy, LOV) and (healthy, LIV) are consistent and are able to separate two class very clearly Captionkey, as shown in figure.

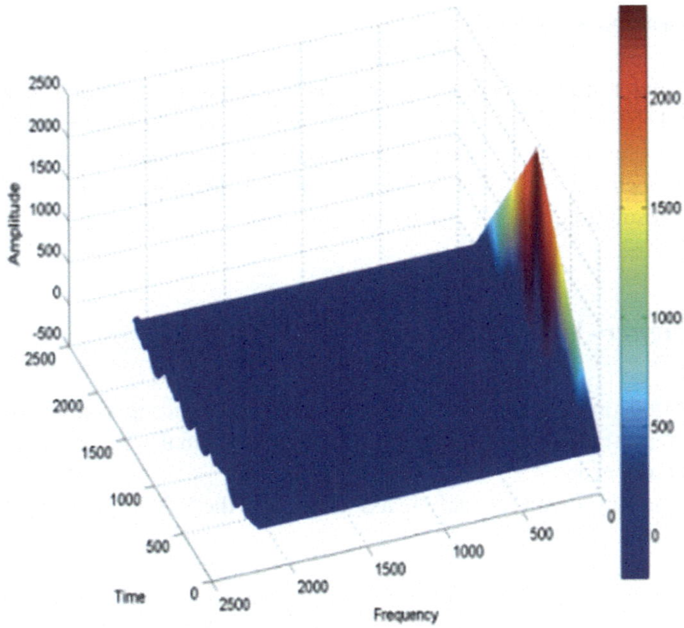

Fig. 4.16 3D plot of healthy versus LOV sample [21]

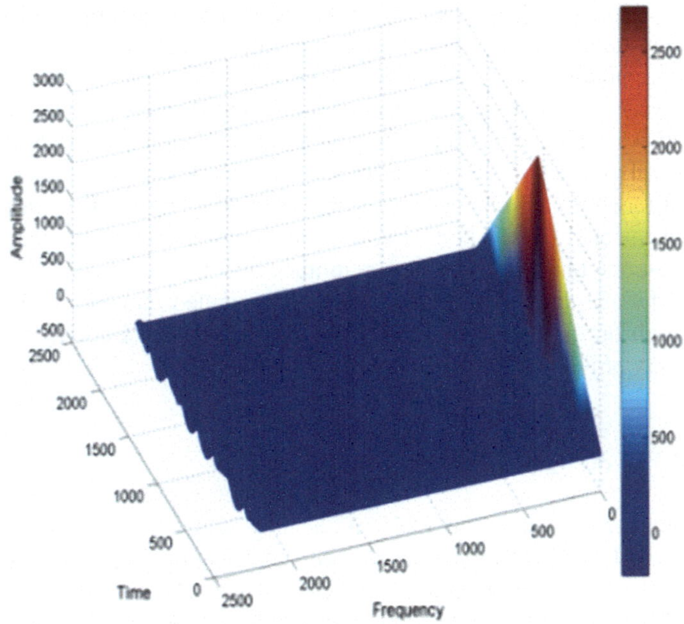

Fig. 4.17 3D plot of healthy versus LIV sample [21]

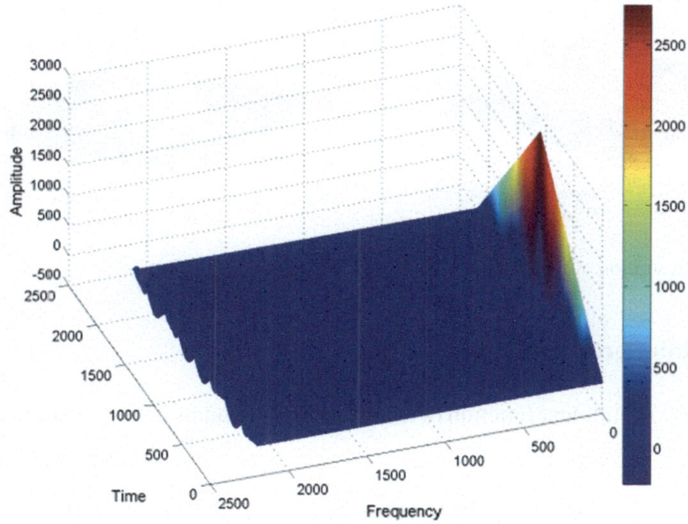

Fig. 4.18 WVD feature for healthy versus LOV sample [21]

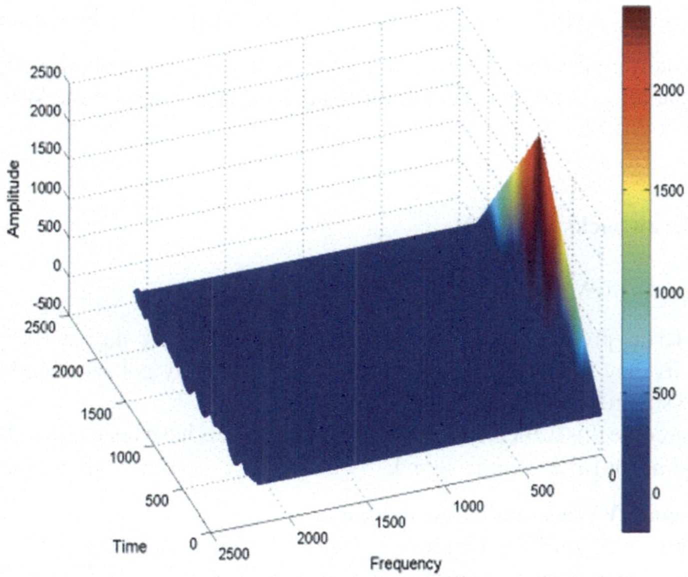

Fig. 4.19 WVD feature for healthy versus LIV sample [21]

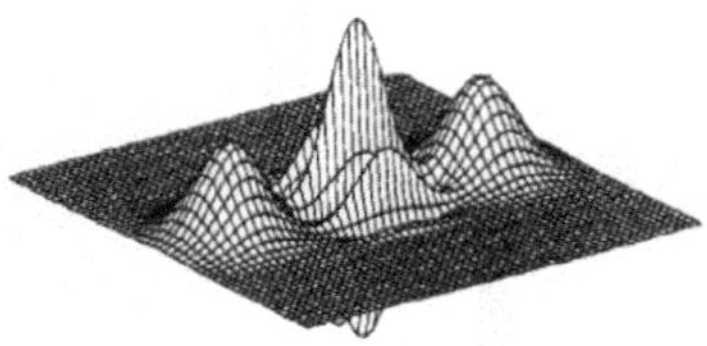

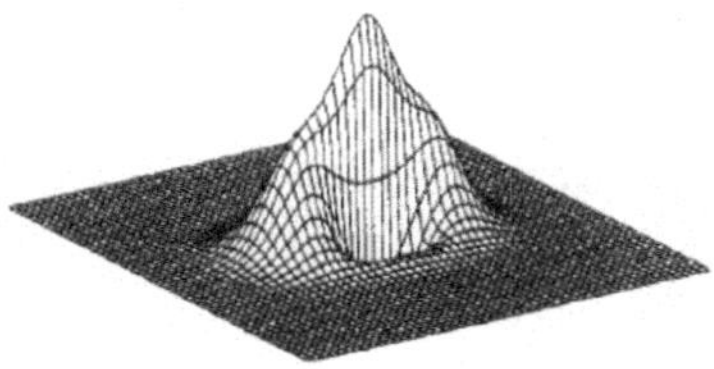

Fig. 4.20 Illustration of interference with WVD [21]

4.4.2.2 Interference in WVD

As the WVD is a bilinear function of the signal s, the quadratic superposition principle applies

$$W_{s+x}(t,\omega) = W_s(t,\omega) + W_x(t,\omega) + 2\Re\{W_{s,x}(t,\omega)\} \tag{4.32}$$

where

$W_{s,x}(t,\omega) = \frac{1}{2\pi}\int s\left(t+\frac{\tau}{2}\right) * s\left(t-\frac{\tau}{2}\right) \mathrm{e}^{-\mathrm{j}\omega\tau}\,\mathrm{d}\tau$ is the cross-WVD of s and x.

These interference terms are troublesome since they may overlap with auto-terms (signal terms) and thus make it difficult to visually interpret the WVD image (Refer Fig. 4.20).

4.4.2.3 Drawbacks of WVD

The drawbacks of WVD are:

i. Auto-terms provide useful information and cross-terms may be harmful in a sense as they obscure the auto-terms. Auto-terms may get sacrificed in a way due to suppression by cross-terms.
ii. To overcome this difficulty, different types of kernels have been utilized to obtain a good result for a specific signal.
    a. Pseudo-Wigner–Ville Distribution
    b. Choi–William Distribution
    c. Born–Jordan Distribution
iii. The WVD spectrum is always real-valued but can very well take on negative values. Therefore, it does not give a “true” density.

4.4.3 *Pseudo-Wigner–Ville Distribution (PWD)*

The pseudo-Wigner distribution obtains the maximum resolution in the time domain with a substantial sacrifice of resolution in frequency direction which is acceptable.

For a signal, $x(t)$ with analytic associate $s(t)$, the PWD ($\mathrm{PW}(t,\omega)$) is defined as

$$\mathrm{PW}(t,\omega)=\frac{1}{2\pi}\int s\left(t+\frac{\tau}{2}\right)*s\left(t-\frac{t}{2}\right)h(t)\,\mathrm{e}^{-\mathrm{j}\omega\tau}\,\mathrm{d}\tau \tag{4.33}$$

where $h(t)$ is known as window function such as Gaussian function and rectangular function. When Gaussian window applied along with WVD, the effect of cross-components gets reduced. PWD maps the signal in two-dimensional functions of time and frequency. Similar procedure as in WVD has been followed here for extracting 72 features as shown in Fig. 4.21.

4.4.3.1 PWD: 3D Visual Analysis

The healthy sample here contains more amplitude range while LOV sample has more outliers along time axis as shown in Fig. 4.22.

Also, in case of healthy and LIV classification, both have differences as healthy data recording has more outlier while LOV has more amplitude range as shown in Fig. 4.23.

4.4.3.2 Drawbacks of PWD

The drawbacks of PWD are:

1. As PWD is obtained by making smoothness along frequency axis, it does not provide considerable reduction in interferences along time axis.
2. It provides poorer resolution along frequency axis.

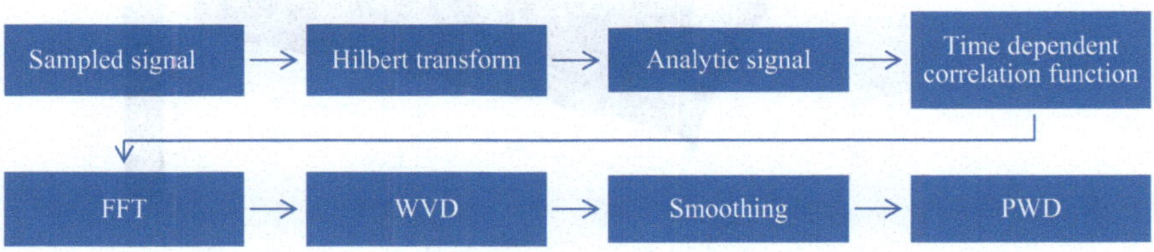

Fig. 4.21 Flow chart of PWD computation [21]

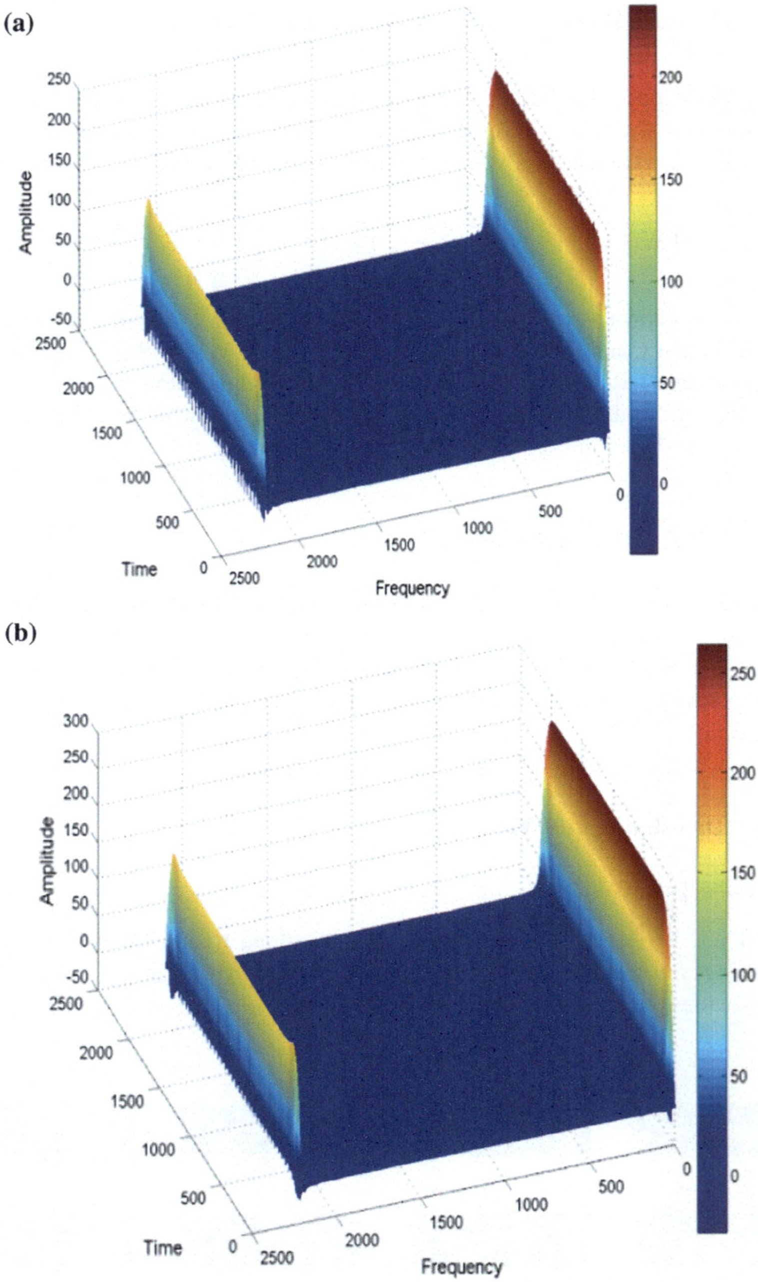

Fig. 4.22 3D plot of **a** healthy and **b** LOV samples after applying PWD [21]

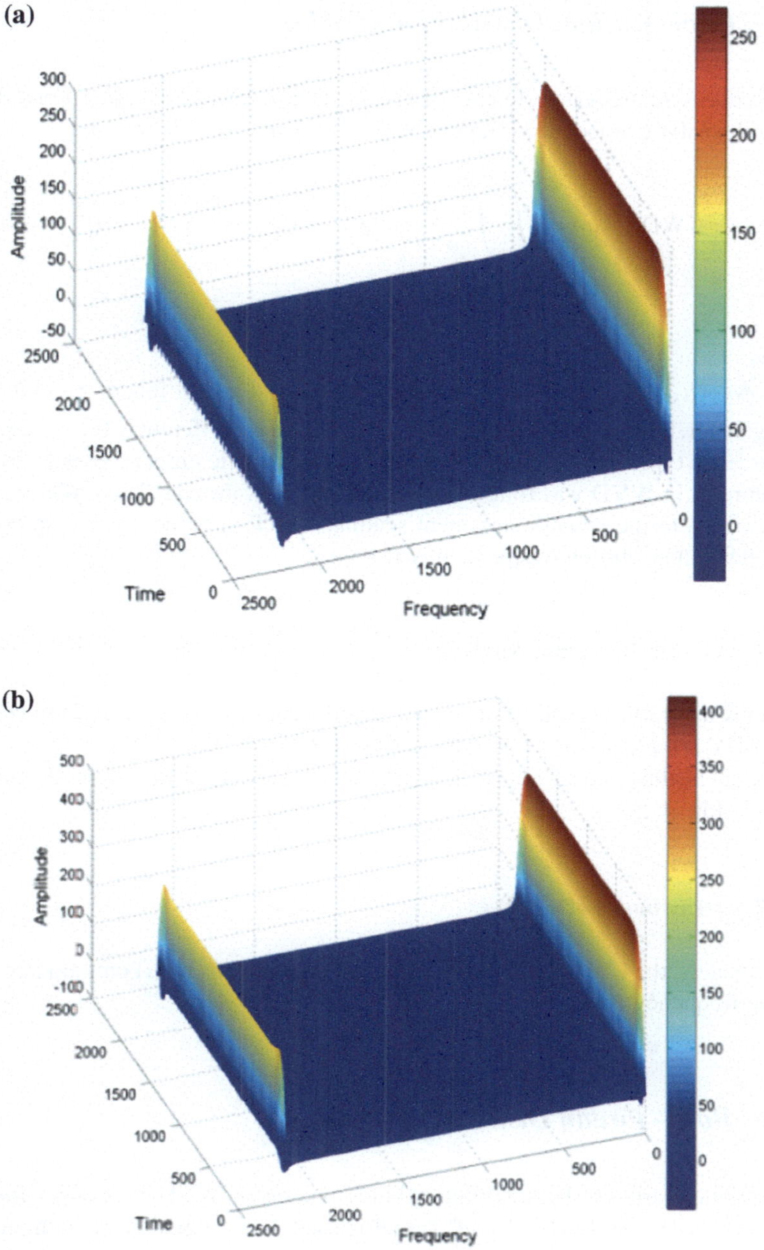

Fig. 4.23 3D plot of **a** healthy and **b** LIV samples after applying PWD [21]

4.4.4 Choi–William Distribution (CWD)

CWD reduces the cross (interference)-terms by using exponential kernel. For a signal, $x(t)$ with analytic associate $s(t)$, the CWD ($\mathrm{CWD}_s(t,\omega)$) is defined as

$$\mathrm{CWD}_s(t,\omega) = \int_{-\infty}^{\infty}\int_{-\infty}^{\infty} \frac{\sigma}{4\pi\tau^2}\exp\left(\frac{-\sigma(\mu-t)^2}{4\tau^2}\right)s\left(\mu+\frac{\tau}{2}\right) * s\left(\mu-\frac{\tau}{2}\right)\mathrm{e}^{-\mathrm{j}\omega\tau}\,\mathrm{d}\mu\,\mathrm{d}\tau \tag{4.34}$$

where $\phi_{\mathrm{CW}}(\mu,\omega) = \exp\left(\frac{-\sigma(\mu-t)^2}{4\tau^2}\right)$ is the exponential kernel function and σ is the scaling factor. It filters out all those cross-terms which are different in time–frequency region. Each type of kernel may be suitable for a specific class of signals. Similar procedure as in WVD was followed to get 36 features instead. The CWD is that it is insensitive to the signal component scaling, therefore an acceptable smoothing parameter can be chosen independently from data.

4.4.4.1 CWD: 3D Visual Analysis

In healthy and LOV classification shown, healthy data covers more amplitude range while LOV sample has more outliers as shown in Fig. 4.24.

Also, in healthy and LIV classification, LIV data covers more amplitude range as shown in Fig. 4.25.

4.4.4.2 Drawbacks of CWD

Unfortunately, the CWD has poor resolution of components with continuously time-varying frequency content.

4.4.5 Born–Jordan Distribution (BJD)

As mentioned earlier by taking different kinds of kernel in WVD, we can get different types of distribution. In BJD "sine" kernel is used. For a signal, $x(t)$ with analytic associate $s(t)$, the BJD ($\mathrm{BJD}(t,\omega)$) is defined as

$$\mathrm{BJD}(t,\omega) = \int_{-\infty}^{\infty}\int_{-\infty}^{\infty} \sin c(u\tau)\, s\left(u+\frac{\tau}{2}\right) * s\left(u-\frac{\tau}{2}\right)\mathrm{e}^{-\mathrm{j}\omega t}\,\mathrm{d}u\,\mathrm{d}\tau \tag{4.35}$$

where $\phi_{\mathrm{BJD}}(\mu,\tau) = \sin c\,(\mu\tau)$ is the sine kernel function.

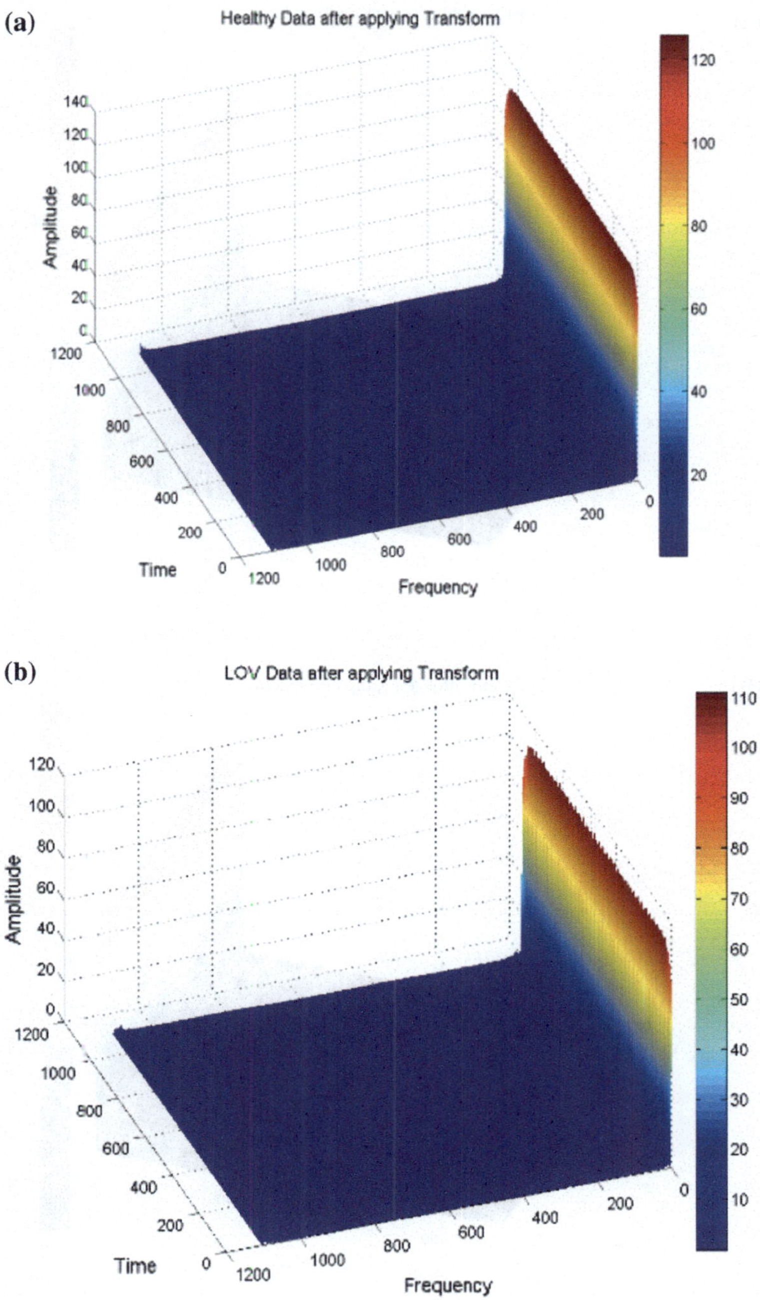

Fig. 4.24 3D plot of **a** healthy and **b** LOV samples after applying CWD [21]

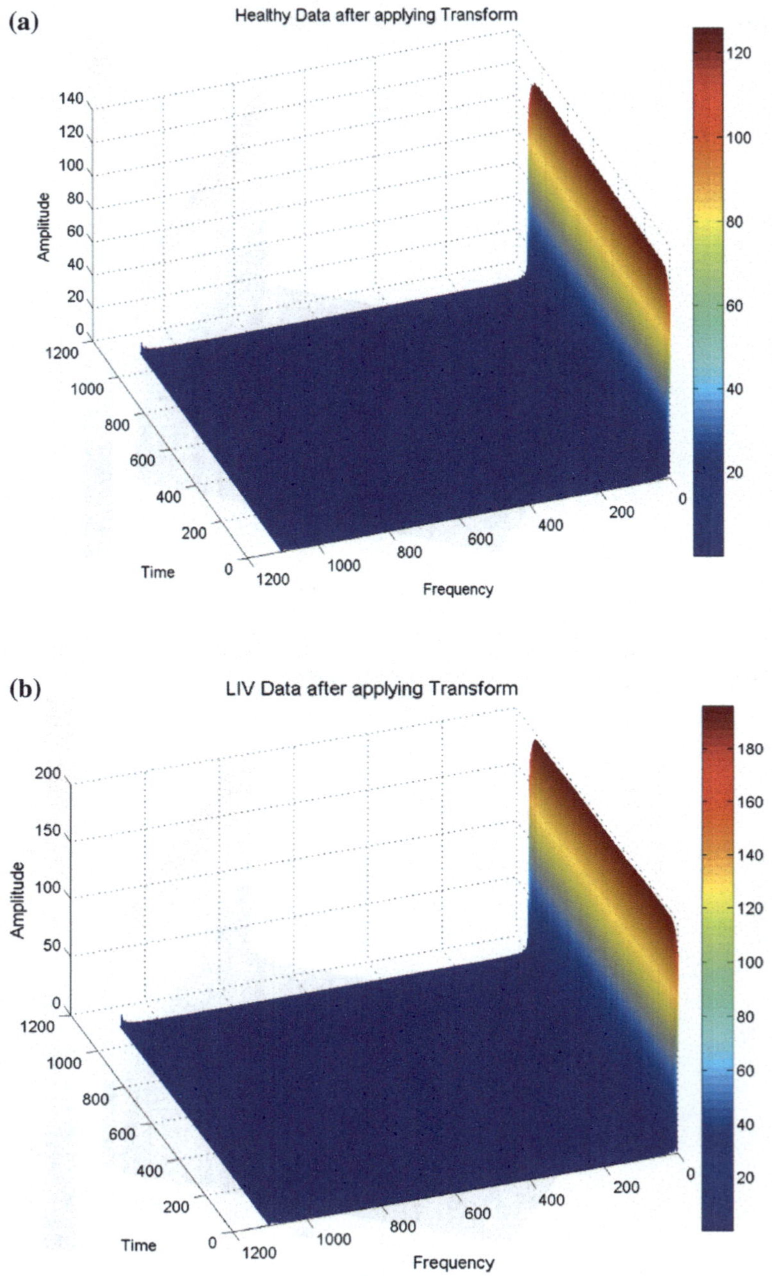

Fig. 4.25 3D plot of **a** healthy and **b** LIV samples after applying CWD [21]

Similar procedure as in WVD has been followed here to get 36 features. Born–Jordan kernel provides the lowest possible variance among the kernels satisfying both marginal and time-support.

4.4.5.1 BJD: 3D Visual Analysis

From the plots of healthy and LOV and healthy and LIV states shown in Figs. 4.26 and 4.27, the healthy sample has large amplitude range while LOV and LIV samples have large outliers.

4.4.5.2 Drawback of BJD

Although BJD reduces the cross-term interferences significantly, it does not provide better resolution along frequency axis.

4.4.6 *S Transform*

The S-transform is an extension of ideas of CWT and is based on a moving and scalable localized Gaussian window.

$$W(t) = \int_{-\infty}^{\infty} h(t)\, w(t - \Gamma, d)\, \mathrm{d}t \tag{4.36}$$

where

$W(\Gamma, d)$ is the scaled replica of the mother wavelet defined by

$$s(\Gamma, f) = \mathrm{e}^{\mathrm{i}2\pi f \Gamma}\, W(\Gamma, d) \tag{4.37}$$

Mother wavelet can be defined as

$$w(t, f) = \frac{f}{2\pi}\, \mathrm{e}^{-\frac{t^2 f^2}{2}}\, \mathrm{e}^{-\mathrm{i}2\pi f t} \tag{4.38}$$

S transform localizes real and imaginary components of the spectrum independently. It simultaneously localizes the phase and amplitude spectrum. It can be expressed as an operation on the FT $H(f)$ of $h(t)$ defined as

$$s(\Gamma, f) = \int H(\alpha + f)\, \mathrm{e}^{\frac{-2\pi\alpha^2}{f^2}}\, \mathrm{e}^{\mathrm{i}2\pi\alpha\Gamma}\, \mathrm{d}\alpha \tag{4.39}$$

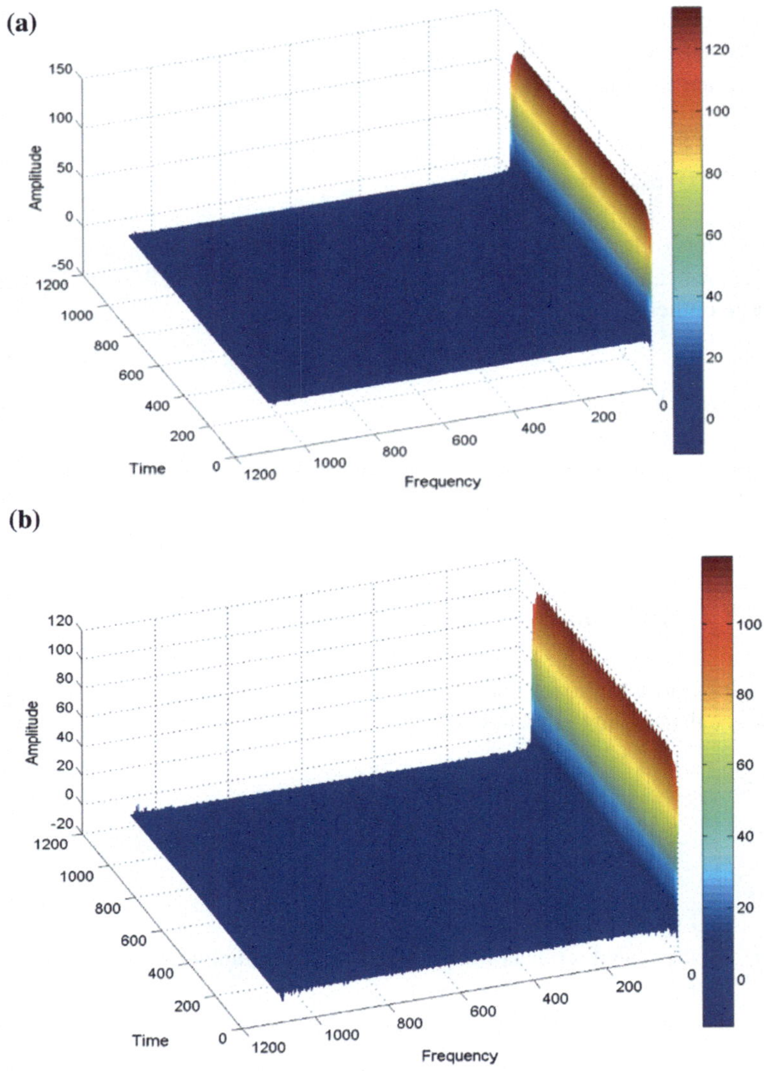

Fig. 4.26 3D plot of **a** healthy and **b** LOV samples after applying BJD [21]

In discrete form, it can be written as

$$s\left[jT, \frac{n}{NT}\right] = \sum_{m=0}^{N-1} H\left[\frac{m+n}{NT}\right] e^{\frac{-2\pi^2 m^2}{n^2}} e^{\frac{i2\pi mj}{N}} \tag{4.40}$$

Similar procedure as in the case of WVD has been followed to get 36 features.

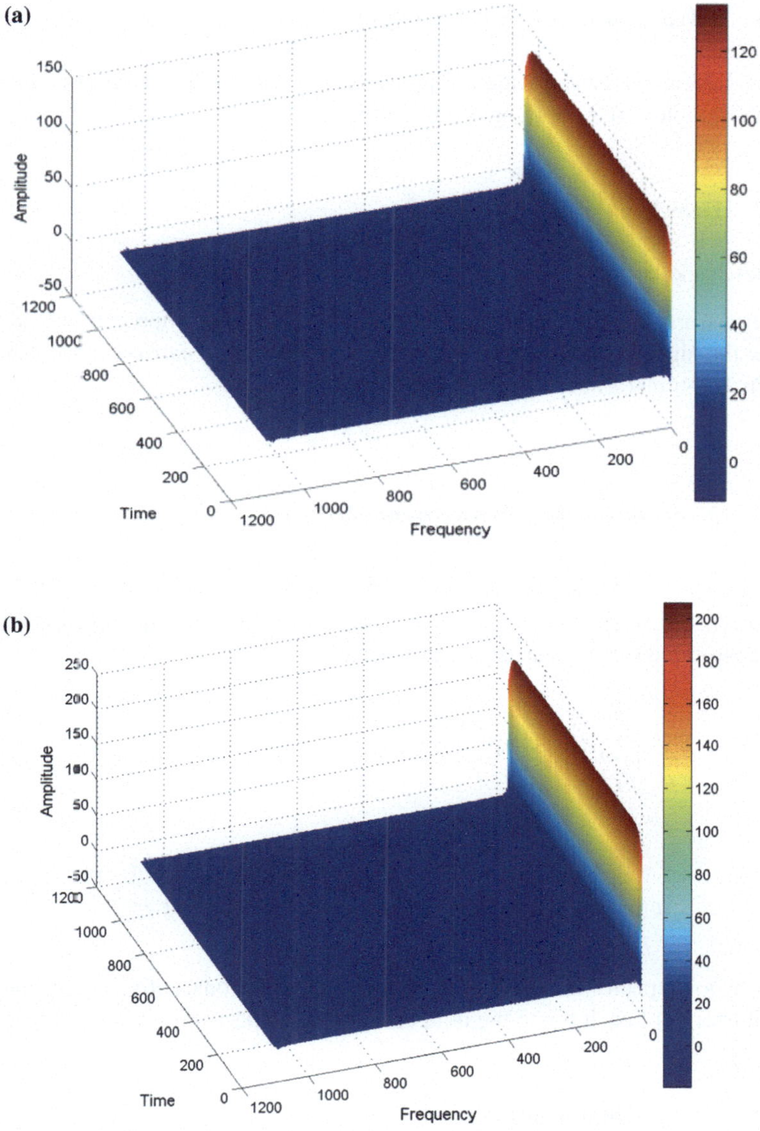

Fig. 4.27 3D plot of **a** healthy and **b** LIV samples after applying BJD [21]

4.4.6.1 S Transform: 3D Visual Analysis

Figures 4.28 and 4.29 show the 3D plot of healthy and LOV samples and healthy and LIV samples after applying S transform.

4.4.6.2 Drawback of S-Transform

The drawbacks of S-transform are:

i. S-transform cannot analyse variation with time in DC component.
ii. For the high frequency component, window will be too narrow, so the points on which S-transform can be applied will be too less.

4.4.7 *Discrete Cosine Transform (DCT)*

DCT expresses a finite sequence of data points in terms of a sum of cosine functions oscillating at different frequencies. It is defined to map the spatial (correlated) data into transformed (uncorrelated) coefficients as

$$y(k) = w(k)\sum_{n=1}^{N} x(n)\cos\left(\frac{\pi(2n-1)(k-1)}{2N}\right) \quad k = 1, 2, 3, \ldots, N \tag{4.41}$$

where

$$w(k) = \frac{1}{\sqrt{N}} \quad k = 1 \text{ and } w(k) = \sqrt{\frac{2}{N}} \quad 2 \le k \le N.$$

The whole spectrum of DCT is divided into eight consecutive equal bins. The ratio of energy of each bin to the total energy of the spectrum gives eight features.

4.4.7.1 DCT: Visual Analysis

After applying DCT, both healthy and LOV samples are quite separable as shown in the plots (Refer Figs. 4.30 and 4.31).

Also, in the case of healthy and LIV, DCT shows difference between these two as given in the DCT spectrum (Refer Figs. 4.32 and 4.33).

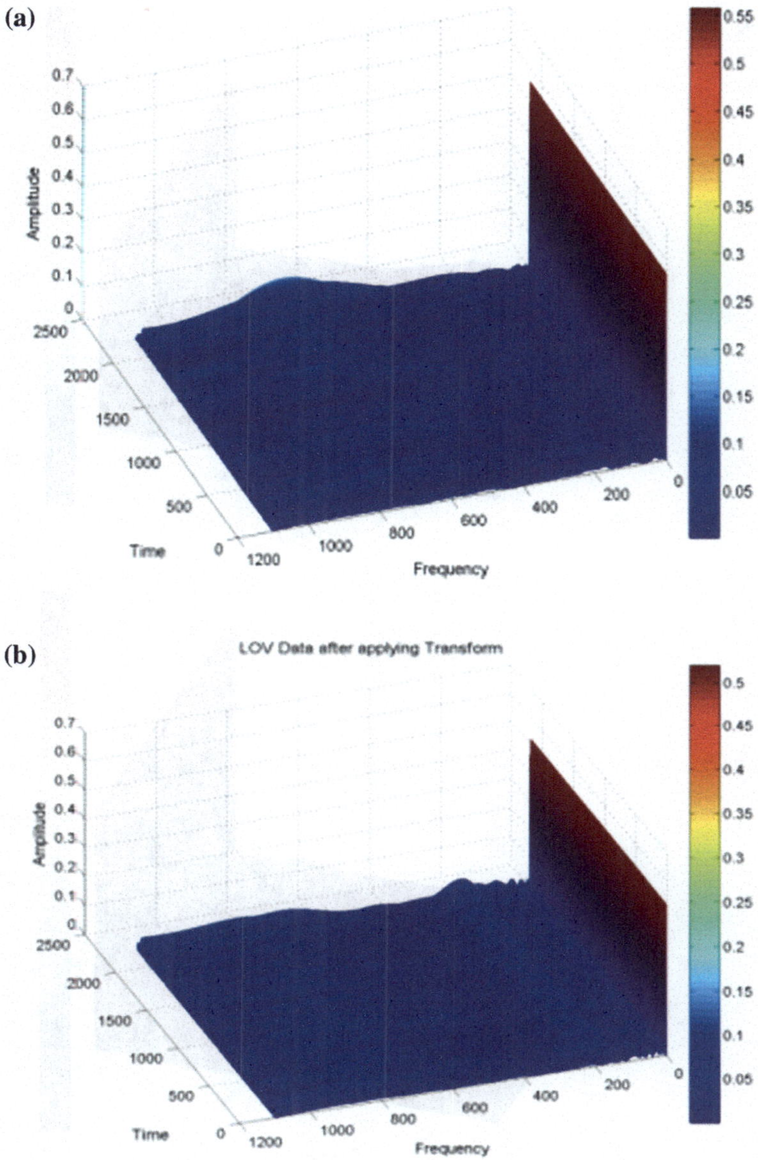

Fig. 4.28 3D plot of **a** healthy and **b** LOV samples after applying S transform [21]

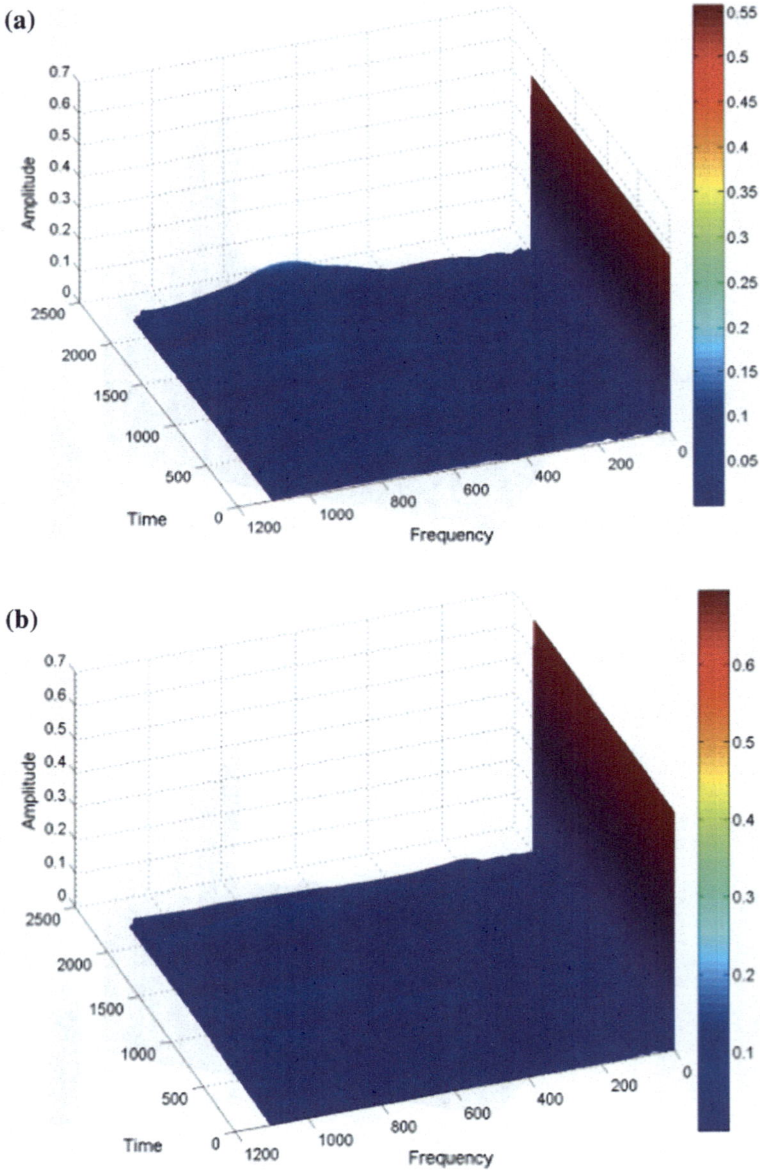

Fig. 4.29 3D plot of **a** healthy and **b** LIV samples after applying S transform [21]

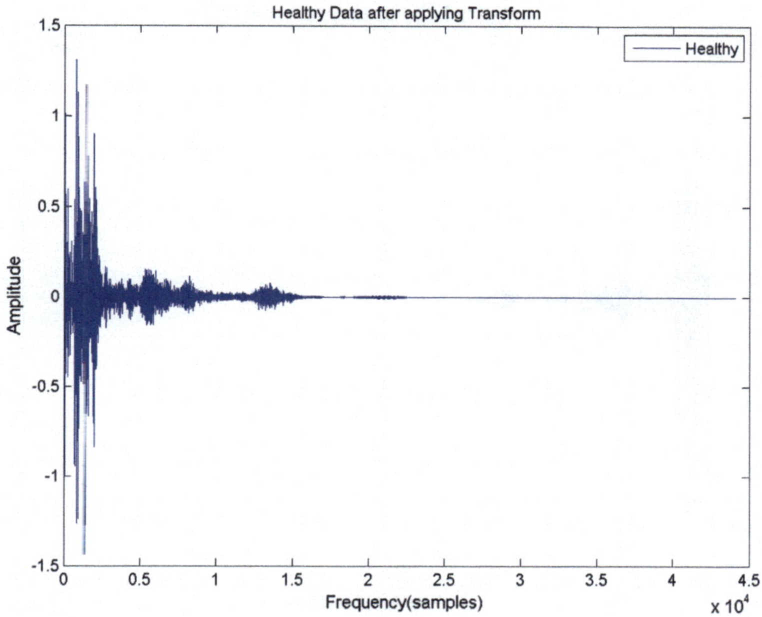

Fig. 4.30 DCT plot of healthy sample [21]

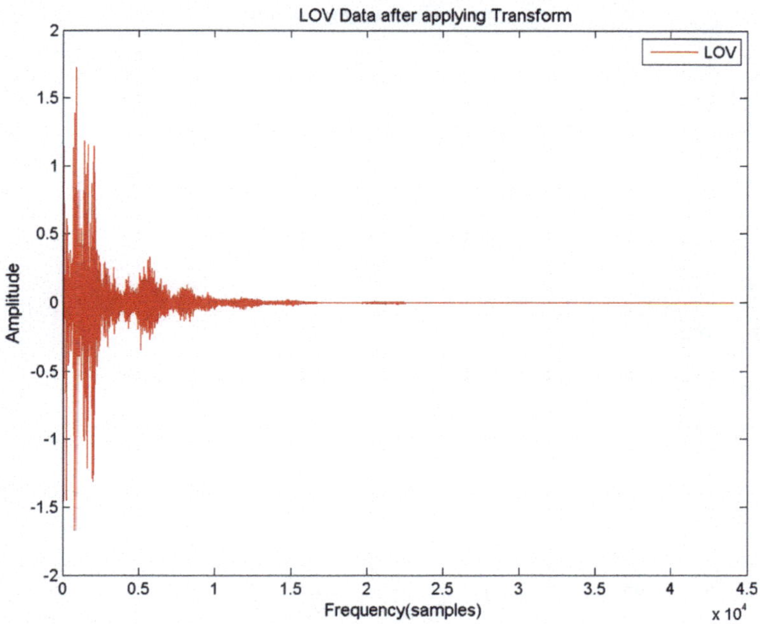

Fig. 4.31 DCT plot of LOV sample [21]

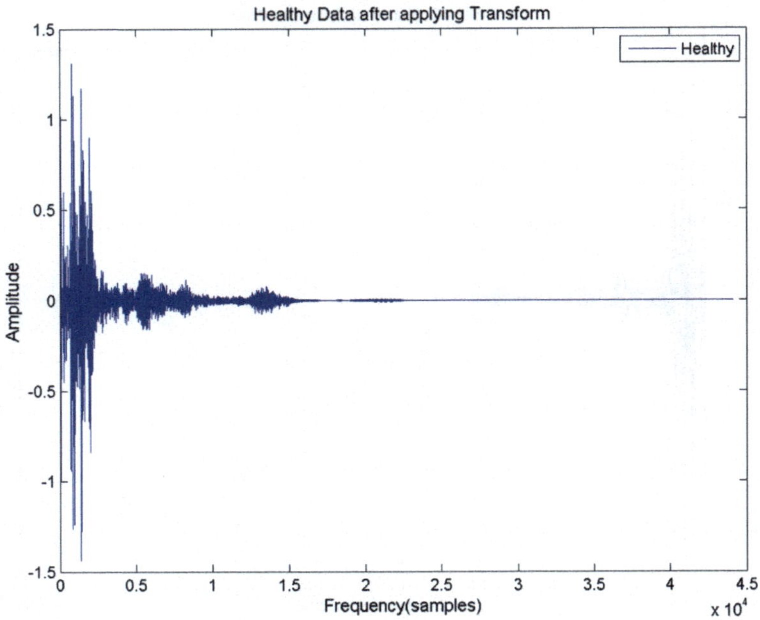

Fig. 4.32 DCT plot of healthy sample [21]

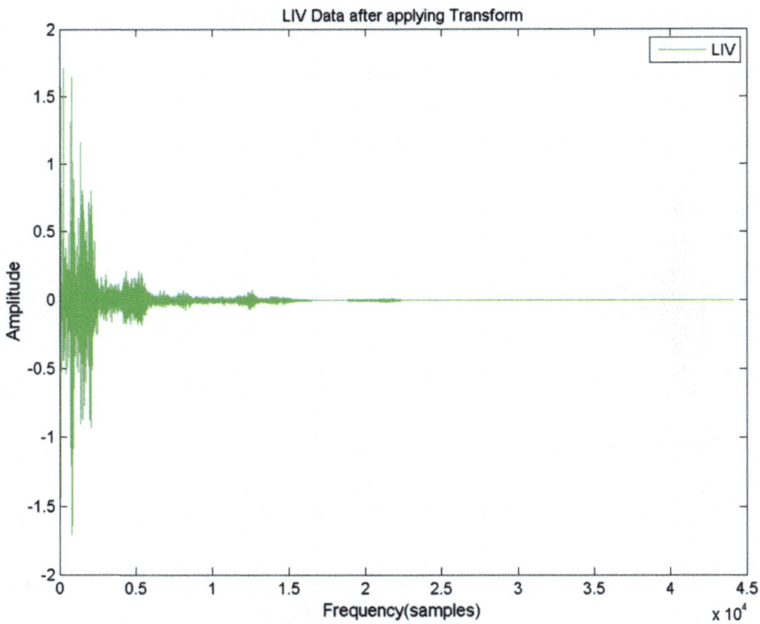

Fig. 4.33 DCT plot of LIV sample [21]

4.4.8 *Autocorrelation*

Autocorrelation is the cross-correlation of a signal with itself. Informally, it is the similarity between observations as a function of time separating them which can be defined by

$$R(\tau) = \int_{-\infty}^{\infty} f(t)^* f(t - \tau)\, \mathrm{d}t \tag{4.42}$$

It is a mathematical tool for finding repeating patterns, such as the presence of a periodic signal obscured by noise or identifying the missing fundamental frequency in a signal implied by its harmonic frequencies. A feature from autocorrelation function is extracted by taking the variance of autocorrelation coefficients.

4.4.8.1 Autocorrelation: Visual Analysis

It is clear from Figs. 4.34, 4.35, 4.36, and 4.37 given that healthy, LOV, and LIV samples are quite separable when one performs their autocorrelation.

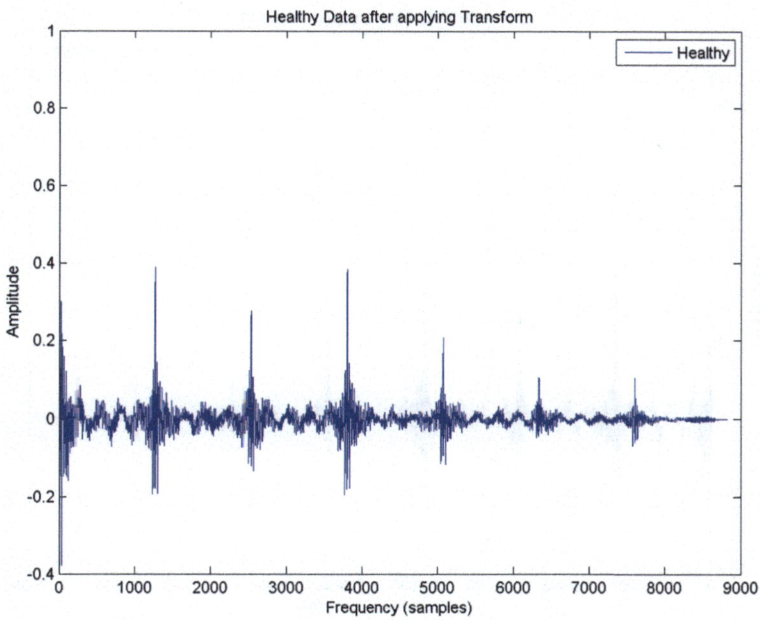

Fig. 4.34 Autocorrelation plot of healthy sample [21]

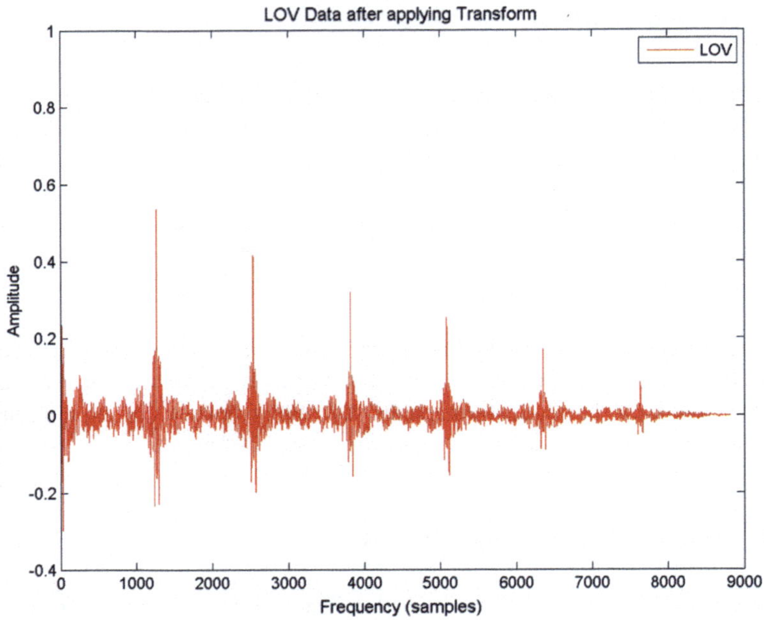

Fig. 4.35 Autocorrelation plot of LOV sample [21]

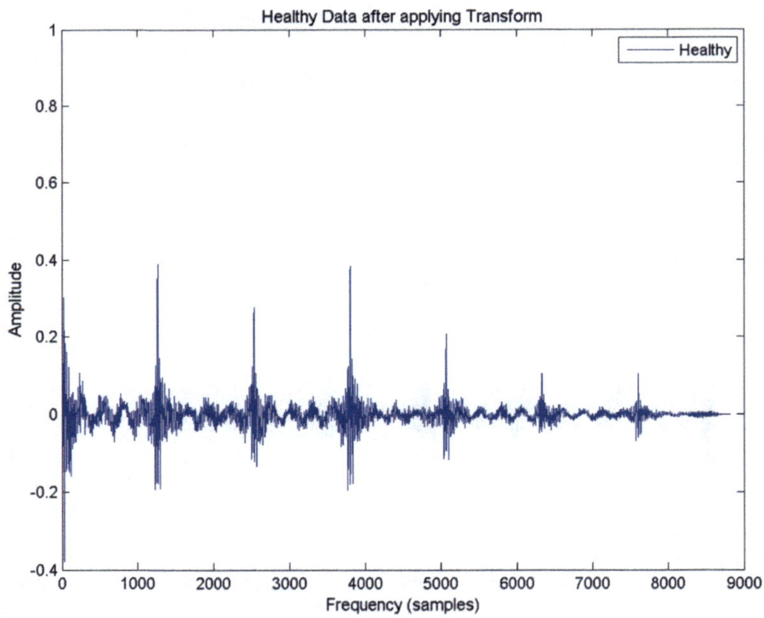

Fig. 4.36 Autocorrelation plot of healthy sample [21]

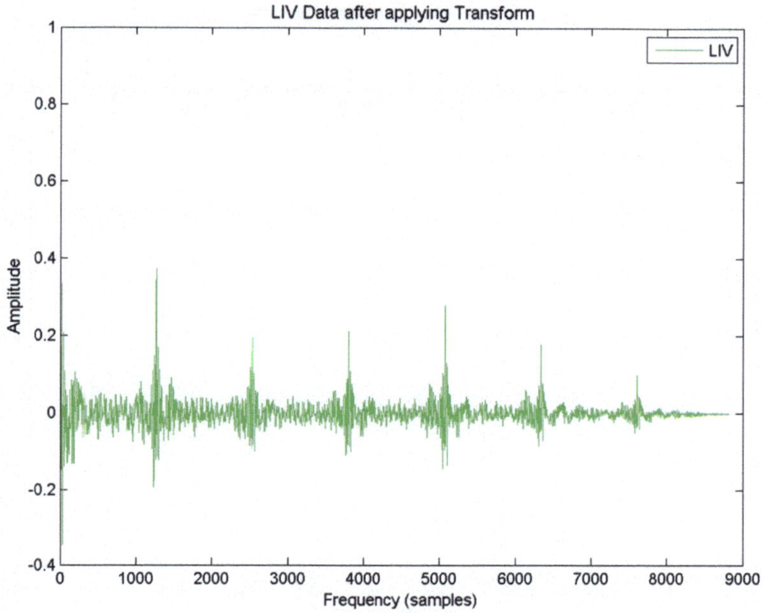

Fig. 4.37 Autocorrelation plot of LIV sample [21]

4.4.9 Updated Morlet Transform (UMT)

The UMT has "sine" function in place of the "cosine" function. Thus, mother wavelet of UMT is given by

$$y_{a,b}(t) = \mathrm{e}^{\frac{-b^2(t-b)^2}{a^2}} \sin\left(\frac{p(t-b)}{a}\right) \tag{4.43}$$

where "a" is scale parameter and "b" is translation parameter. By changing them, multiple son wavelets can be obtained. Five features have been considered from the resultant output. They are as follows

- Standard deviation of output function
- Entropy of output function
- Kurtosis of output function
- Skewness of output function
- Variance of the output function.

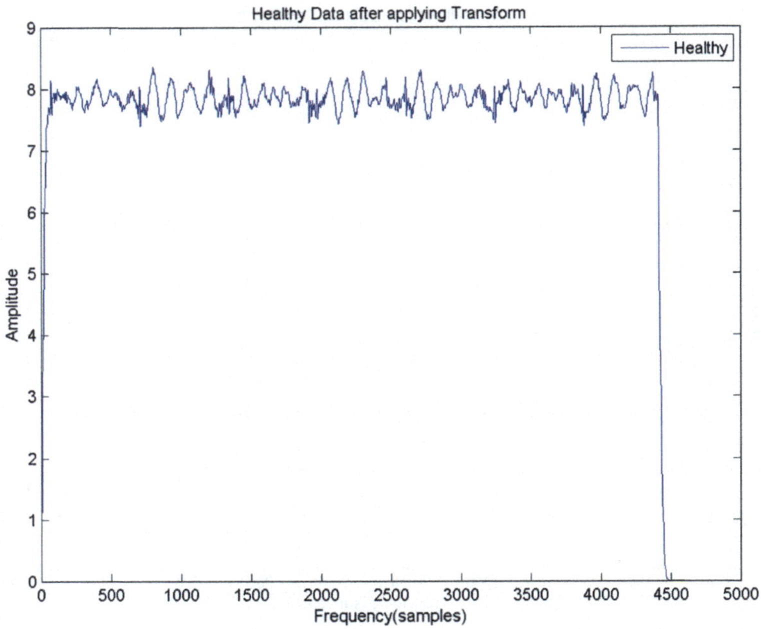

Fig. 4.38 UMT plot of healthy sample [21]

4.4.10 UMT: Visual Analysis

The plots of healthy and LOV data shown in Figs. 4.38 and 4.39 show that both the states have separation to classify them based on unknown test data.

The plots of healthy and LIV data shown in Figs. 4.40 and 4.41 show that both the states have separation to classify them based on unknown test data.

4.4.11 Convolution with Sine Wave

The recorded signal $x(t)$ is convolved with sine wave $p(t)$ which gives a third function $y(t)$ that is typically viewed as a modified version of the original signal. The obtained signal can be presented as

$$p(t) = \begin{cases} 0 & \text{if } t < 0 \\ \sin(t) & \text{otherwise} \end{cases} \tag{4.44.1}$$

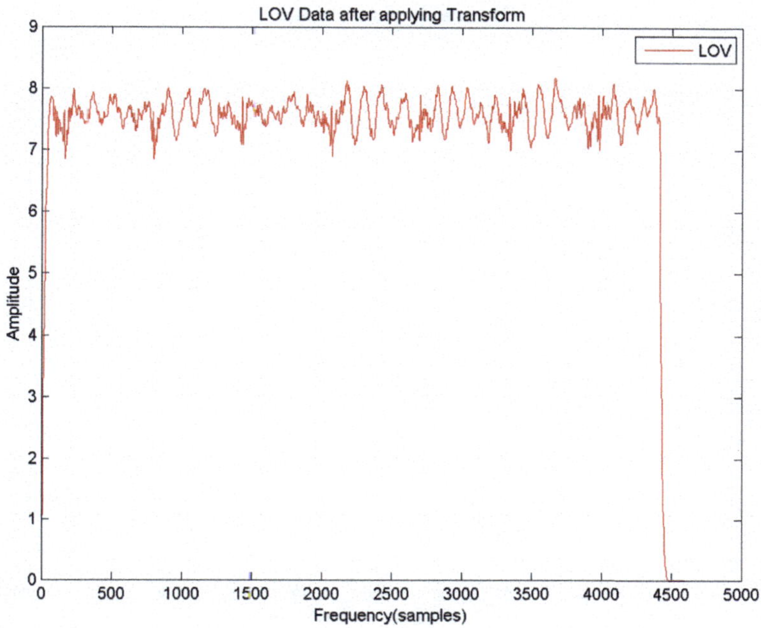

Fig. 4.39 UMT plot of LOV sample [21]

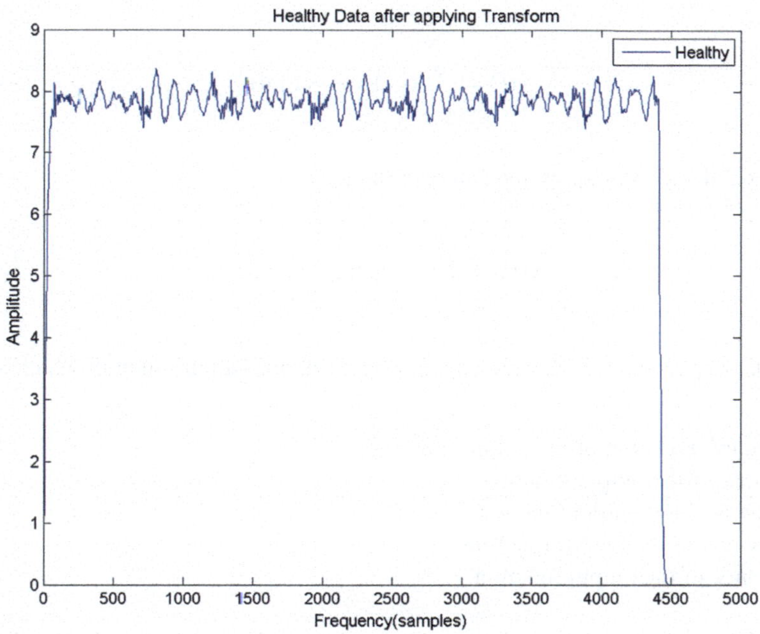

Fig. 4.40 UMT plot of healthy sample [21]

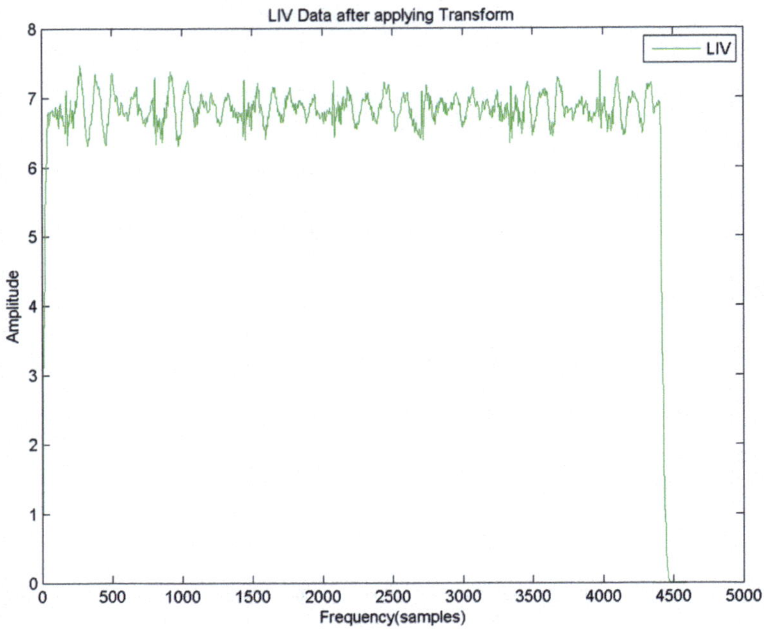

Fig. 4.41 UMT plot of LIV sample [21]

$$y(t) = \int_{0}^{\infty} x(\tau)p(t-\tau)\,\mathrm{d}\tau \tag{4.44.2}$$

Hence, the above expression can be written as

$$y(t) = \int_{0}^{\infty} x(\tau)\sin(t-\tau)\,\mathrm{d}\tau \tag{4.45}$$

Thus, five features are extracted from the resultant output function which are as follows

- Standard deviation of the output function
- Entropy of the output function
- Kurtosis of the output function
- Skewness of the output function
- Variance of the output function.

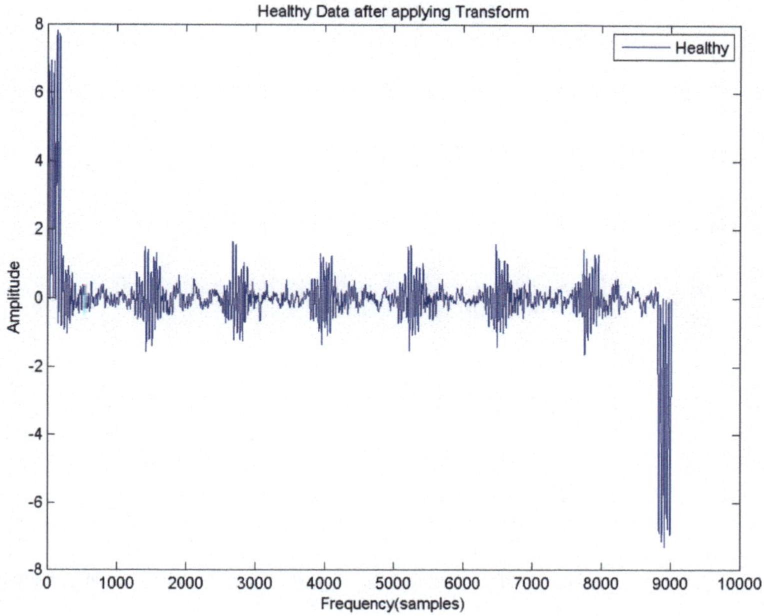

Fig. 4.42 Convolution using sine plot of healthy sample [21]

4.4.11.1 Convolution Using Sine: Visual Analysis

From the plots of healthy and LOVshown in Figs. 4.42 and 4.43, respectively and healthy and LIV data shown in Figs. 4.44 and 4.45, respectively, it is quite clear that both healthy and faulty states are separable. This is the reason to use convolution using sine in the feature extraction process.

4.5 Feature Extraction Tool

4.5.1 Feature Extraction Tool for Android Platform

The steps to use feature extraction tool for Android platform are following.

Step 1 As the user clicks on "FeatureExtractionApp.apk", welcome page appears as shown in Fig. 4.46. Here, user is asked to tap anywhere on the screen to move on to the main activity page.

Step 2 The first activity page of the app appears as shown in Fig. 4.47. This app page is mainly meant for taking input from the user. At this page, various inputs are taken from the user.

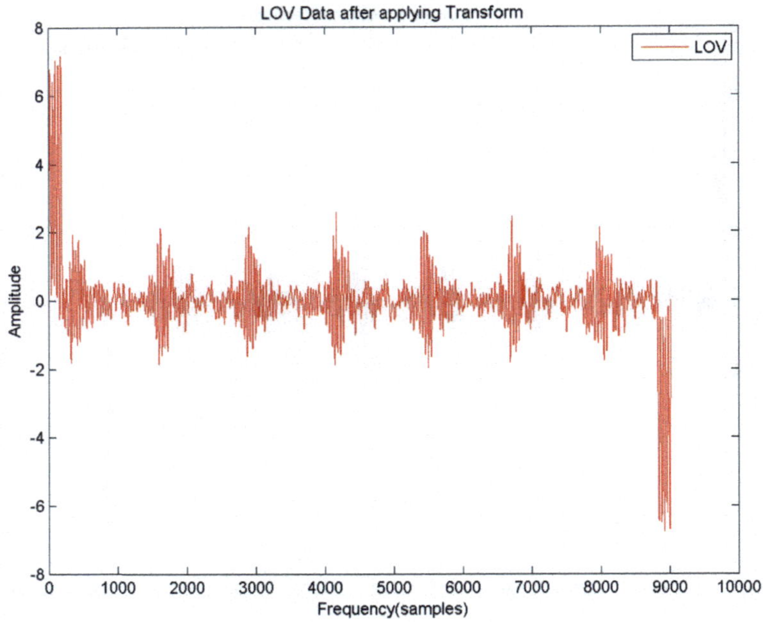

Fig. 4.43 Convolution using sine plot of LOV sample [21]

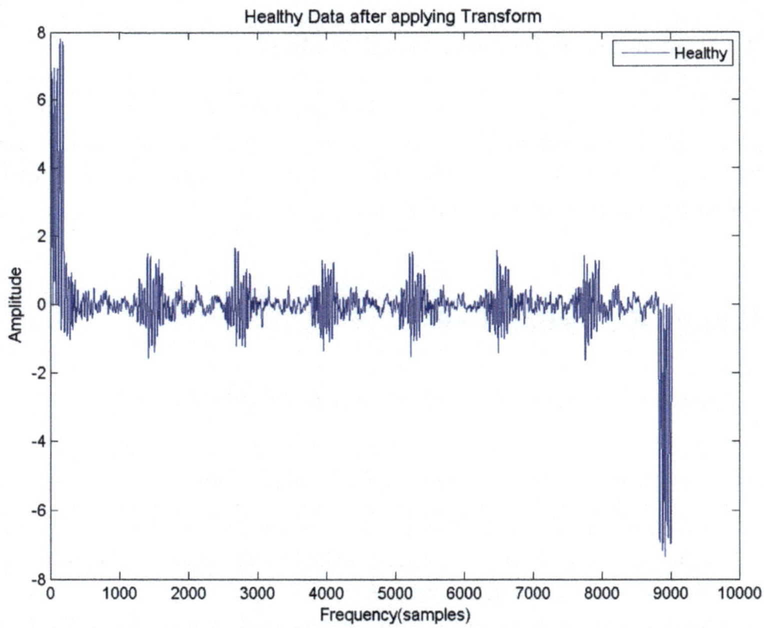

Fig. 4.44 Convolution using sine plot of heathy sample [21]

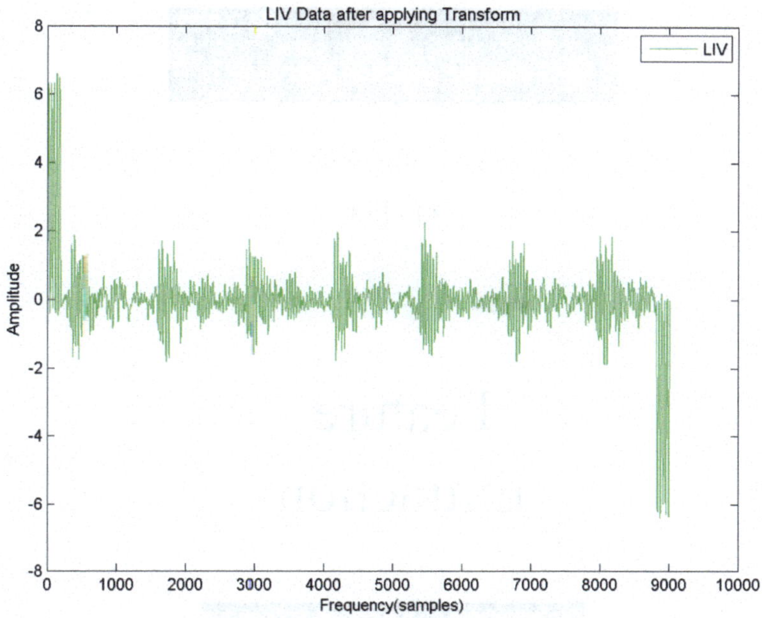

Fig. 4.45 Convolution using sine plot of LIV sample [21]

Step 3 As the user clicks on "Browse Data File" button, browsing option from SD card opens as shown in Fig. 4.48.

Step 4 The data file is saved inside the databank folder, so the user have to choose the databank folder. It will open the directory and list all files inside it as shown in Fig. 4.49.

Step 5 On selecting any file, it first checks whether selected file format is ".dat" or ".txt". If it is in the chosen format, the file gets selected and a message appears on the screen as shown in Fig. 4.50.

Step 6 If the user clicks on "YES" option, the file gets selected and returns to the first page. The selected file name appears in a text box as shown in Fig. 4.51. In case the user selects NO option, the activity page returns to the first page without selecting any file and leaves the box blank.

Step 7 The next step is to select a domain from choices in which the data should be analyzed. As shown in Fig. 4.52, the user can select any of the domains, namely time, frequency, and wavelet domains. These are being multiple choice boxes, so more than one domain can be selected at a moment.

Step 8 Once all the parameters are selected, click on Extract Features button which calls the next activity page.

Fig. 4.46 Feature extraction welcome page

Step 9 The next activity page is the processing page where domain analysis on the selected data takes place. This activity page does not provide any element in the user interface for the user to interact with app. Only a progress spinner is displayed as shown in Fig. 4.53, which indicates that the application is busy in extracting features. Once all the features are extracted, several feature vectors are obtained and written in a text file in .CSV format. The next activity page is then called.

Step 10 As shown in Fig. 4.54, once feature extraction is done, the result activity page appears. In this page, a text message displays "Extracted features have been stored in Reading_Features.txt in the same directory" to indicate that features have been saved in a text file. The user is given a choice to perform the feature extraction task again or want to exit the application.

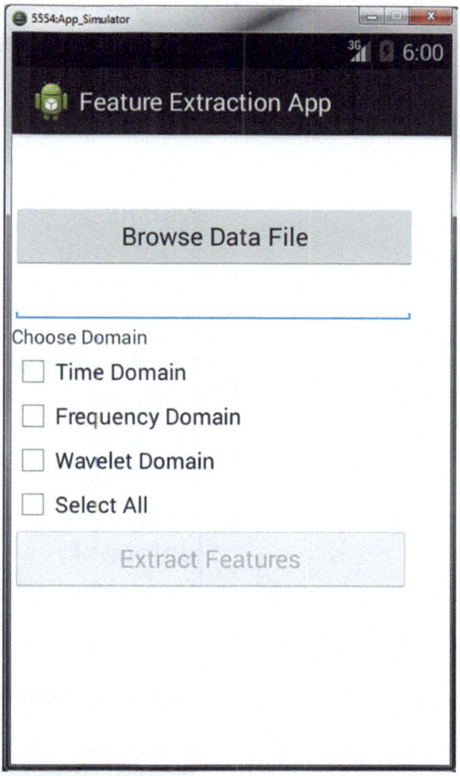

Fig. 4.47 Feature extraction activity page 1 [10]

4.5.2 Feature Extraction Tool for Windows Mobile

Fig. 4.55 shows snapshots of feature extraction tool for windows mobile platform.

4.5.3 Windows Tablet Platform

Fig. 4.56 shows snapshots of feature extraction tool for windows tablet platform.

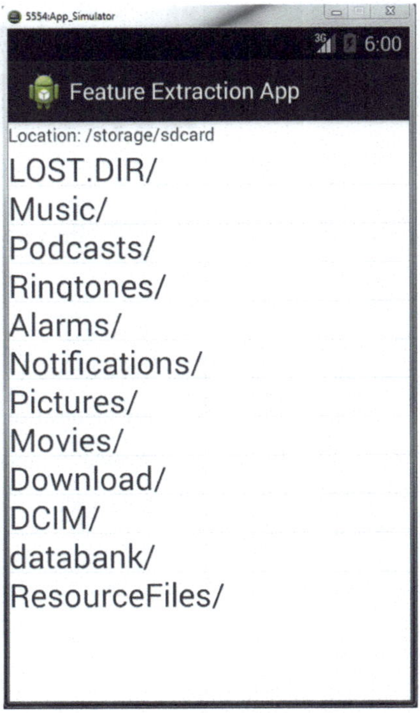

Fig. 4.48 File browse activity page 1 [10]

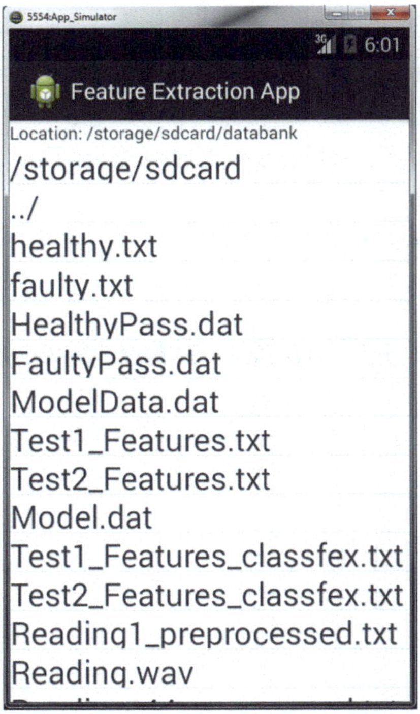

Fig. 4.49 File browse activity page 2 [10]

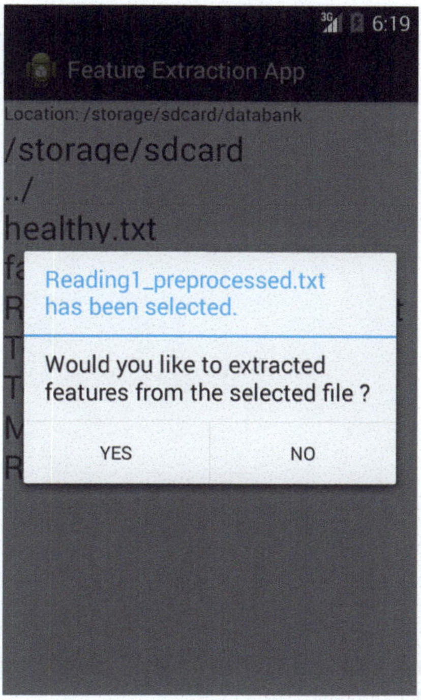

Fig. 4.50 Confirmation activity page [10]

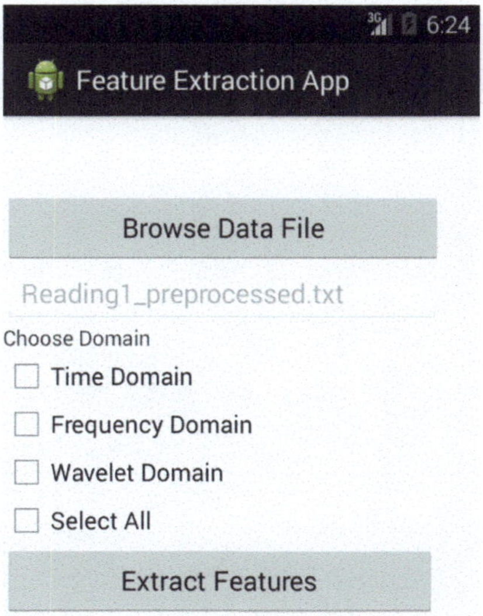

Fig. 4.51 Selected data file name in textbox [10]

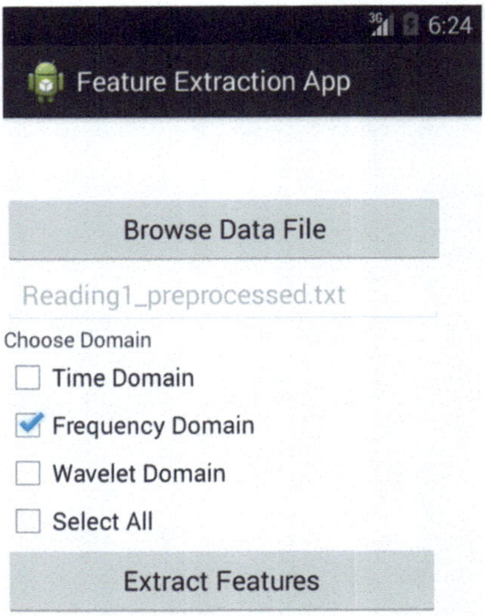

Fig. 4.52 Select feature domain page [10]

Fig. 4.53 Processing activity page [10]

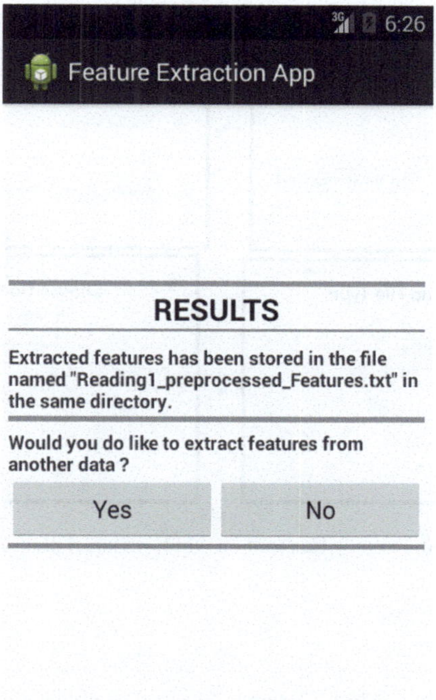

Fig. 4.54 Feature extraction result activity page [10]

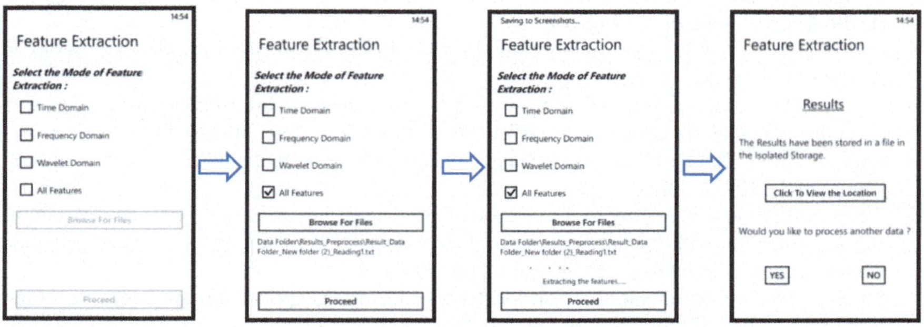

Fig. 4.55 Snapshots of feature extraction tool on windows mobile platform [10]

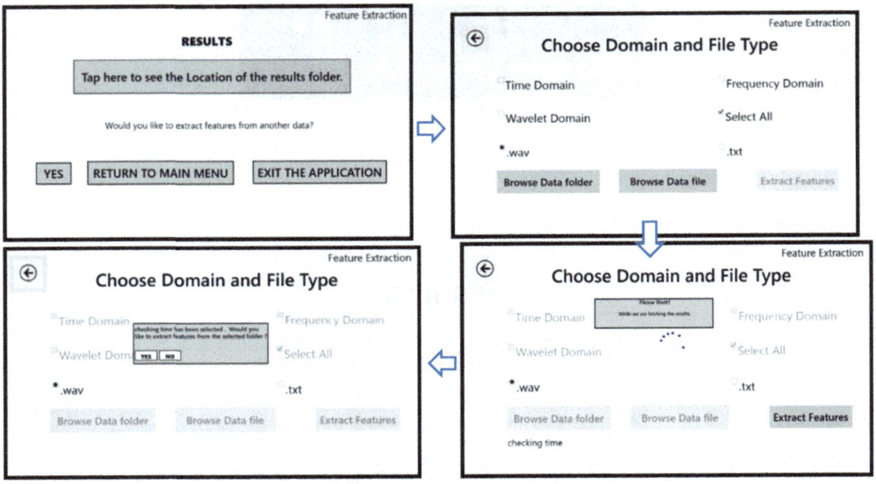

Fig. 4.56 Snapshots of feature extraction tool on windows tablet platform [10]

References

1. Khorshidtalab, A., Salami, M.J.E., Hamedi, M.: Evaluating the effectiveness of time-domain features for motor imagery movements using SVM. In: IEEE International Conference on Computer and Communication Engineering, pp. 909–913 (2012)
2. He, Q., Kong, F., Yan, R.: Subspace-based gearbox condition monitoring by kernel principal component analysis. Int. J. Mech. Syst. Sig. Process. **21**(4), 1755–1772 (2007)
3. Verma, N.K., Gupta, V.K., Sharma, M., Sevakula, R.K.: Intelligent condition based monitoring of rotating machines using sparse auto-encoders. In: IEEE Conference on Prognostics and Health Management, pp. 1–7 (2013)
4. Daubechies, I.: Ten lectures on wavelets. In: SIAM (ed.) CBMS-NSF Conference Series in Applied Mathematics (1992)
5. Verma, N.K., Sevakula, R.K., Dixit, S., Salour, A.: Intelligent condition based monitoring using acoustic signals for air compressors. IEEE Trans. Rel. **65**(1), 291–309 (2016)
6. Lajos, T., Tóth, T.: On finding better wavelet basis for bearing fault detection. Acta Polytech. Hung. **10**(3), 17–35 (2013)
7. Wang, S.B., Zhu, Z.K., Wang, A.Z.: Gearbox fault feature detection based on adaptive parameter identification with Morlet wavelet. In: Proceedings of IEEE International Conference on Wavelet Analysis and Pattern Recognition, pp. 409–414 (2010)
8. Ming, J.: Feature extraction based on Morlet wavelet and its application for mechanical fault diagnosis. J. Sound Vib. **234**, 135–148 (2000)
9. Stens, J.L.: Butterworth Low Pass Filter. Lecture Notes. Available at http://www.ece.uah.edu/courses/ee426/Butterworth.pdf. Accessed on June 2014
10. Verma, N.K., Singh, S., Gupta, J.K., Sevakula, R.K., Dixit, S., Salour, A.: Smartphone application for fault recognition. In: 6th International Conference on Sensing Technology, Kolkata, India, pp. 1–6 (2012)
11. Goumas, S.K., Zervakis, M.E., Stavrakakis, G.S.: Classification of washing machines vibration signals using discrete wavelet analysis for feature extraction. IEEE Trans. Instrum. Meas. **51**(3), 497–508 (2002)
12. Verma, N.K., Sarkar, S., Dixit, S., Sevakula, R.K., Salour, A.: Android app for intelligent CBM. In: 2013 IEEE International Symposium on Industrial Electronics, Taipei, Taiwan, pp. 1–6 (2013)

13. Liu, B., Lingand, S.-F., Meng, Q.F.: Machinery diagnostics based on wavelet packets. J. Vib. Control **3**(1), 5–17 (1997)
14. Yang, D.M.: Induction motor bearing fault detection with non-stationary signal analysis. In: IEEE International Conference on Mechatronics, pp. 1–6 (2007)
15. Bruce, A., Donoho, D., Gao, H.Y.: Wavelet analysis for signal processing. IEEE Spectr. **33**(10), 26–35 (1996)
16. Maurits, M., Roose, D.: Wavelet-based image denoising using a Markov random field a priori model. IEEE Trans. Image Process. **6**(4), 549–565 (1997)
17. Sevakula, R.K., Verma, N.K.: Wavelet transforms for fault detection using SVM in power systems. In: IEEE International Conference on Power Electronics, Drives and Energy Systems, Bengaluru, India, pp. 1–6 (2012)
18. Verma, N.K., Gupta, J.K., Singh, S., Sevakula, R.K., Dixit, S., Salour, A.: Feature level analysis. In: IEEE Workshop on Computational Intelligence: Theories, Applications and Future Directions, IIT Kanpur, India, pp. 148–152 (2013)
19. Verma, N.K., Gupta, R., Sevakula, R.K., Salour, A.: Signal transforms for feature extraction from vibration signal for air compressor monitoring. In: IEEE Region 10 TENCON, Bangkok, Thailand, pp. 1–6 (2014)
20. Saraswat, G., Singh, V., Verma, N.K., Salour, A., Liu, J.: Prognosis of diesel engine (MBT) using feature extraction techniques: a comparative study. In: IEEE International Conference on Prognostics and Health Management, Washington, USA, pp. 11–13 (2018)
21. Verma, N.K., Sevakula, R.K., Goel, S.: Study of transforms for their comparison. In: International Conference on Industrial and Information Systems, Gwalior, India, pp. 1–6 (2014)

Chapter 5
Feature Selection

Abstract As detailed in Chap. 4, features have been extracted from the pre-processed data. Too many features may lead to the curse of dimensionality issues. This chapter explains how to obtain a set of relevant features which is the process of feature selection. Feature selection is the process in which most informative variables are selected for the generation of the model. It helps to remove the redundant data and contributes to proper classification. While minimizing the redundancy, we should keep in mind that the predicted information must be preserved as much as possible. This chapter describes various feature selection methods such as Principal Component Analysis (PCA)-based approach, Mutual Information (MI), Bhattacharyya Distance (BD), and Independent Component Analysis (ICA). A novel feature selection based on graphical review has also been elaborated later in this chapter.

5.1 Introduction

Feature selection is the process of selecting a subset of relevant features for building robust learning models. By removing irrelevant and redundant features from the data, we can improve the performance of learning models [1–3]. The obvious reason for the selection of a subset of features, instead of complete feature set, is to increase the computational efficiency. Feature selection serves the following purposes:

- Removing the curse of dimensionality,
- Speeding up the learning process, and
- Reducing the storage space.

Features represent the characteristics of a machine. These characteristics are regularity, sensitivity, and classification, which vary from device to device. Therefore, it is desirable to use more effective features for machine monitoring, which are not directly available from the raw data [4–7]. The following approaches can do feature selection:

(1) PCA-based approach
(2) MI-based approach
(3) BD-based approach

N. K. Verma and A. Salour, *Intelligent Condition Based Monitoring*, Studies in Systems, Decision and Control 256, https://doi.org/10.1007/978-981-15-0512-6_5

(4) ICA-based approach
(5) GA-based approach

5.2 Principal Component Analysis Based Approach

Principal Component Analysis (PCA) is one of the most popular methods for dimensionality reduction [8]. It is a simple and non-parametric method of extracting relevant information from the raw data. PCA reduces the complex data to a lower dimension, which reveals the hidden structure with minimal additional efforts that often underly these structures. PCA is efficient in compressing information and eliminating correlations between variables. The goal of PCA is to compute the most meaningful basis to express a noisy dataset [9]. The procedure of applying this technique begins by first computing the covariance matrix of input data variables. Unit Eigenvectors (EVs) of the covariance matrix are arranged in descending order of their eigenvalues which in this case are known as singular values. These values capture the amount of variation present in the direction of corresponding EVs of the data. Better features are expected to have higher variance; hence, transforming the top EVs is selected to be the new feature axes. It is also to be referred to as Principal Components (PCs). For bringing out the input data in the space of new feature axes, their projections across each of the primary components are obtained. These projection values are found by performing the dot product of the input data with each of the principal components. Note that the features are not selected here; instead, they are transformed into a new orthogonal space where the principal components form the basis functions.

5.2.1 *PCA as Dimension Reduction Tool for Classification*

Initially Feature Vector (FV) matrix may not be of full rank. Hence, it may have high amount of redundancy. Let, feature set for any dataset is "X"

$$X = \begin{bmatrix} x_{1,1} & \cdots & x_{1,d} \\ \vdots & \ddots & \vdots \\ x_{N,1} & \cdots & x_{N,d} \end{bmatrix} \tag{5.1}$$

Here, N is the total number of the samples available for each class and d is the dimension of the feature set. The next step involves the calculation of Covariance Matrix COV X.

$$\text{COV } X = \sum_{i=1}^{N} (x_i - \mu_x)^{\mathrm{T}} (x_i - \mu_x) \tag{5.2}$$

where μ_x is the mean of the FV, which is arithmetic mean of all the samples of FVs.

$$\mu_x = \frac{1}{N}\sum_{i=1}^{N} x_i \tag{5.3}$$

The eigenvalues and EVs are calculated as follows:

$$\lambda \times V = \text{COV } X \times V \tag{5.4}$$

where λ is the eigenvalue and V is the EV of COV X.

For d dimension feature set, maximum d distinct eigenvalues can be obtained. Arrange the eigenvalues in descending order such that $\lambda_1 > \lambda_2 > \lambda_3 > \cdots > \lambda_d$ and their corresponding EVs are arranged in such order as $V_1 > V_2 > V_3 > \cdots > V_d$. Higher the eigenvalue more is the variance in the direction of the corresponding EV . Select only the most representative principal components using Accumulative Contribution Rate (ACR). It is a traditional method of selecting PCs. Using ACR, let us call matrix of first "m" selected EVs as

$$\text{EV}_m = \begin{bmatrix} V_{11} & \cdots & V_{im} \\ \vdots & \ddots & \vdots \\ V_{d1} & \cdots & V_{dm} \end{bmatrix} \tag{5.5}$$

where "d" is a dimension of feature matrix. Size of EV_m is $d \times m$. Projection of original mean adjusted FVs over EV_m gives first m PC as,

$$\text{PCs} = X \times \text{EV}_m \tag{5.6}$$

5.2.2 *Procedure for Feature Selection Using PCA*

Step 1 Get the data matrix.
Step 2 Subtract the mean of each dimension. This produces a dataset whose mean is zero.
Step 3 Calculate the covariance matrix.
Step 4 Find EVs and eigenvalues for the covariance matrix.
Step 5 Arrange eigenvalues in descending order with their corresponding EVs. EV with the highest eigenvalue is the PC of the dataset.
Step 6 Construct FV by selecting only p EVs.
Step 7 Take the transpose of FV and multiply it on the left of the original transposed dataset.

5.3 Mutual Information (MI)-Based Feature Selection

This feature selection method is based on mutual information which is a measure of relevance and redundancy among the other features. R. Battiti [10] defined the feature reduction problem as the process of selecting the most relevant k features from an initial set of features and proposed a greedy selection method. Ideally, the feature selection problem can be solved by maximizing joint MI information $I(C;\, S)$ between the class variable C and the subset of selected features S. However, computing Shannon's MI between high-dimensional vectors is impractical because the number of samples and computational time required becomes prohibitive. To overcome these limitations, Battiti adopted a heuristic criterion for approximating the ideal solution. Instead of calculating the joint MI between the selected feature set and the class variable, only $I(C;\, f_i)$ and $I\left(f_i;\, f_j\right)$ are computed, where f_i and f_j are individual features. Battiti's Mutual Information Feature Selector (MIFS) selects features, which maximize the information about the class, corrected by subtracting a quantity proportional to the average MI with the previously selected features.

Entropy It is a measure of uncertainty in a variable which is defined as follows:

$$H(X) = -\sum P(x_i) \log P(x_i) \quad \text{where } i = 1, \ldots, n \tag{5.7}$$

$P(x_i)$ is the probability density function for n variables.

Conditional Entropy It quantifies the probability of an event when another event has already occurred. Mathematically, it can be defined as follows:

Let "X" and "Y" are two variables where $X = x_1, \ldots, x_n$ and $Y = y_1, \ldots, y_n$, respectively. The conditional entropy between "X" and "Y" can be defined as

$$H(X|Y) = -\sum P(x_i, y_i) \log P(x_i, y_i) \quad i = 1, \ldots, n \tag{5.8}$$

Mutual Information It is defined as the information gained about random variable with the evidence of another random variable. Since it is a probability-based measure, it is free from normalization and coordinate system. Mutual information "I" between "X" and "Y" can be written as,

$$\begin{aligned} I(X, Y) &= H(X) - H(X|Y) \\ &= H(X) - H(Y|X) \\ &= H(X) + H(Y) - H(X, Y) \\ &= \sum_{x,y} P(x, y) \log \frac{P(x, y)}{P(x) \times P(y)} \end{aligned} \tag{5.9}$$

Classification If "C" represents a class and "X_i", "X_j" are FVs then,

Case 1: If $I(C, X) = \text{high} \Rightarrow$ X is a relevant feature
Case 2: If $I(X_1, X_2) = \text{high} \Rightarrow$ X redundancy between "X" *and* "Y".

5.3.1 MIFS Algorithm

Given an initial set F with n features, find subset $S \subset F$ with k features that maximize the $I(C; S)$, i.e., mutual information between the class variable C and the subset of selected features S. Battiti's MIFS is a heuristic incremental search solution to the above-defined problem. The algorithm is as follows:

Step 1 *Initialization*: F be the "initial set of n features," and S be the "empty set".
Step 2 *Computation of MI with the output class*:
For each, $f_i \in F$ compute $I(C; f_i)$.
Step 3 *Selection of the first feature*: Find the feature f_i that maximizes $I(C; f_i)$; set $F \leftarrow F \backslash \{f_i\}$ and $S \leftarrow \{f_i\}$.
Step 4 *Greedy selection*: Repeat until $|S| = K$,
where $|S|$: cardinality of S.
(a) *Computation of the MI between variables*: For all pairs $(f_i; f_s)$ with $f_i \in F$ and $f_s \in S$,
Compute $I(f_i; f_s)$ if it is not yet available.
(b) *Selection of the next feature:* Choose the feature $f_i \in F$ that maximizes

$$I(C, f_i) - \beta \sum_{f_s \in S} I(f_s, f_i)$$

Set $F \leftarrow F \backslash \{f_i\}$ and $S \leftarrow \{f_i\}$.
Step 5: Output the set containing the selected features.

The parameter β is a user-defined parameter that regulates the relative importance of the redundancy between the candidate feature and the set of selected features. Based on the selection criteria of Step 4 (b), three more variants of MIFS have been introduced as mentioned below.

5.3.2 Minimum Redundancy, Maximum Relevance (MRMR)

It was proposed by Peng et al. [11] where instead of assigning a fixed value to β, it is being given the following value:

$$I(C; f_i) - \frac{1}{|s|} \sum_{f_s \in S} I(f_s; f_i) \tag{5.10}$$

where

$|S|$ is the cardinality of set S.

5.3.3 Mutual Information Feature Selection Under Uniform Information Distribution (MIFS-U)

MIFS-U assumes that the information about the input data is distributed uniformly [12]. The criteria function for MIFS-U is as follows:

$$I(C; f_i) - \beta \sum_{f_s \in S} \frac{I(C; f_s)}{H(f_s)} I(f_s; f_i) \tag{5.11}$$

5.3.4 Normalized Mutual Information Feature Selection (NMIFS)

NMIFS measures redundancy among the features by using normalized MI. P. A. Estevez [13] introduced NMIFS. The criteria function of NMIFS is defined as follows:

$$I(C; f_i) - \frac{1}{S} \sum_{f_s \in S} \mathrm{NI}(f_i; f_s) \tag{5.12}$$

5.4 Bhattacharyya Distance (BD) Based Feature Selection

In statistics, BD measures the similarity of two discrete or continuous probability distributions [14]. It is closely related to the Bhattacharyya coefficient, which is the measure of the amount of overlap between two statistical samples or populations. For discrete probability distributions p and q, the Bhattacharyya coefficient is defined as:

$$\mathrm{BC}(p, q) = \sum_{x \in X} \sqrt{p(x)q(x)} \tag{5.13}$$

For continuous probability distributions p and q, the Bhattacharyya coefficient is defined as:

$$\mathrm{BC}(p, q) = \int \sqrt{p(x)q(x)}\,\mathrm{d}x \tag{5.14}$$

BD between any two distributions P_k and P_l is given by:

$$\mathrm{BD} = \frac{1}{8}(m_k - m_l)^{\mathrm{T}} p^{-1}(m_k - m_l) + \frac{1}{2}\left(\frac{\det P}{\sqrt{\det P_k \det P_l}}\right) \tag{5.15}$$

where m_i and P_i are mean vector and covariance matrix of class I and P and can be defined as:

$$P = \frac{(P_k + P_l)}{2} \tag{5.16}$$

Firstly, compute BD for each feature. Then, features get ranked based on the this value. The feature with the maximum value of BD is ranked as one.

The algorithm to select K features out of N given features using BD is as follows:

Step 1 Let us consider a set F be the initial set of N features and set S the initial set of selected features.

Step 2 Calculate the BD as mentioned in Eq. (5.15) for each feature $f_i \in F$ with all other features $f_k \in F \backslash f_i$, for $k = 1$ to N and $k \neq i$.

Step 3 Find $f_i \in F$ that maximizes mean BD where $j \neq i$.

$$\frac{1}{n}\sum_{j=1}^{n} \mathrm{BD}(f_i, f_j)$$

where n is the number of features.

Step 4 Set $S \leftarrow \{f_i\}$ and set $F \leftarrow \{f_i\}$.

Step 5 Select next f_j that maximizes,

$$\frac{1}{l}\sum_{j=1}^{l} \mathrm{BD}(f_i, f_j)$$

$\forall f_i \in F$, where $f_i \in F$ and l is the number of features in S.

Step 6 Set $S \leftarrow S \cup \{f_i\}$ and set $F \leftarrow F \backslash f_i$.

Step 7 Go to step 5 if $|S| \neq K$, otherwise exit.

5.5 Independent Component Analysis (ICA) Based Feature Selection

This method [15] of feature selection is more useful with those data that follows non-Gaussian distributions. The main goal is to find a linear representation of non-Gaussian data so that the components are statistically independent, or as independent as possible. Components of observed random vector x are the sum of independent components and mixing weights. It is represented as,

$$x_j = a_{j1}s_1 + a_{j2}s_2 + \cdots + a_{jn}s_n \quad (5.17)$$

where

x_j is the observed random vector we can also write the above equation as $x = As$,
x is the observed random vector,
s is the independent component vector, and
A is the mixing matrix.

Compute mixing matrix A, and find the inverse of A denoted by W. Compute the independent components s as,

$$s = Wx \quad (5.18)$$

Compute the kurtosis of each independent component. Degree of independence is indicated by the absolute value of kurtosis, and ICA features are ranked in order of their degree of independence. For Gaussian distributed independent components, kurtosis is zero. For feature classification, we need strongly independent features. So, select those features, which have the maximum absolute value of kurtosis.

Fast ICA algorithm is implemented for ICA-based feature selection. It is based on a fixed-point iteration scheme for finding a maximum of the non-Gaussian of $\boldsymbol{w}^{\mathrm{T}}x$, where $\boldsymbol{w}$ is a weight vector that the neuron is able to update by a learning rule. The basic form of the Fast ICA algorithm is as follows:

Step 1: Choose an initial (e.g. random) weight vector $\boldsymbol{w}$
Step 2: Compute

$$W^{+} = E\{xg(\boldsymbol{w}^{\mathrm{T}}x)\} - E\{g^{l}(\boldsymbol{w}^{\mathrm{T}}x)\}\boldsymbol{w}$$

Step 3: Let, $W = \frac{W^{+}}{\|W^{+}\|}$
Step 4: If not converged, go back to Step 2.

5.6 Graphical Analysis (GA)-Based Feature Selection

Apart from the well-defined feature selection methods, this section explains the new method for feature selection [16]. Since the foremost task is to identify parameters, which depict the performance of the features by showing consistency and ability to distinguish over a prolonged time. Once these parameters are decided, a threshold value is estimated to filter out the good features. Initially, the method was developed based on intuition where the all features were plotted manually, and then based on their distinguishing performance, good and bad features were selected.

As shown in Fig. 5.1, "Sample Feature 1" has both the conditions mixed up, whereas "Sample Feature 2" can differentiate between the two conditions. Intuitively, "Sample Feature 2" is a better choice than "Sample Feature 1" in terms of classification accuracy. Based on this observation, a good feature is defined to be the one, which is consistent for the healthy as well as the faulty plots. To utilize this observation at large scale, it was needed to automate this process.

5.6.1 *Identification of Performance Parameters of Features*

A good feature is defined to be the one, which is consistent for a healthy as well as the faulty state of a machine. Here, some characteristics of a good feature and bad feature plots are shown in Fig. 5.2.

- The plots should not intersect too much, i.e., they must have minimum common values for all the samples throughout the plot. As shown in Fig. 5.2b, the plot shows a bad feature because the healthy and the faulty plots are intersecting each other many times while the plot shown in Fig. 5.2a is a good feature because there are no intersections at all.
- The plots in Fig. 5.3 should not vary abnormally from their average variation throughout the plot. Fig. 5.3a shows the normal variation of the plot so it can be

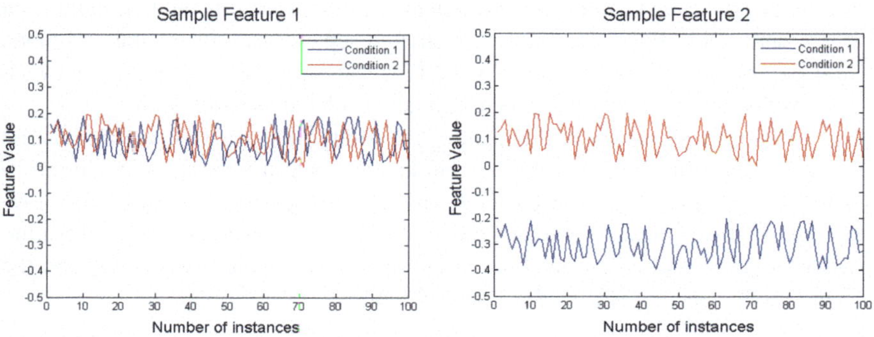

Fig. 5.1 Comparison between two features discriminating capability [16]

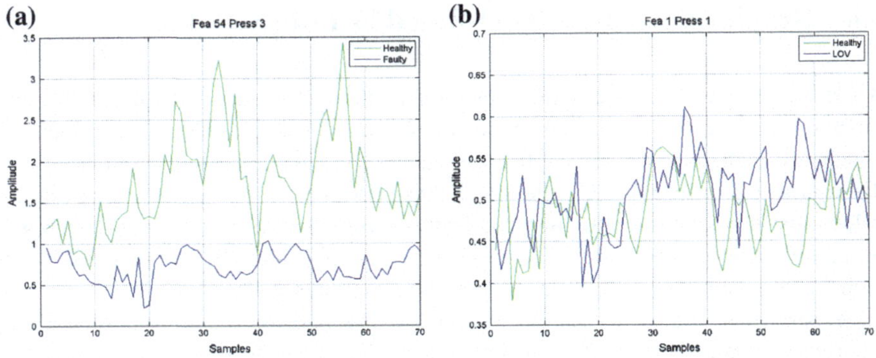

Fig. 5.2 **a** Good feature and **b** Bad feature [16]

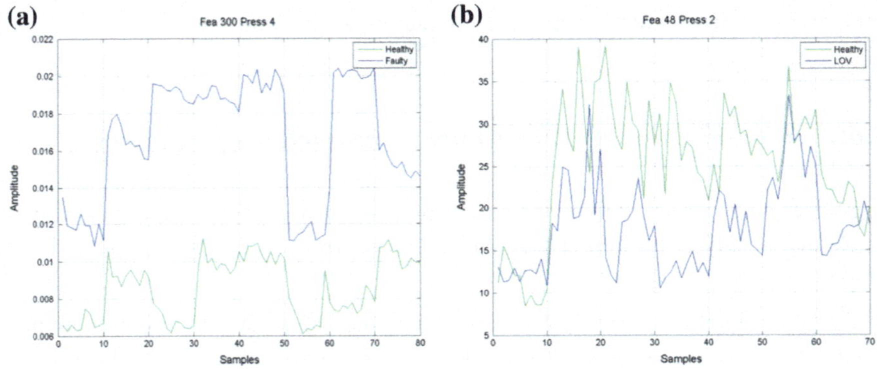

Fig. 5.3 **a** Good feature and **b** Bad feature [16]

considered as the good feature while Fig. 5.3b shows the abnormal variation of the plot so it is considered as the bad feature.

- The plots shall be separable, i.e., there shall be an imaginary line which should mostly lie above the healthy plot, without the loss of generality, if it lies below then it is a faulty plot. Fig. 5.4a shows the enough separation between the two plots, so it is considered as good feature while Fig. 5.4b shows less separation between the healthy and the faulty plots; hence, it is considered as bad feature.
- The ratio of mean values of the healthy and faulty plots (or vice versa) must lie above a certain threshold. It's physical implication measures the amount of improbability that a faulty feature must resemble its healthy counterpart. Figure 5.5a shows optimum ratio of means between the two plots, so it is considered as good feature while Fig. 5.5b shows the much less ratio of means between the healthy and the faulty plots; hence, it is considered as bad feature.

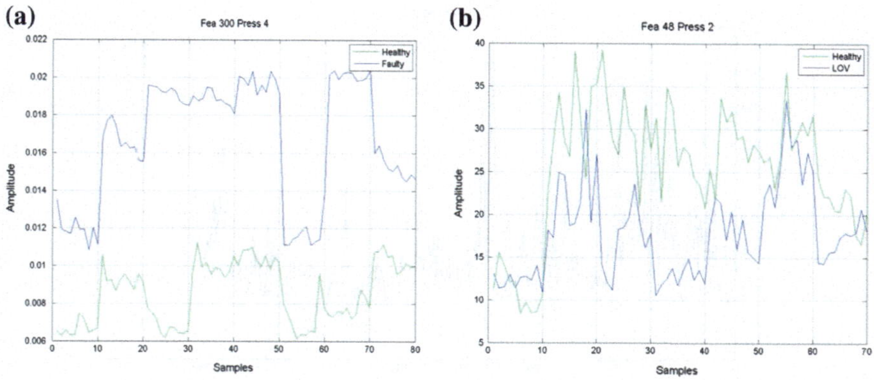

Fig. 5.4 **a** Good feature and **b** Bad feature [16]

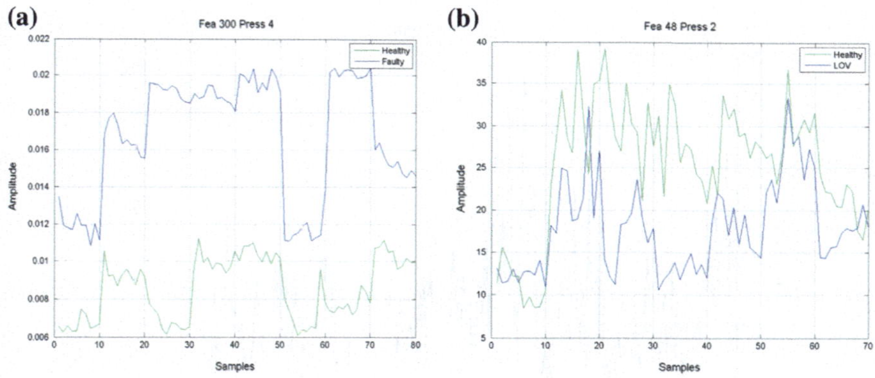

Fig. 5.5 **a** Good feature and **b** Bad feature [16]

Four parameters were identified to capture features' ability to distinguish two states effectively and with reliability. These parameters are as follows:

- **Zero-Crossing Rate (ZCR)**: this is a measure of how many times the plot of healthy and faulty data intersects each other. Mathematically, it is observed that several times the difference of healthy plot and faulty plot attains zero value. Low ZCR implies a better feature as shown in Fig. 5.6.
- **Separation**: this measures plot's ability to have two of its components (i.e., healthy data and faulty data) separated by an imaginary line. Mathematically, it counts the occurrences of healthy and faulty plots crossing the imaginary line of separation. As shown in Fig. 5.7, upper line indicates minimum of faulty while lower line indicates maximum of healthy data. Low separation count implies a better feature.
- **Ratio**: ratio implies the amount of uncertainty a faulty feature resembles its healthy counterpart. It is the ratio of the mean of lower plot (here, it is healthy) to the mean of upper plot (here, it is faulty) for over all datasets. As shown in Fig. 5.8, upper line

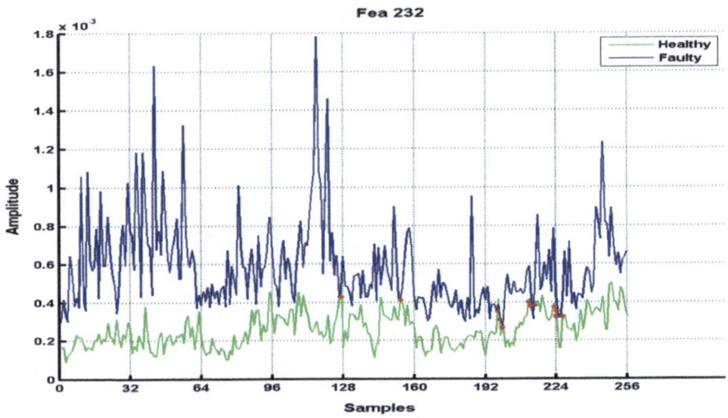

Fig. 5.6 ZCR value for healthy and faulty datasets [16]

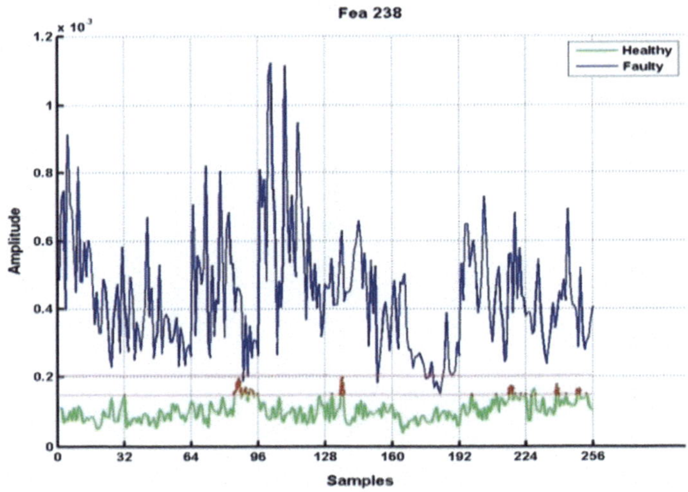

Fig. 5.7 Separation count for healthy and faulty datasets [16]

indicates the faulty datasets while lower line indicates the healthy datasets. Lower ratio implies better ability to distinguish the features, i.e., better feature.

– **Standard Deviation**: this measures the abnormal variation in the plot of a feature. Mathematically, it computes the deviation of the mean value of the plot divided into datasets of equal data width. This is calculated after normalizing the feature values. Low standard deviation implies better feature as shown in Fig. 5.9.

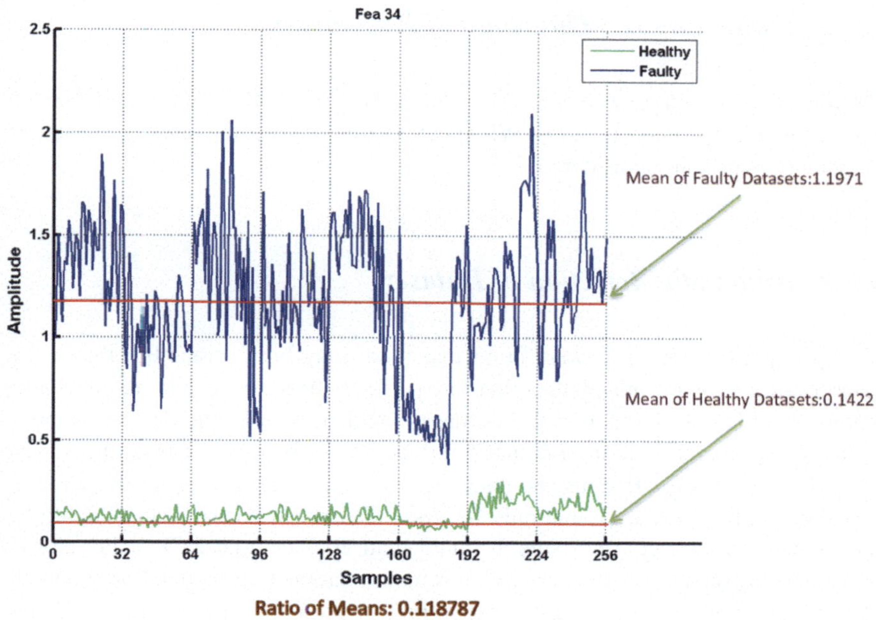

Fig. 5.8 Ratio of means for healthy and faulty datasets [16]

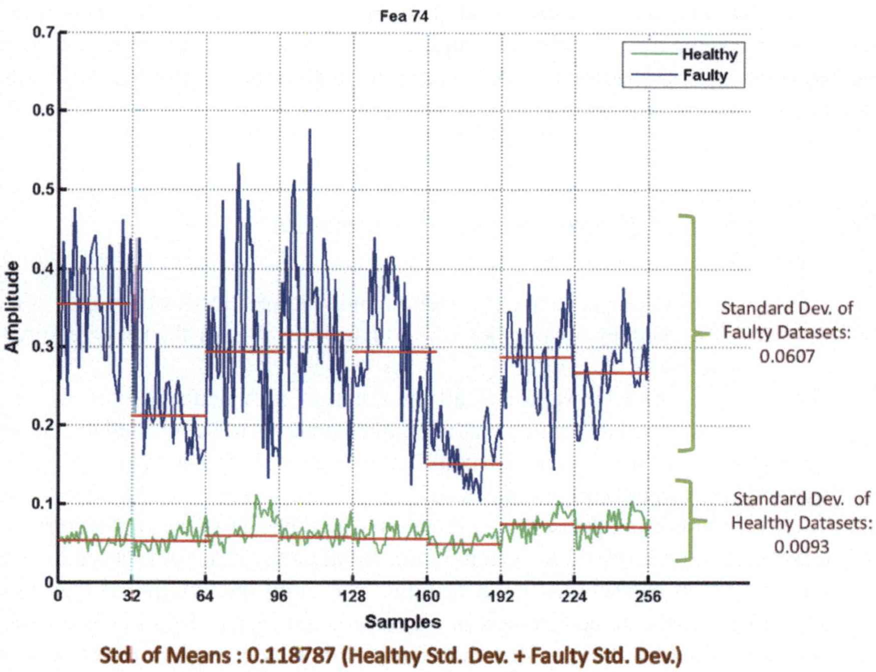

Fig. 5.9 Standard deviation for healthy and faulty datasets [16]

5.6.2 Estimating the Threshold of Parameters

The threshold of each parameter has been decided by finding the average of parameter values for selected good features. While calculating this, all the datasets present in the feature matrix are considered.

5.6.3 Automatic Rejection of Dataset

During experimentation, it was found that there may be a possibility that a few datasets are corrupted, and all the others are perfectly fine. For such a case, the entire feature would be declared as bad. In order to tackle this situation, dataset rejection is employed wherein a limited number of datasets can be rejected from the feature if they are adversely affecting the entire feature set. Dataset rejection is based on four thresholds which are found only for good features. A dataset is considered for rejection when dataset contributes to more than 80% of total ZCR value and its average ZCR, separation and ratio value is more than their corresponding threshold values. A dataset is rejected if more than half of the good features suggest being rejected.

One more question raises what to do first: selection of good features or rejection of corrupt datasets. As we know good features may get rejected due to corrupt dataset and good dataset also seems corrupted due to bad features. Thus, both dataset rejection and feature selection need to go on simultaneously to give the best features based on consistent datasets.

5.6.4 Automated Feature Selection Algorithm

i. To consider simultaneous dataset rejection and feature selection, a loop is constructed until execute till number of good features is less than the required number of features.
ii. The average parameter values of all the datasets are calculated only for the good features which determine that the if any dataset is to be rejected or not. In the beginning, all the features are considered as good. It should be noted that in one loop, only one dataset is rejected.
iii. Based on non-corrupt datasets after rejection, half the value of the total features is selected as good features. In each iteration based on the value of four parameter averages of all datasets for good features, if corrupt datasets present, it gets rejected. It must be noted that there is a maximum limit (given by user) for dataset rejection till which datasets that can be rejected. This process of rejection and selection continues till the count of good features goes lower than the required number of good features. Figure 5.10 depicts the flow chart of feature selection algorithm.

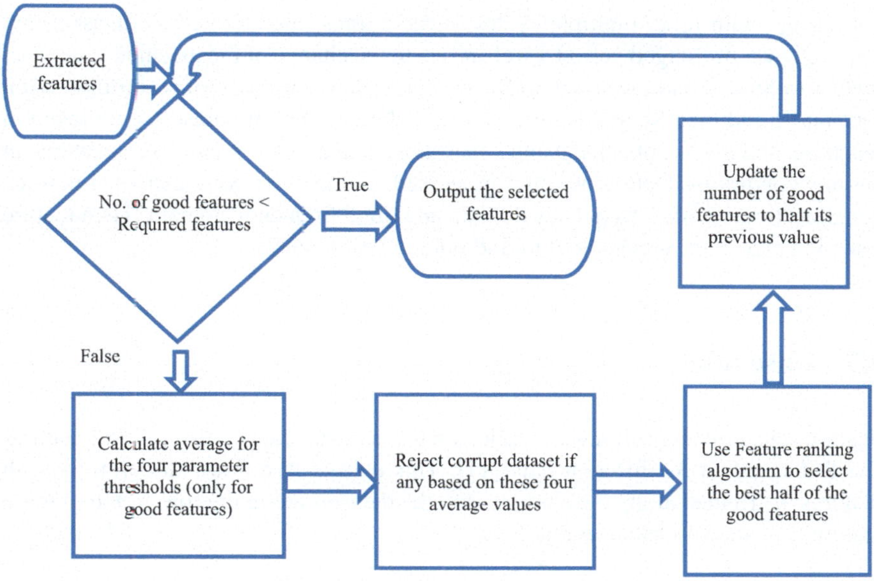

Fig. 5.10 Flow chart of feature selection algorithm [16]

The proposed algorithm needs the following parameters:

- Two feature matrices, for healthy (state 1) and faulty (state 2) state, with number of rows equal to the number of samples, i.e., (number of datasets × number of recordings in each dataset) and number of columns equal to the total number of features.
- Number of datasets are needed to judge the sample width (number of samples) of each dataset.
- Number of datasets that can be rejected at max.
- Number of good features desired (default is 30).

5.6.5 *Limitations*

However, there are some limitations to this method. Equal weightage is given to all the four parameters for judging the good features, which may not always be true. The user needs to decide how many datasets can be rejected. The algorithm does not returns the desired number of good features; instead, it is a number that is always lower than the desired number. During the development of GA-based feature selection, we found that machine data should be taken over a prolonged time because air compressor over long periods of time gradually changes its characteristics, even though it lies in the same state. Hence, to capture those trends in data which do not change with time, multiple cycles of recordings (also termed as datasets) are acquired over prolonged period. Cycle refers to the change of the machine state from State 1 to State 2 and then back to State 1. One complete cycle of recordings taken for State 1 and then State 2 is termed as one dataset. No. of datasets were taken in real time, and it was collected using smartphone and accelerometer from machine in varying conditions. Dataset was pre-processed, and features were extracted in three domains: time domain, frequency domain and time–frequency domain. Then feature selection algorithm was applied to find out the best features.

5.7 Case Study

Feature selection algorithm was applied on the acoustic data consist of 286 features and 8 datasets of 32 different data [17]. This data is stored in ".dat" format with features distributed along the columns and the datasets along the row. Some of these features are shown further in Fig. 5.11.

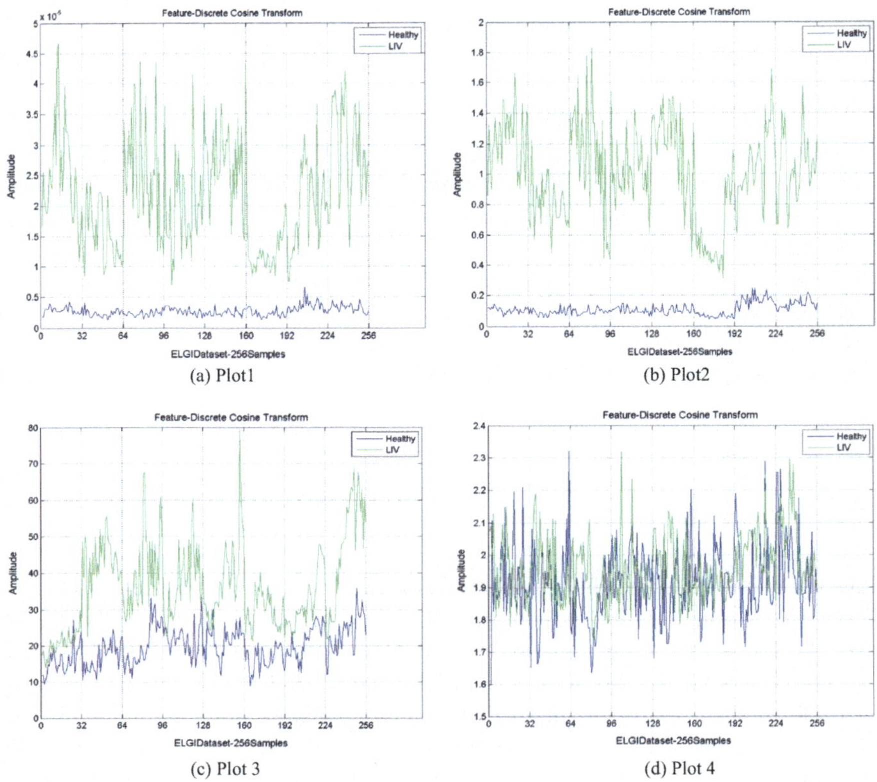

(a) Plot1

(b) Plot2

(c) Plot 3

(d) Plot 4

Fig. 5.11 Feature plot [16]

(a) it can be seen that the faulty data is varying abruptly while the feature plot can be considered as a good feature as shown in the Fig. 5.11.
(b) the mean ratio between the healthy (blue) and the faulty (green) data is very small that required, whereas in the feature plot, the healthy data and faulty data are not easily distinguishable shown in Fig. 5.11.

The test of each threshold value has been implemented as a function which takes arguments as healthy data, faulty data, and number of datasets, threshold values and an array score (286×1) to indicate which feature(s) have been rejected so far by other tests. This function then returns the variable score with updates of rejected feature(s) by the test. This function is made for removing dataset which performs poorly for all the features and also causes removal of some good features. This function has to be called in the starting of the program. It uses the given values of thresholds to determine good features as well as simultaneously find the datasets which adversely impact the good features. Since there are a vast number of random bad features, they would incorrectly indicate a dataset to be bad, whereas the reason for this might be simply the bad feature instead of dataset. This would result in an approximately equal probability of each dataset being rejected, which is obviously wrong. Hence, only good features are used to find bad features. The rejected function takes arguments as healthy data, faulty data, number of dataset, and threshold values. It returns the rejected number of dataset number w.r.t. the initial dataset numbering.

5.8 Development of Feature Selection Tool

5.8.1 Feature Selection Tool for Android Platform

Step 1 As the user runs the Feature Selection App, a home page appears as shown in Fig. 5.12. Here, the user is asked to tap anywhere on the screen to redirect on the main activity page.

Step 2 As the user taps on the home page, feature selection activity page opens as shown in Fig. 5.13. From here, the user can select the input files and parameters required for the feature selection process.

Step 3 On clicking "Browse Healthy Data/Browse Faulty Data" button(s), a screen appears which shows the databank folder containing required healthy and faulty files. On clicking the databank folder, the required files, i.e., "Healthy-Data.txt" and "FaultyData.txt" appear as shown in Fig. 5.14.

Step 4 On clicking the respective files ("HealthyData.txt" and "FaultyData.txt"), a confirmation screen appears as shown in Fig. 5.15.

Step 5 On clicking the respective data file, a confirmation screen pops up as shown in Fig. 5.15. On clicking "YES", the screen Activity Page 1 with both the selected files in respective text boxes appears. While on clicking "NO", it returns to the Activity Page 1. Now, enter the no. of dataset and max. no. of dataset rejected as shown in Fig. 5.16.

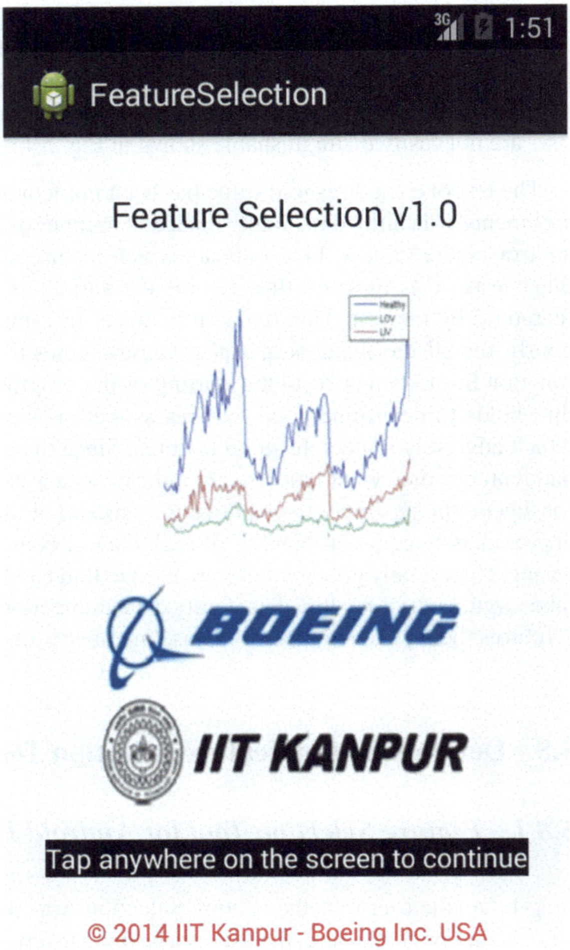

Fig. 5.12 Home page [16]

Step 6 After clicking "Select Best Features" button, a screen showing the processing of data appears as shown in Fig. 5.17.

After the processing is done, the result page appears showing that results are stored in a model file ("model.dat"). This model file contains the following information about the selected features is shown in Fig. 5.18. (1) Indices of the selected features. (2) Mean of the selected healthy features. (3) Mean of the selected faulty features. (4) Standard deviation of the selected healthy features. (5) Standard deviation of the selected faulty features.

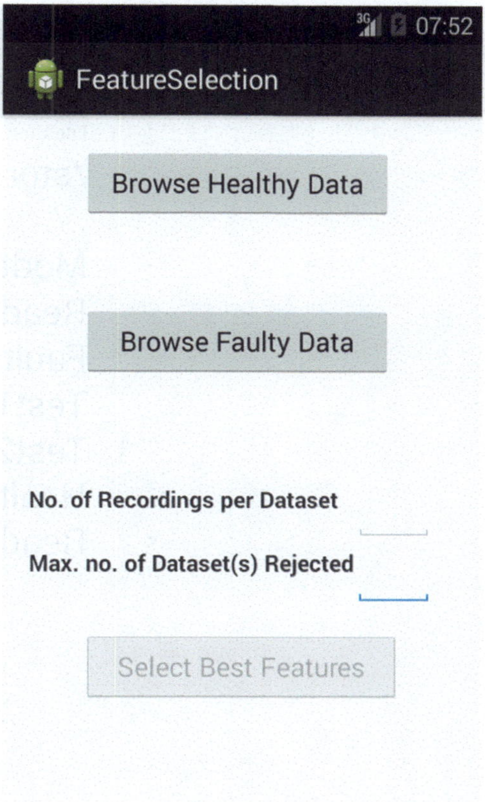

Fig. 5.13 Feature selection Activity page 1 [18]

5.8.2 Feature Selection Tool for Windows Mobile Phone

Figure 5.19 show snapshot of feature selection tool on windows mobile phone.

5.8.3 Feature Selection Tool Windows Tablet Platform

Figure 5.20 shows snapshots of feature selection tool on windows tablet platform.

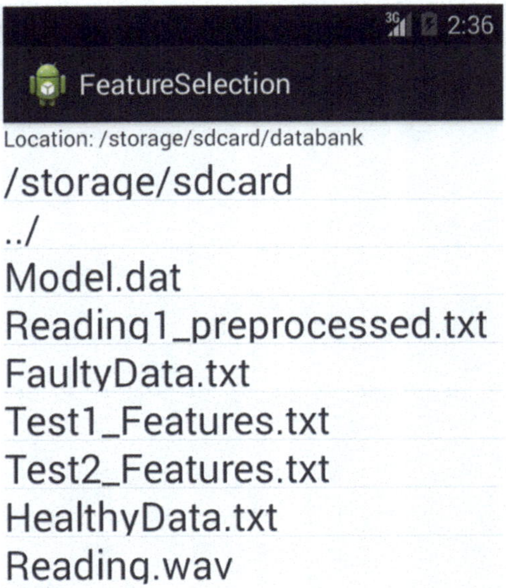

Fig. 5.14 Browse Activity page 2 [18]

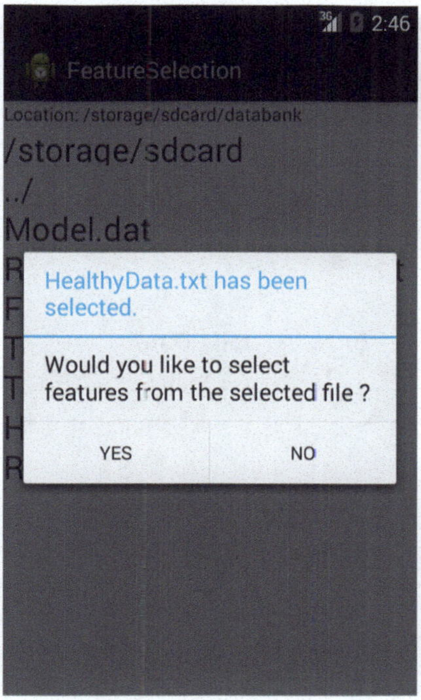

(a) File confirmation page1

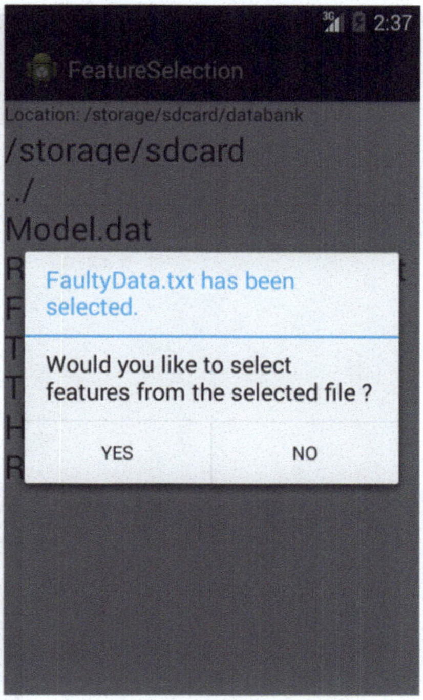

(b) File confirmation page2

Fig. 5.15 File confirmation page [18]

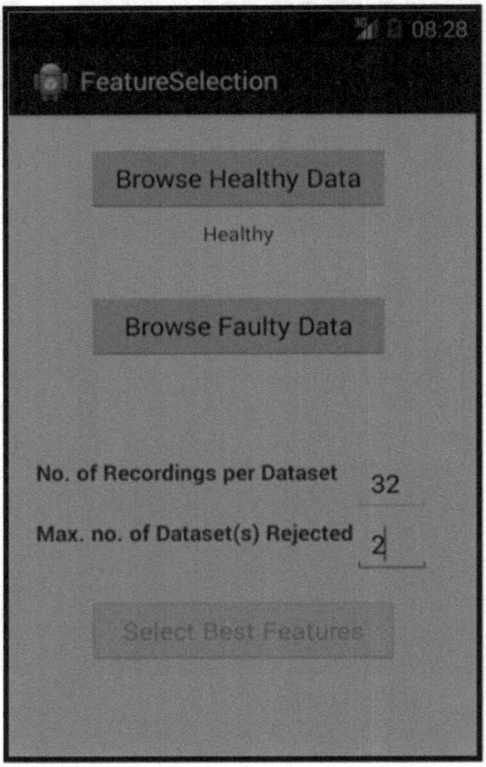

Fig. 5.16 Selected file activity page [18]

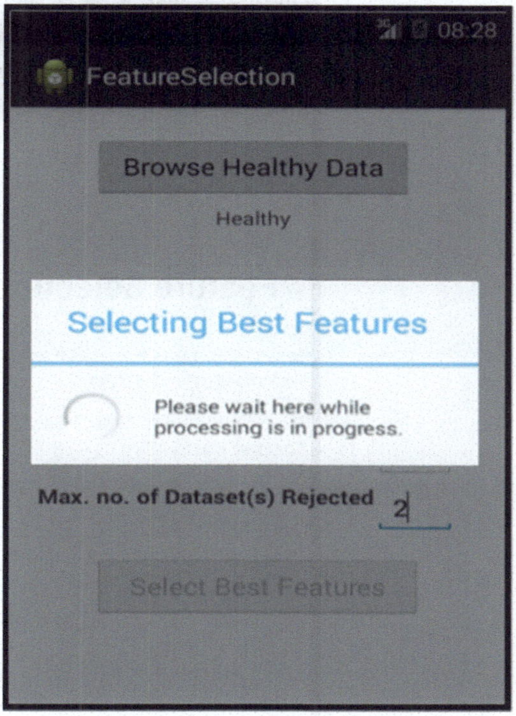

Fig. 5.17 Feature selection processing page [18]

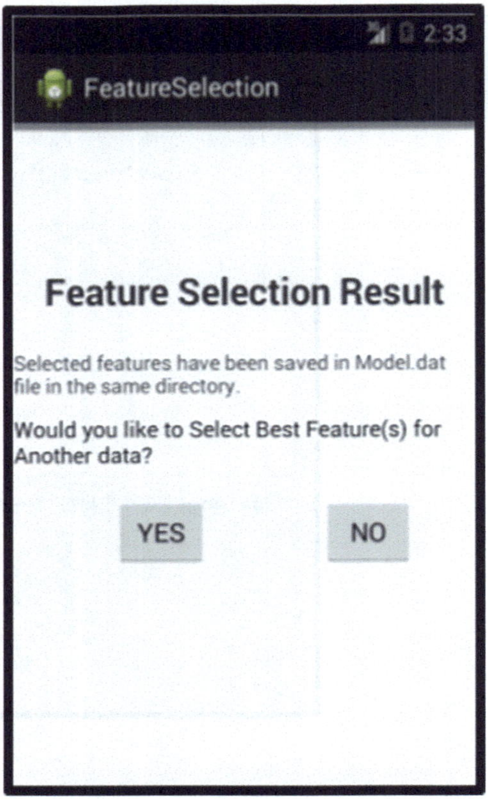

Fig. 5.18 Result activity page [18]

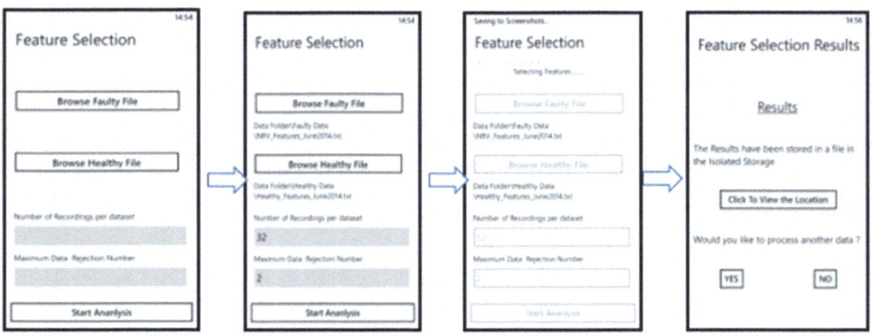

Fig. 5.19 Snapshots of feature selection tool on windows mobile phone [19]

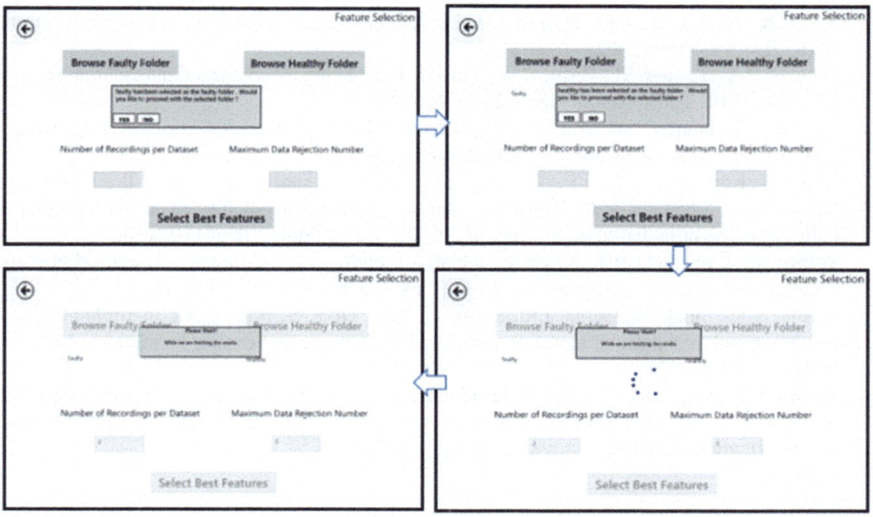

Fig. 5.20 Snapshots of feature selection tool on windows tablet platform [19]

References

1. Verma, N.K., Maini, T., Salour, A.: Acoustic signature based intelligent health monitoring of air compressors with selected features. In: 9th International Conference on Information Technology: New Generations, Nevada, USA, pp. 839–845 (2012)
2. Shlens, J.: A Tutorial on Principal Component Analysis. University of California, Systems Neurobiology Lab., California, USA (2005)
3. Thirukovalluru, R., Sevakula, R.K., Dixit, S., Verma, N.K.: Generating optimum feature sets for fault diagnosis using denoising stacked auto-encode. In: IEEE International Conference on Prognostics and Health Management, Canada, USA, pp. 1–7 (2016)
4. Verma, N.K, Gupta, V.K., Sharma, M., Sevakula, R.K.: Intelligent condition based monitoring of rotating machines using sparse auto-encoders. In: IEEE Conference on Prognostics and Health Management, Maryland, USA, pp. 1–7 (2013)
5. Maurya, S., Singh, V., Dixit, S., Verma, N.K., Salour, A., Liu, J.: Fusion of low-level features with stacked autoencoder for condition based monitoring of machines. In: IEEE International Conference on Prognostics and Health Management, Washington, USA, pp. 11–13 (2018)
6. Sharma, A.K., Singh, V., Verma, N.K., Liu, J.: Condition based monitoring of machine using Mamdani fuzzy network. In: Prognostics and System Health Management Conference, Chongqing, China, pp. 26–28 (2018)
7. Verma, N.K., Dixit, S., Sevakula, R.K., Salour, A.: Computational framework for machine fault diagnosis with autoencoder variants. In: 2018 International Conference on Sensing, Diagnostics, Prognostics, and Control, Xi'an, China, pp. 15–17 (2018)
8. Jolliffe, I.T.: Principal Component Analysis. Springer, New York (1986)
9. He, Q., Yan, R., Kong, F., Du, R.: Machine condition monitoring using principal component representations. J. Mech. Syst. Signal Pr. **23**(2), 446–466 (2009)
10. Battiti, R.: Using mutual information for selecting features in supervised neural net learning. IEEE Trans. Neural Netw. **5**(4), 537–550 (1994)
11. Peng, H., Long, F., Ding, C.: Feature selection based on mutual information: criteria of max dependency, max-relevance and min-redundancy. IEEE Trans. Pattern Anal. Mach. Intell. **27**(8), 1226–1238 (2005)

12. Kwak, N., Choi, C.H.: Input feature selection for classification problems. IEEE Trans. Neural Netw. **3**(1), 143–159 (2002)
13. Estevez, P.A., Tesmer, M., Perez, C.A., Zurada, J.M.: Normalized mutual information feature selection. IEEE Trans. Neural Netw. **20**(2), 189–201 (2009)
14. Guorong, X., Peiqi, C., Minhui, W.: Bhattacharyya distance feature selection. In: Proceedings 13th IEEE International Conference on Pattern Recognition, **2**, pp. 195–199 (1996)
15. Aapo, H., Karhunen, J., Oja, E.: Independent Component Analysis, vol. 46. Wiley (2004)
16. Verma, N.K., Sevakula, R.K., Thirukovalluru, R.: Pattern analysis framework with graphical indices for condition based monitoring. IEEE Trans. Rel. **66**(4), 1085–1100 (2017)
17. Verma, N.K., Sevakula, R.K., Dixit, S., Salour, A.: Intelligent condition based monitoring using acoustic signals for air compressors. IEEE Trans. Rel. **65**(1), 291–309 (2016)
18. Verma, N.K., Singh, S., Gupta, J.K., Sevakula, R.K., Dixit, S., Salour, A.: Smartphone application for fault recognition. In: 6th International Conference on Sensing Technology, Kolkata, India, pp. 1–6 (2012)
19. Verma, N.K., Singh, J.V., Gupta, M., Sevakula, R.K., Dixit, S.: Windows mobile and tablet app for acoustic signature machine health monitoring. In: International Conference on Industrial and Information Systems, Gwalior, India, pp. 1–6 (2014)

Chapter 6
Fault Recognition

Abstract This chapter details the last module of the fault diagnosis framework, i.e., fault recognition. After the collection of relevant subsets of the features, as detailed in the Chap. 5, classification is performed to decide the machine state. Chapter 2 provides details about different kinds of machine faults. Classification is the task to categorize the given object by learning the relationship between the selected set of features and their class label. Since fault recognition is also treated as a classification process; here, the description of different classification techniques, such as *k*-means clustering, *k*-nearest neighbour (*k*-NN), Naive Bayes classifier, Support Vector Machine (SVM), Multiclass classification algorithms, etc., is provided in this chapter.

6.1 Classification

To recognize an object, it is essential to learn the distinctive characteristics of that object so that it can be distinguished from other objects. Classification is a similar process where the relationship between training samples and their class data is learned. Fault recognition is also a classification problem where fault classes need to be identified by analyzing collected features. Here, training samples refer to the sets of relevant features obtained after the feature selection process. As mentioned in Chap. 3, data is acquired in each state of a machine, and therefore, samples from each fault class are included in training samples. This phase is called the training phase and it gives a trained model. Once you have a training model, you can use this model later for predicting the class of unknown given sample. The focus of this work is to identify the faults in the compressor [1–3]. Some popular algorithms for classification are *k*-means clustering [4], *k*-NN classifier, decision tree, naïve Bayes classifier, SVM classifier [5], etc. For real-time fault diagnosis systems [6, 7], a highly generalized classifier is needed; hence, SVM has been chosen. The next subsections discuss a brief introduction to these classifiers.

N. K. Verma and A. Salour, *Intelligent Condition Based Monitoring*, Studies in Systems, Decision and Control 256, https://doi.org/10.1007/978-981-15-0512-6_6

6.2 *k*-means Clustering

k-means clustering is an algorithm to classify or to group objects based on attributes/features into *k* number of groups where *k* is a hyper-parameter. The grouping is done by minimizing the sum of squares of distances between data and the corresponding cluster centroid. Thus, the purpose of the *k*-means clustering is to cluster the data. A plot of the *k*-means clustering is shown in Fig. 6.1. The basic steps of *k*-means clustering are simple and are presented below.

6.2.1 *Steps for* k*-means Clustering Algorithm*

Step 1 Place *k* points in the space, which represents initial centroids.
Step 2 Assign each point to the group that has the closest centroid.
Step 3 When all objects have been assigned, recalculate the positions of the *k* centroid.
Step 4 Repeat steps 2 and 3 until the centroids no longer move. This produces a separation of the objects to form groups from which the metric is to be minimized and calculated.

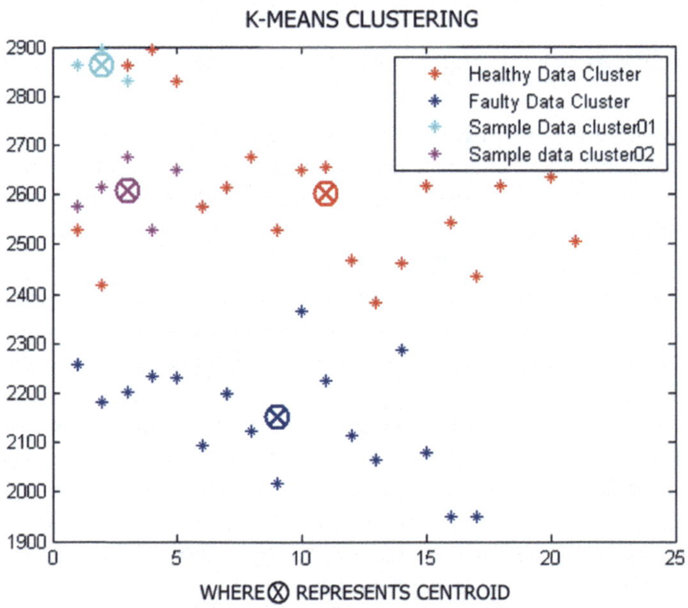

Fig. 6.1 Plot of *k*-means clustering [4]

6.3 *k*-nearest Neighbour (*k*-NN) Classifier

k-nearest neighbour (*k*-NN) classifier is a sample-based classifier. It operates on the premise that the classification of unknown samples can be done by relating the unknown to the known based on some distance. The intuition is that two samples far apart in the space defined by an appropriate distance function are less likely to belong to the same class than two closely situated samples to belong to the same class. Classification using a sample-based classifier can be understood as locating the nearest neighbour in sample space and labelling the unknown sample with the same class label as that of the located (known) neighbour. This approach is often referred to as the nearest neighbour classifier. The downside of this simple approach is the lack of robustness that characterizes the resulting classifiers. The high degree of local sensitivity makes nearest neighbour classifiers highly susceptible to noise in the training data.

More robust models can be achieved by locating k, where $k > 1$, neighbours and letting the majority vote [8, 9] decide the outcome of class labelling. A higher value of k results in a smoother and less locally sensitive function. The nearest neighbour classifier can be regarded as a special case of the more general *k*-NN classifier, hereafter referred to as a *k*-NN classifier. The drawback of increasing the value of k is that as k approaches n, where n is the size of the sample base, the performance of classifier will approach to that of the most straightforward statistical baseline, with the assumption that all unknown samples belong to the class most frequently represented in training data. This problem can be avoided by limiting the influence of distant samples. One way of doing so is to assign a weight to each vote, where the weight is a function of the distance between unknown and known samples. By letting each weight be defined as the inverse of squared distance between the known and unknown samples, votes cast by distant samples will have very little influence on the decision process compared to samples in the near neighbourhood. The distance-weighted voting usually serves as good middle ground as far as local sensitivity is concerned. A plot of the *k*-NN classifier is shown in Fig. 6.2.

6.4 Naïve Bayes Classifier

Naïve Bayes classifier is a probabilistic classifier based on Bayes theorem which assumes all the features to be linearly independent. Using prior information of individual classes and likelihood of samples, the classifier for a test sample computes a posteriori probability of each class. The class having maximum posteriori probability will become the assigned class for the sample. When priori probabilities and likelihood are not available, the same is learned from training samples. Priori probability is based on deductive reasoning and not on past behaviour, whereas posteriori probability is given by accounting relevant evidence and background.

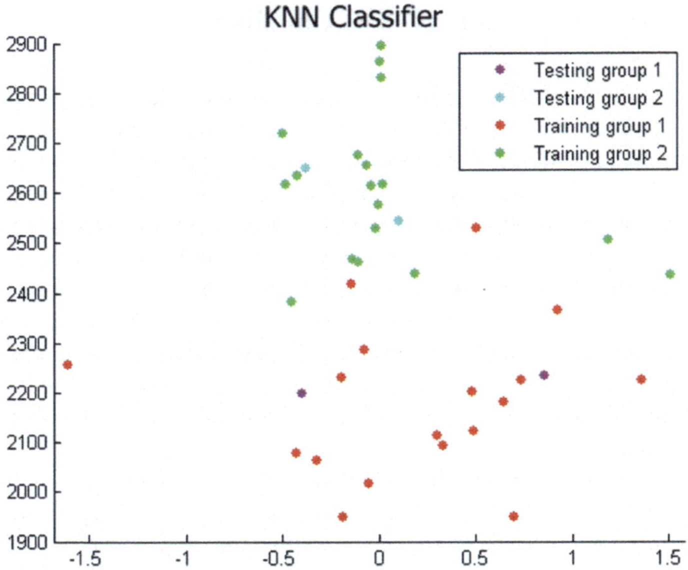

Fig. 6.2 Plot of k-NN classifier [4]

$$P(c_i|x_w, X) = \frac{p(x_w|c_i, X)P(c_i|X)}{\sum_{c'=1}^{L} p(x_w|c', X)P(c'|X)} \tag{6.1}$$

where c_i refers to the class under consideration for finding posteriori probability and L refers to the number of classes. X refers to training data used for getting prior information, and $\boldsymbol{x_w}$ refers to the test sample under consideration. Bayes classifier acts as an important benchmark in classification and works best for large training data.

6.5 SVM Classifier

SVM was introduced as a binary classifier, in which a linearly separable case finds the hyperplane which separates two classes with maximum possible margin. It comes under the supervised learning category. SVM constructs a hyperplane or set of hyperplanes in a high or infinite-dimensional space, which can be used for classification, regression, or other tasks. Intuitively, a good separation is achieved by the hyperplane which has the largest distance to the nearest training data points of any class (so-called functional margin), as in general the larger the margin, the lower the generalization error of the classifier.

Classification of data is a common task in machine learning. Suppose some given data points each belong to one of the two classes, and the goal is to decide which

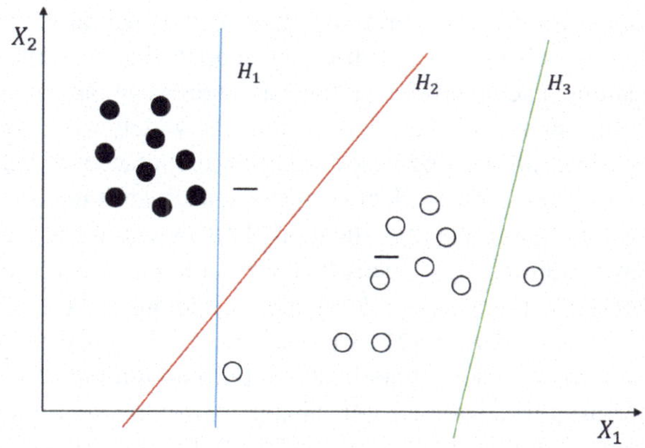

Fig. 6.3 Plot for SVM classifier [10]

class a new data point will fall. In the case of SVM, a data point is viewed as a *p*-dimensional vector, and the objective to know whether we can separate such points with a uniform straight hyperplane. This is called a linear classifier. There are many hyperplanes that might classify the data.

However, maximum separation (margin) between the two classes is usually desired. So, a hyperplane is chosen such that the distance from it to the nearest data point on each side is maximized. If such a hyperplane exists, it is clearly of interest and is known as the maximum margin hyperplane, and such a linear classifier is known as a maximum margin classifier. Figure 6.3 has been plotted between two features X_1 and X_2. The H_3 (Green) does not separate the two classes, $H1$ (Blue) does with a small margin, and $H2$ (Red) with the maximum margin.

6.5.1 *Introduction to SVM-Based Classification*

The value of a supervised classification is usually a function of its accuracy. One of the main objectives in classification is to achieve high accuracy; if possible, a small number of training samples will make the classification process as useful and economical as possible. One attractive classifier for this application is a SVM. SVM classifications may be more accurate than the widely used alternatives such as classification by maximum likelihood, decision tree, and neural network-based approaches. In the past years, SVMs have been proposed for pattern recognition and function approximation problems [11].

SVM tries to fit an optimal separating hyperplane between classes by focusing on the training samples that lie at the edge of the class distributions, the support vectors. The optimal separating hyperplane is oriented such that it is placed at the maximum distance between the sets of support vectors. It is because of this orientation that SVM is expected to generalize more accurately on unseen cases relative to classifiers that aim to minimize the training error such as neural networks. So, with SVM classification, only some of the training samples that lie at the edge of class distributions in feature space (support vectors) are needed in the establishment of decision surface unlike statistical classifiers such as the widely used maximum-likelihood classifiers in which all training cases are used to characterize the classes. Also, it may sometimes be possible to identify the most useful training sites for the provision of support vectors before the classification. This potential for accurate classification based on small training sets means that the adoption of SVM classification can provide the analyst with considerable savings in training data acquisition.

The standard SVMs were originally designed for binary classifications. However, many practical applications fall into multiclass classification problems, which are usually converted into binary ones. One such problem is the classification of Condition-Based Monitoring (CBM) datasets. These datasets are taken while monitoring via computer for specific applications. Here, the CBM dataset is of a compressor, which was put under surveillance regarding its fault diagnosis and health monitoring. The solution to the classification problem is generally achieved by decomposing the multiclass problem into a series of binary analyses, which can be addressed with a binary SVM by applying algorithms like one-against-one, one-against-all, etc. Here, the aim is to evaluate the classification of CBM data via SVM multiclass classification. The SVM classification, here, has been implemented through various existing widely used algorithms which are described later here. Also, a new SVM-based technique is proposed, which is observed to be giving significantly better classification results than other existing algorithms results. The result is tabulated, compared, and shown later in the results section of this chapter. At this point, it becomes important to describe and elucidate the fundamentals of SVM classification. So, the next section deals with the description of SVM, which is followed by related well-known and widely used other SVM-based methods. The CBM dataset taken here is a standard dataset and its description is also presented in the later section.

6.5.2 Support Vector Machine (SVM)

SVM is an important concept and its algorithm can be better understood with a mathematical explanation and example. Let $S = \{(x_1, y_1), (x_2, y_2), \ldots, (x_i, y_i)\}$ be a training set where x_i are m-dimensional attribute vectors, $y_i \in \{-1, +1\}$, $y_i = -1$, and $y_i = +1$ for class 1 and class 2, respectively.

According to, the SVM classifier is defined as follows:

$$D(x) = w^{\mathrm{T}}\Phi(x) + b = 0 \tag{6.2}$$

where $\Phi(x)$ is a mapping function,w^{T} is a vector in the feature space, and b is a scalar.

To classify the data as linearly separable in the feature space, the decision function satisfies the following condition:

$$y_i\left(w^{\mathrm{T}}\Phi(x)+b\right)\geq 1 \quad \text{for } i=1,2,\ldots l. \tag{6.3}$$

Among all the separating hyperplanes, the optimal separating hyperplane has the maximal margin between two classes. It can be formed as follows:

$$\min_{w,b} J(w,b)=\frac{1}{2}w^{\mathrm{T}}w. \tag{6.4}$$

Subject to Eq. (6.3), if the training data is nonlinearly separable, slack variables are introduced in Eq. (6.4) to relax the hard margin constraints as follows:

$$y_i(w^{\mathrm{T}}\Phi(x_i)+b)\geq 1-\zeta_i \quad \text{for } i=1,2,\ldots l, \text{ and} \tag{6.5}$$

$$\xi_i\geq 0 \quad \text{for } i=1,2,\ldots l. \tag{6.6}$$

This technique allows the possibility of having examples that violate Eq. (3.3). In order to obtain the optimal separating hyperplane, one should minimize

$$\min J_{w,b,\zeta_i}(w,b,\zeta_i)=\frac{1}{2}w^{\mathrm{T}}w+\gamma\frac{1}{2}\sum_{i=1}^{l}\zeta_i \tag{6.7}$$

Subject to Eqs. (6.5) and (6.6), where γ is a parameter that determines the trade-off between the maximum margin and the minimum classification error. The optimization problem of Eq. (6.7) is a convex quadratic program that can be solved using Lagrange multiplier method. By introducing Lagrange multipliers α_i and $\beta_i(i=1,2,\ldots l)$, one can construct the Lagrangian function as follows:

$$L(w,b,\alpha_i,\zeta_i,\beta_i)=J_{w,b,\zeta i}-\sum_{i=1}^{l}\alpha_i\left\{y_i\left[w^{\mathrm{T}}\Phi(x_i)+b\right]-1+\zeta_i\right\}-\sum_{i=1}^{l}\beta_i\zeta_i. \tag{6.8}$$

According to the Kuhn–Tucker theorem, the solution of optimization problem is given by the saddle point of Lagrangian function and can be shown to have an expression

$$w=\sum_{i=1}^{l}\alpha_i y_i\Phi(x_i). \tag{6.9}$$

The training examples (x_i, y_i) with nonzero Lagrangian coefficients α_i are called support vectors. The coefficients can be found by solving the following convex quadratic programming problem.

$$\max_\alpha\left(\sum_{i=1}^{l}\alpha_i-\frac{1}{2}\sum_{i=1}^{l}\sum_{j=1}^{l}y_iy_j\left(j(x_i)^{\mathrm{T}}\times j\left(x_j\right)\right)\alpha_i\alpha_j\right) \tag{6.10}$$

subject to

$$\sum_{i=1}^{l}\alpha_iy_i=0,\text{ and }0\le\alpha_i\le\gamma, i=1,\ldots,l. \tag{6.11}$$

By substituting Eq. (6.9) into Eq. (6.2), the classifier can be obtained. Given a new input x, $f(x)$ can be estimated using Eq. (6.12). If $f(x) > 0$, the sample belongs to class 1; otherwise, it belongs to class 2.

$$f(x)=\operatorname{sgn}\left\{\sum_{i=1}^{l}\alpha_iy_i\cdot\left(\Phi(x)^{\mathrm{T}}\cdot\Phi(x)\right)+b\right\} \tag{6.12}$$

where $\operatorname{sgn}(x)=\begin{cases}1, x>0\\0, x\le 0\end{cases}$

In Eq. (6.13), the pair-wise inner product in the feature space can be computed from the original data items using a kernel function. The kernel function can be denoted as,

$$K(x,x_i)=\Phi(x)^{\mathrm{T}}\cdot\Phi(x_i). \tag{6.13}$$

The typical kernel functions include polynomial kernel functions and radial basis functions (RBFs). In this way, $f(x)$ can be rewritten as follows:

$$f(x)=\operatorname{sgn}\left\{\sum_{i=1}^{l}\alpha_iy_i\cdot K(x_i,x)+b\right\}. \tag{6.14}$$

6.6 Multiclass Classification Algorithms

The multiclass classification problem refers to assigning each of the observations into one of C classes. In this section, one-against-one (OAO), Decision-Directed Acyclic Graph (DDAG) method, Fuzzy Decision Function (FDF) methods, and one-against-all (OAA) methods are briefly introduced. At first, the OAO scheme is introduced.

Assume $S = \{(x_1,y_1),(x_2,y_2),\ldots,(x_i,y_i)\}$ is a training set, where $x_i \in R^m$ and $y_i \in (1, 2, \ldots C)$. For the OAO scheme, one needs to determine $C(C-1)/2$ classifiers for the C-class problem. The optimal hyperplane with SVMs for class i against class j is,

$$D_{ij}(x) = w_{ij}^{\mathrm{T}}\Phi(x) + b_{ij} = 0 \quad i < j, 1 < j \leq C, 1 \leq i < C \tag{6.15}$$

where w_{ij}^{T} is a vector in the feature space, $\Phi(x)$ is a mapping function, and b_{ij} is a scalar.

Here, the orientation of optimal hyperplane is defined via following equation:

$$D_{ij}(x) = -D_{ji}(x). \tag{6.16}$$

Owing to different decision functions, three algorithms can be derived based on OAO scheme.

6.6.1 One-Against-One Decision Function Method

For the input vector, one computes

$$D_i(x) = \sum_{j \neq i, j=1}^{C} \mathrm{sgn}\big(D_{ij}(x)\big), \tag{6.17}$$

and classifies x into the class

$$C^* = \arg\max(D_i(x)), \quad i = 1, \ldots C. \tag{6.18}$$

However, if i is not unique in Eq. (6.18), then x is unclassifiable. To visualize the unclassifiable region, a three-class problem is taken as an example.

As illustrated in Fig. 6.4, arrowheads denote the plus orientation of each hyperplane. Assume x_i falls in the shaded region, and then one obtains the following inequalities:

$$D_{12}(x_i) > 0, \quad D_{13}(x_i) < 0, \text{ and } \quad D_{23}(x_i) > 0. \tag{6.19}$$

According to Eq. (6.16), one can gain

$$D_{21}(x_i) < 0, \quad D_{31}(x_i) > 0, \text{ and } \quad D_{32}(x_i) < 0. \tag{6.20}$$

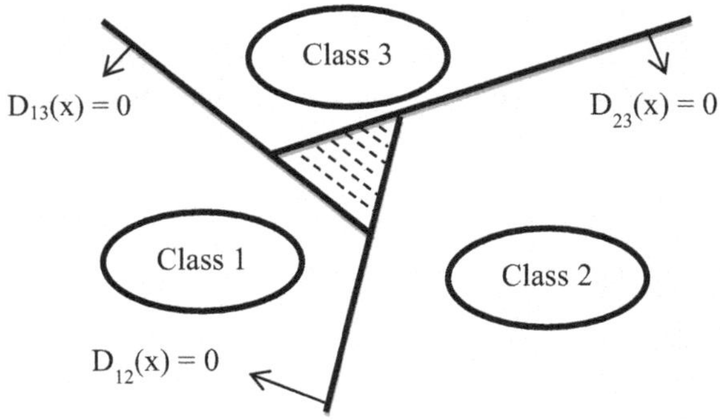

Fig. 6.4 Representation of one-against-one decision function [10]

On implementing Eq. (6.17), following equations can be achieved:

$$\begin{aligned}D_1 &= sgn(D_{12}(x_i)) + sgn(D_{13}(x_i)) = 1 + 0 = 1,\\ D_2 &= sgn(D_{21}(x_i)) + sgn(D_{23}(x_i)) = 0 + 1 = 1, and\\ D_3 &= sgn(D_{31}(x_i)) + sgn(D_{32}(x_i)) = 1 + 0 = 1.\end{aligned} \tag{6.21}$$

Hence, based on these equations,

$$D_1(x_i) = D_2(x_i) = D_3(x_i). \tag{6.22}$$

It can be observed that one cannot utilize Eq. (6.18) to classify the input sample x_i. Therefore, there exists an unclassifiable region in the OAO method. In the numerical experiments, for the samples in the unclassifiable region, they are usually randomly assigned to one class.

6.6.2 Decision-Directed Acyclic Graph Method

To resolve the unclassifiable region, the DDAG method was developed based on the OAO scheme.

Figure 6.5 shows one case of the decision trees for the three classes shown in Fig. 6.4. In Fig. 6.5, $\bar{i}$ ($i = 1, 2,$ and 3) means that x does not belong to class i. At the top-level classification, one can choose any pair of class, except for the leaf node if $D_{ij} > 0$, then one can regard x not to belonging to class j. If $D_{12}(x) > 0$, it means x does not belong to class 2. Thus it belongs to either class 1 or class 3 and therefore the next classification pair is class 1 and class 3. Thus, the unclassifiable region is resolved but it depends on the tree information. By this formulation, the decision boundaries of DDAG are the same as those in Eq. (6.18) for classifiable regions. In addition, the unclassifiable region is assigned to the classes associated with the leaf nodes of the decision tree.

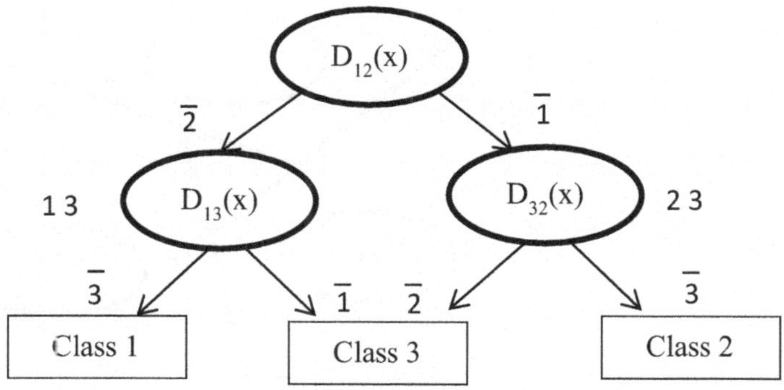

Fig. 6.5 DDAG decision function method based on OAO scheme [10]

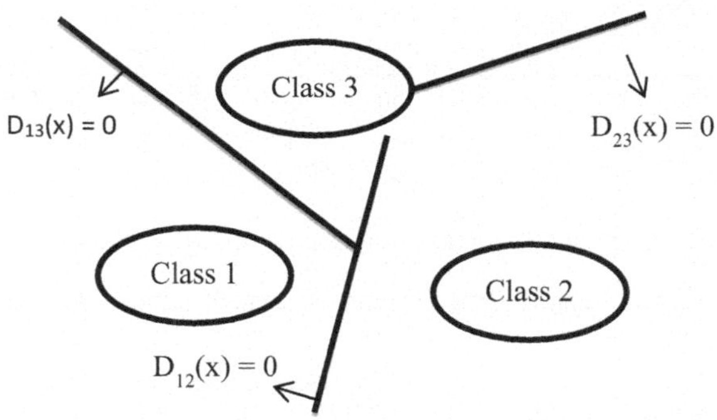

Fig. 6.6 Resolved unclassifiable regions based on DDAG decision function method [10]

As shown in Fig. 6.6, the unclassifiable region is assigned to class 3 if one utilizes a decision function method.

6.6.3 *Fuzzy Decision Function Method*

To overcome the unclassifiable region, another method known as FDF method based on OAO scheme is introduced. For the input vector x, the 1-D membership function $m_{ij}(x)(i, j = 1, 2, \ldots, C)$ in the directions orthogonal to the optimal separating hyperplanes $D_{ij}(x) = 0$ is defined as follows:

$$m_{ij}(x) = \begin{cases} 1, & 1 \leq D_{ij}(x) \\ D_{ij}, & \text{otherwise} \end{cases} \tag{6.23}$$

The membership functions $m_i(x)$ are given by

$$m_i(x) = \min\bigl(m_{ij}(x)\bigr) \quad j = 1, 2, \ldots C. \tag{6.24}$$

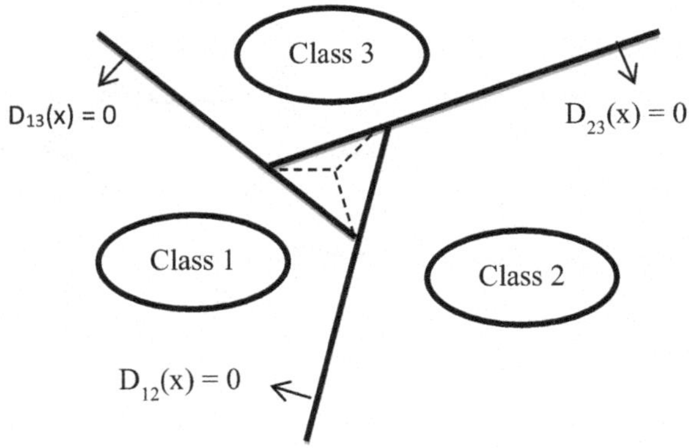

Fig. 6.7 FDF method based on OAO [10]

Table 6.1 Number of samples present in the dataset

	Class 1	Class 2	Class 3	Class 4	Class 5	Class 6	Class 7	Total
Initial-data	225	350	350	350	350	350	350	1975
Training	113	175	175	175	175	175	175	1163
Testing	112	175	175	175	175	175	175	1162

Using Eq. (6.24), sample x is classified into the class

$$C^* = \arg\max(m_i(x)) \quad i = 1, 2, \ldots C. \tag{6.25}$$

Using Eq. (6.25), the unclassifiable region is resolved as shown in Fig. 6.7.

Here, rows describe the instances at which data is acquired and the columns describe the features. The number of samples in the dataset is reported in Table 6.1.

6.6.4 One-Against-All Method

For a C class problem, the OAA method constructs C SVM models. The ith SVM is trained with all the training examples in ith class with positive labels and all other examples with negative labels. The final output of the OAA method is the class which corresponds to the SVM with the highest output value.

6.6.4.1 Using Radial Basis Function (RBF) Kernel Function

This method uses the basic concept of OAO algorithm. On using the RBF kernel function during the implementation of the SVM algorithm, the choice of sigma plays a very vital role while obtaining convergence and obtaining high accuracy value. Instead of fixing a sigma value for the calculation of decision function in all possible cases, the sigma value is optimized, and the optimal value is chosen for every case in the calculation of decision function. During the implementation of SVM for the optimization of the case, the sigma value is assigned, which best suits that case on an individual basis. Thus, with each classification, the net convergence becomes more effective and better in terms of accuracy rate percentage. This can be easily inferred from the tables shown in the result section.

6.7 Datasets and Methods

The CBM dataset taken here is collected at the workshop, Department of Electrical Engineering, IIT Kanpur under "Health Monitoring for Rotating Machine" project. The experiments were conducted on a personal computer with a 3.0 GHz CPU and 3 GB of RAM. All the SVM-based methods used were trained by half of the dataset chosen fairly from the main dataset, ensuring representation of all classes present in the required percentage. The remaining half of the main dataset was used for testing and analysis purpose. The SVM-based methods were applied using the Gaussian kernel. The kernel parameter sigma (σ) and the regularization parameter lambda (λ) were empirically optimized by minimizing the error rate on the validation dataset [12].

After applying pre-processing techniques [13] and implementing feature extraction [9–15] and selection algorithms [7] on the raw dataset, the final CBM dataset contains 1975 rows and 93 columns.

The sample dataset taken is the pressure reading of compressor (in lb/in^2). Once the compressor starts, its pressure is set to increase incrementally. The extent of pressure value up to which it is expected to rise from zero value is set as 150 lb/in^2. In range of $0 - 150$ lb/in^2 pressure, a total of 75 readings are expected to be taken. These readings are taken via data acquisition (DAQ) hardware and sensors. Here, four accelerometers/transducers/mics are installed at four different positions of compressor, which helps in obtaining data of compressor in digital form. Now, this ($0 - 150$ lb/in^2) range of pressure is divided into 15 intervals of 10l b/in^2 each and in each interval five readings are taken for running state of the compressor. The features considered in different domains for classification are illustrated in Table 6.2.

In the CBM dataset, one more column is added, which represents the class of the dataset. "Class" here basically means the state of compressor system during which the readings are taken. As the collected data comprises readings of four different conditions, the data is categorized into four different classes which are illustrated in the Table 6.3.

In the application of SVM-based methods, RBF kernels have been used throughout for calculating the accuracy rate. For each problem, the accuracy rate is estimated using different kernel parameters, sigma (σ) and lambda (λ) where

$$\sigma = \left[2^{-4}, 2^{-3}, 2^{-2}, \ldots, 2^{7}, 2^{8}\right] \text{ and}$$
$$\lambda = \left[2^{-4}, 2^{-3}, 2^{-2}, \ldots, 2^{7}, 2^{8}\right].$$

Table 6.2 Feature description

Features in different domains		
Time domain (eight features)	F_1	Absolute mean
	F_2	Maximum peak
	F_3	Root mean square
	F_4	Variance
	F_5	Kurtosis
	F_6	Crest factor
	F_7	Shape factor
	F_8	Skewness
Frequency domain (eight features)	F_9	Bin 1/total spectral energy
	F_{10}	Bin 2/total spectral energy
	F_{11}	Bin 3/total spectral energy
	F_{12}	Bin 4/total spectral energy
	F_{13}	Bin 5/total spectral energy
	F_{14}	Bin 6/total spectral energy
	F_{15}	Bin 7/total spectral energy
	F_{16}	Bin 8/total spectral energy
Continuous wavelet transform, (seven features)	F_{17}	Standard deviation
	F_{18}	Entropy
	F_{19}	Kurtosis
	F_{20}	Skewness
	F_{21}	Standard variance
	F_{22}	Sum of peak
	F_{23}	Zero-crossing rate
Discrete wavelet transform (nine features)	$F_{24}-F_{26}$	Variance of coefficients at level-1,2 and 3
	$F_{27}-F_{29}$	Mean of coefficients at level-1,2 and 3
	$F_{30}-F_{32}$	Variance of autocorrelated coefficients at level-4,5 and 6
Wavelet packet transform, (254 features)	$F_{33}-F_{286}$	Wavelet packet node energy

So, each problem is run with 13 different values of sigma. The problem is rerun for 13 different values of lambda for the best value of sigma.

Table 6.3 Class description of data

Class	State
Class 1	Healthy condition
Class 2	Leakage Inlet valve (LIV)
Class 3	Leakage Outlet valve (LOV)
Class 4	Non-Returning Valve fault (NRV)
Class 5	Rider belt fault
Class 6	Bearing fault on opposite flywheel side
Class 7	Bearing fault on flywheel side

6.8 Results and Discussion

In this section, along with experimental results, an implementation of SVM-based methods on CBM dataset along with the performance of proposed method with OAO, OAA, and FDF methods have been presented in Tables 6.4, 6.5, 6.6 and 6.7.

i. For all the features present in CBM dataset, RBF kernel function-based algorithm gives better classification result (classification accuracy of 98.3939%)
ii. Using the feature selection algorithm, Mutual Information Feature Selection Under Uniform Information Distribution (MIFS-U) (beta parameter, $B = 0.5$ and features, $N = 10$), the RBF kernel function-based algorithm again proved to be superior to others with classification accuracy of 99.0148%). Figure 6.8 shows a bar chart for comparison of classification accuracies obtained using different algorithms.

Table 6.4 Accuracy of OAO based method

One-Against-One method										
No. of features	PCA	MIFS				mRMR	MIFS-U			NMIFS
		B = 0	B = 0.3	B = 0.5	B = 1		B = 0.3	B = 0.5	B = 1	
5	43.5961	52.9557	73.1527	75.6158	59.2365	73.8916	70.4433	51.6010	75.6158	55.2956
10	43.1034	63.9163	74.3842	65.1478	40.6404	74.8768	82.6355	**63.5468**	65.1478	72.0443
26	44.4581	**89.1626**	72.9064	66.1330	60.7143	93.4729	82.6355	60.9606	66.1330	91.9951
35	64.6552	93.9655	74.1379	68.7192	49.2611	93.3498	76.4778	64.6552	**67.7340**	81.0345
50	57.5123	89.6552	73.1527	71.5517	73.8916	86.6995	79.8030	**77.4631**	66.3793	97.2906
Without Feature Selection			**97.783251231527090**							

Table 6.5 Accuracy of OAA based method

One-Against-All method										
No. of features	PCA	MIFS				mRMR	MIFS-U			NMIFS
		B = 0	B = 0.3	B = 0.5	B = 1		B = 0.3	B = 0.5	B = 1	
5	13.7931	13.7931	13.7931	13.7931	13.7931	13.7931	13.7931	13.7931	13.7931	13.7931
10	13.7931	13.7931	13.7931	13.7931	13.7931	13.7931	13.7931	**13.7931**	13.7931	13.7931
26	13.7931	**13.7931**	13.7931	13.7931	13.7931	13.7931	13.7931	13.7931	13.7931	13.7931
35	13.7931	13.7931	13.7931	13.7931	13.7931	13.7931	13.7931	137931	**13.7931**	13.7931
50	20.4433	13.7931	21.6749	20.9360	20.5665	13.7931	19.3350	**18.5961**	27.7094	13.7931
Without Feature selection			**95.320197044334980**							

Table 6.6 Accuracy of FDF based method

Fuzzy Decision Function method										
No. of features	PCA	MIFS				mRMR	MIFS-U			NMIFS
		B = 0	B = 0.3	B = 0.5	B = 1		B = 0.3	B = 0.5	B = 1	
5	75.4926	78.2020	91.3793	89.0394	87.5616	92.3645	88.5468	92.8571	89.0394	81.7734
10	83.3744	65.8867	81.0345	75.4926	84.8522	75.7389	77.2167	**71.5517**	75.4926	69.8276
26	13.7931	**28.4483**	73.6453	69.9507	75.9852	86.2069	77.2167	71.5517	69.9507	44.4581
35	13.7931	65.2709	97.6601	92.8571	48.8916	74.3842	31.6502	35.2217	**94.4581**	68.3498
50	20.4433	59.3596	97.0443	93.5961	93.5961	49.5074	96.9212	**96.9212**	93.5961	98.0296
Without Feature selection		**98.152709359605910**								

Table 6.7 Accuracy of RBF kernel function-based method

Using RBF kernel function										
No. of features	PCA	MIFS				mRMR	MIFS-U			NMIFS
		B = 0	B = 0.3	B = 0.5	B = 1		B = 0.3	B = 0.5	B = 1	
5	73.7685	34.4828	56.6502	35.2217	77.4631	35.8374	57.0197	73.1527	35.2217	50
10	56.4039	50.4926	56.5271	56.1576	35.5911	35.3448	78.2020	**99.0148**	56.1576	56.8966
26	54.4335	**99.8768**	56.6502	34.8522	35.2217	82.6355	78.2020	77.0936	56.1576	77.9557
35	74.7537	86.8227	35.3448	35.3448	34.3596	90.3941	77.7094	76.8473	**94.9507**	77.3399
50	74.1379	78.0788	76.7241	73.2759	94.5813	78.0788	77.2167	**97.4138**	95.0739	76.8473
Without Feature Selection		**98.399014778325120**								

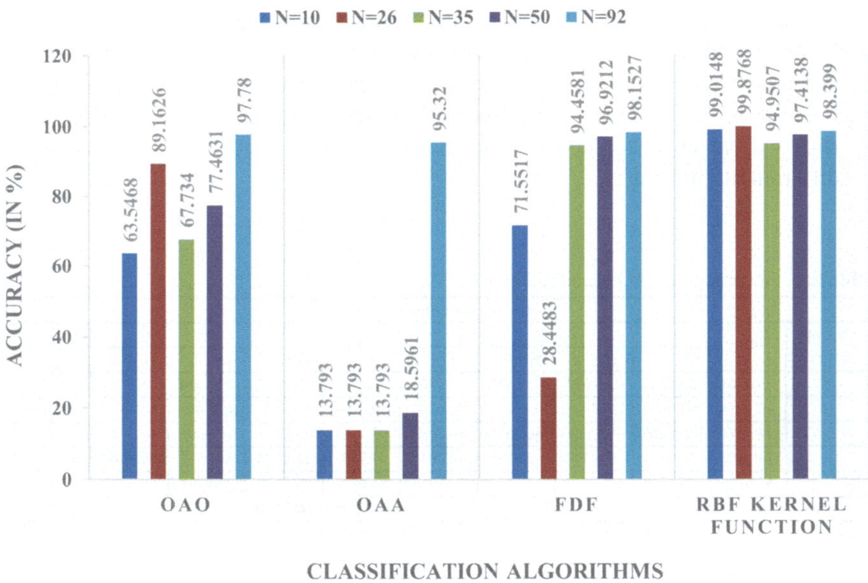

Fig. 6.8 Bar chart for comparison of classification accuracies of different algorithms [1]

6.9 Conclusions

Here, the classification of CBM dataset is done through various well-known SVM-based methods, and the results obtained through the RBF kernel function-based method is found to be dominant over the results of all the other methods compared here. Though DAG method also gives good results, it has a disadvantage of very high time complexity. Thus, with the better selection of sigma parameter while calculating kernels in the RBF kernel function-based SVM method, it could be preferred over all the other methods for classification of CBM dataset.

6.10 Classification Tool

6.10.1 Classification Tool for Android Platform

Step 1 As soon as the user clicks on "ClassificationApp.apk", welcome page of classification app appears as shown in Fig. 6.9. The user is asked to tap on screen to move on to the next activity page.

Step 2 Once the user taps, next activity page appears as shown in Fig. 6.10. This app page is mainly meant for taking input from user.

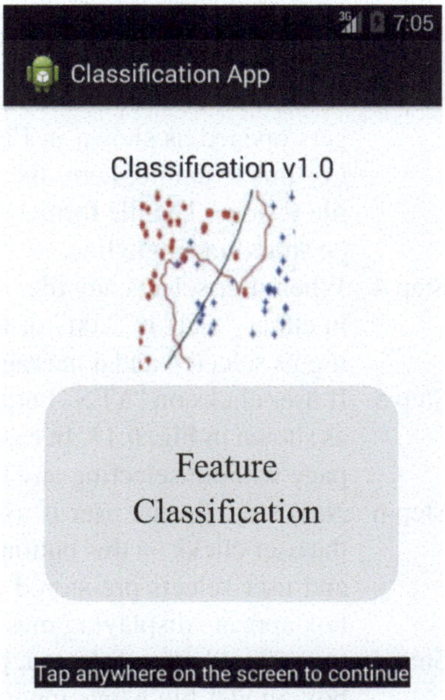

Fig. 6.9 Classification app welcome page [16]

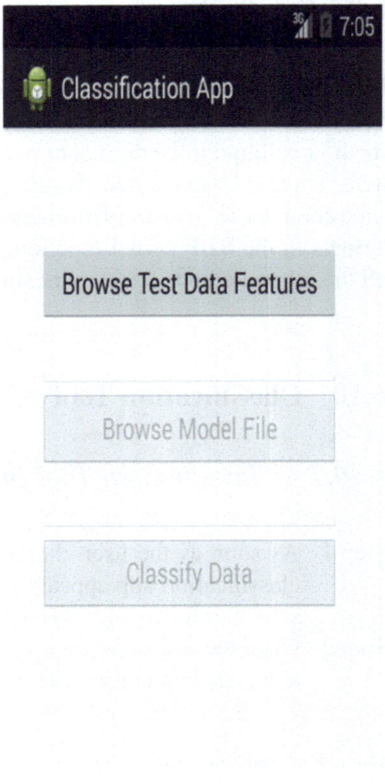

Fig. 6.10 Classification activity input page [16]

Step 3 As the user clicks on "Browse Data File", Secure Digital (SD) card directory gets opened as shown in Fig. 6.11 where user can select the test data file. For this example, here, the databank folder is chosen where test1_feature file resides. The file format should be either .txt or .dat and all the data must be stored in single line.

Step 4 When user selects any file, it first checks whether the selected file format is in either ".dat" or ".txt" or not. If the file is in one of the required formats, it gets selected and a message appears on screen as shown in Fig. 6.12.

Step 5 If user clicks on "YES" button, the file gets selected and returns to first page as shown in Fig. 6.13. In case the user selects "NO" button, it returns to first page without selecting any file.

Step 6 Next step, for the user is to click on "Browse Model File" tab. As soon as the user clicks on this button, SD card directory opens as shown in Fig. 6.14 and user selects pre-stored data file. As mentioned earlier, again dialogue box appears displaying message file has been selected.

Step 7 Once the file gets selected, page returns to first activity page wherein a text box, model file name appears as shown Fig. 6.15. Now, the user clicks on "Classify Data" button to call next activity page.

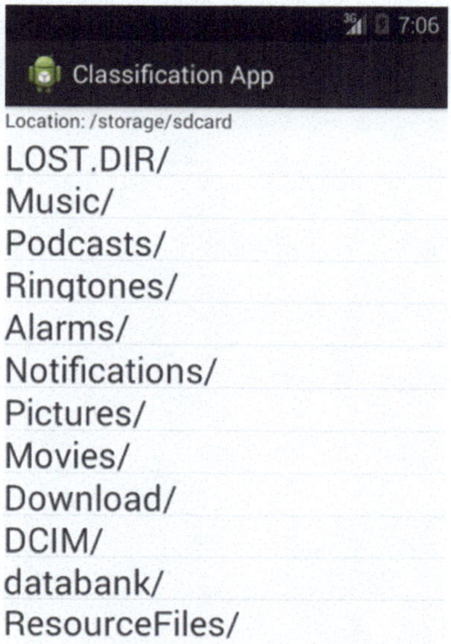

Fig. 6.11 Browse activity page [16]

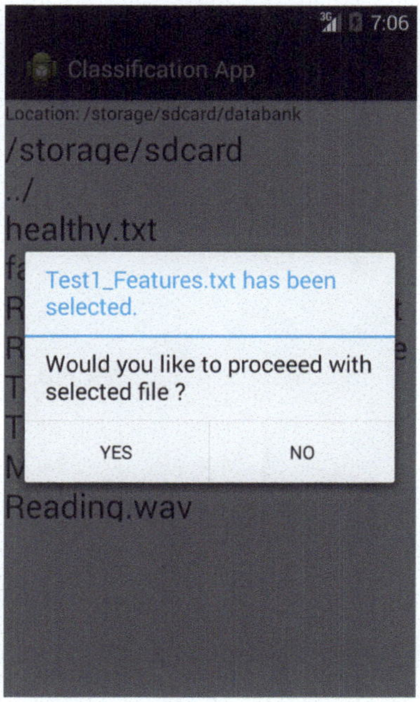

Fig. 6.12 Confirmation activity page [16]

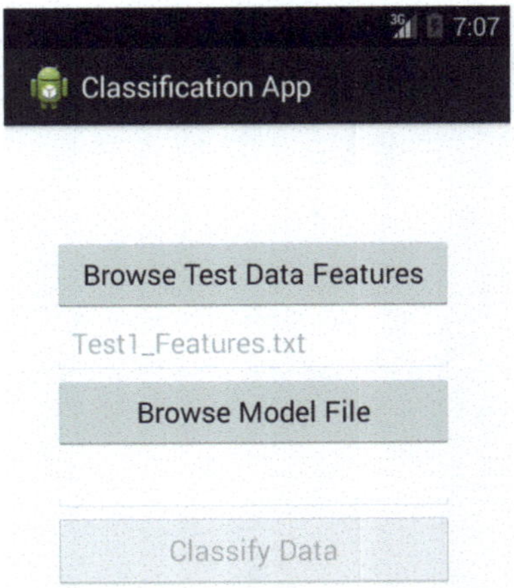

Fig. 6.13 Test file name shown in textbox [16]

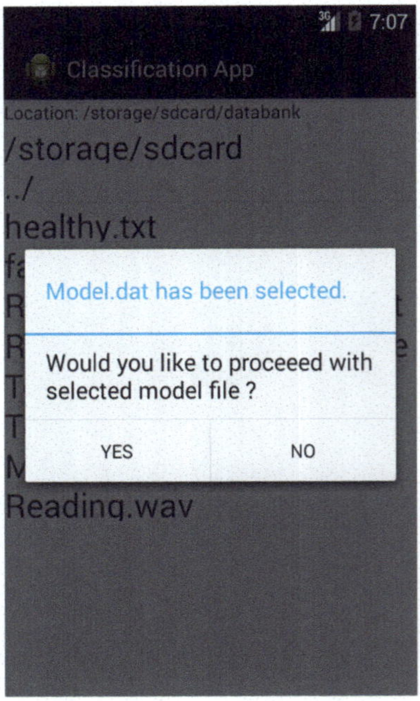

Fig. 6.14 Message displayed on selection of file [16]

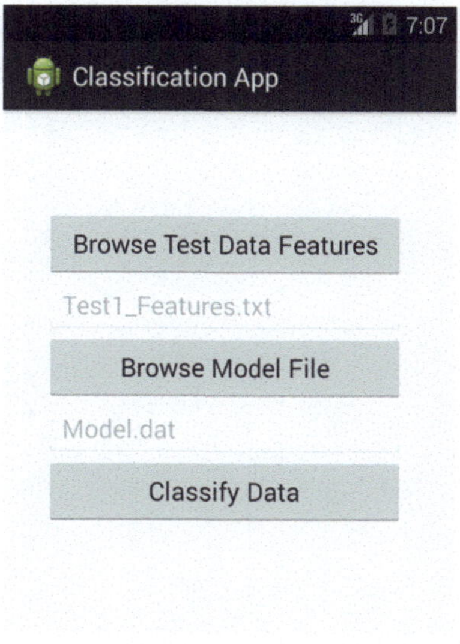

Fig. 6.15 Activity page with selected file name [16]

Step 8 This activity page does not provide any element in the user interface for user to interact with application. Only a progress spinner is displayed as shown in Fig. 6.16 which shows that the application is busy in processing the data.

Step 9 Once the data classification process completes, result is saved, and next activity page is called which displays the result of fault diagnosis test. In the

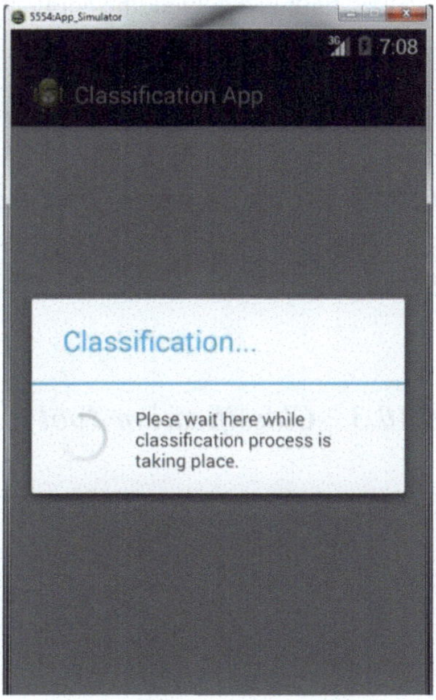

Fig. 6.16 Processing activity page [16]

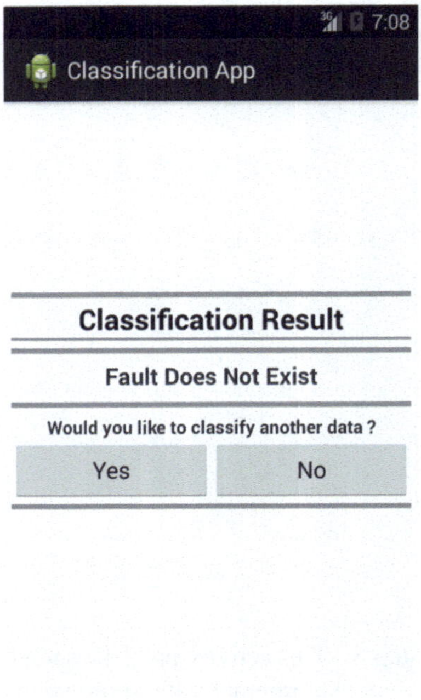

Fig. 6.17 Classification result activity page [16]

backend, activity page starts; the first thing done is collecting information of results from previous activity page. In case a fault is detected, the notification area will display a red coloured text "Fault Detected in Machine" and if no fault is detected, the application will display "Fault does not exist" as shown in Fig. 6.17.

6.10.2 Classification Tool for Windows Mobile

See Fig. 6.18.

6.10.3 Classification Tool for Windows Tablet Platform

See Fig. 6.19.

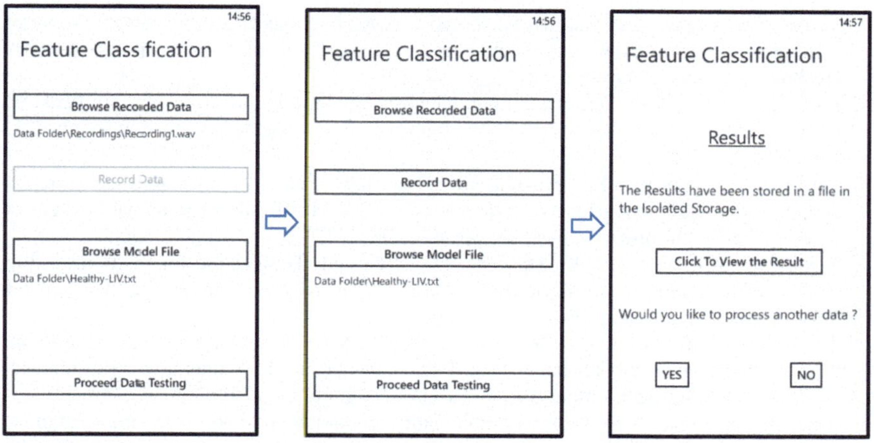

Fig. 6.18 Snapshots of feature selection tool on windows mobile phone [17]

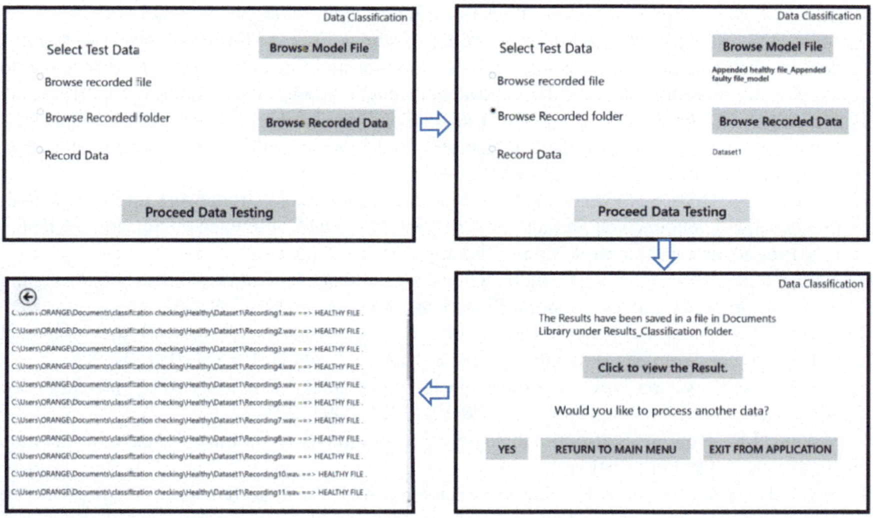

Fig. 6.19 Snapshots of feature selection tool on windows tablet [17]

References

1. Verma, N.K., Roy, A., Salour, A.: An optimized fault diagnosis method for reciprocating air compressors based on SVM. In: Proceedings of the IEEE Control and System Graduate Research Colloquium Incorporating 2011 IEEE International Conference on System Engineering and Technology, Selangor, Malaysia, pp. 65–69 (2011)
2. Sevakula, R.K., Verma, N.K.: Wavelet transforms for fault detection using SVM in power systems. In: IEEE International Conference on Power Electronics, Drives and Energy Systems, Bengaluru, India, pp. 1–6 (2012)

3. Verma, N.K., Gupta, V.K., Sharma, M., Sevakula, R.K.: Intelligent condition based monitoring of rotating machines using sparse auto-encoders. In: IEEE Conference on Prognostics and Health Management, Maryland, USA, pp. 1–7 (2013)
4. Jamal, A., Verma, N.K.: Automatic fault diagnosis system using acoustic data. In: IEEE 8th International Conference on Industrial and Information Systems, Kandy, Sri Lanka, pp. 421–426 (2013)
5. Verma, N.K., Agrawal, A.K., Sevakula, R.K., Prakash, D., Salour, A.: Improved signal preprocessing techniques for machine fault diagnosis. In: 2013 IEEE 8th International Conference on Industrial and Information Systems, pp. 403–408 (2013)
6. Ramkumar, A.J., Verma, N.K., Dixit, S.: Detection and classification for faults in drilling process using vibration analysis. In: IEEE Conference on Prognostics and Health Management, Washington, USA, pp. 1–6 (2014)
7. Thirukovalluru, R., Sevakula, R.K., Dixit, S., Verma, N.K.: Generating optimum feature sets for fault diagnosis using denoising stacked auto-encoder. In: IEEE International Conference on Prognostics and Health Management, Canada, USA, pp. 1–7 (2016)
8. Agarwal, A., Verma, N.K.: Generalization ability of majority vote point classifiers for motor fault diagnosis. In: IEEE International Conference on Industrial and Information Systems (ICIIS), IIT Roorkee, India, pp. 844–849 (2016)
9. Saraswat, G., Maurya, S., Verma, N.K.: Health monitoring of main battle tank engine using Mamdani type fuzzy model. In 2017 International Conference on Computational Intelligence: Theories, Applications and Future Directions, IIT Kanpur, India, pp. 403–414 (2017)
10. Maurya, S., Singh, V., Dixit, S., Verma, N.K., Salour, A., Liu, J.: Fusion of low-level features with stacked autoencoder for condition based monitoring of machines. In: IEEE International Conference on Prognostics and Health Management, Washington, USA, pp. 1–8 (2018)
11. Sharma, A.K., Singh, V., Verma, NK., Liu, J.: Condition based monitoring of machine using Mamdani fuzzy network. In: IEEE Prognostics and System Health Management Conference, Chongqing, China, pp. 1159–1163 (2018)
12. Verma, N.K., Dixit, S., Sevakula, R.K., Salour, A.: Computational framework for machine fault diagnosis with autoencoder variants. In: 2018 International Conference on Sensing, Diagnostics, Prognostics, and Control, Xi'an, China, pp. 353–358 (2018)
13. Verma, N.K., Sevakula, R.K., Dixit, S., Salour, A.: Intelligent condition based monitoring using acoustic signals for air compressors. IEEE Trans. Rel. **65**(1), 291–309 (2016)
14. Verma, N.K., Sevakula, R.K., Thirukovalluru, R.: Pattern analysis framework with graphical indices for condition based monitoring. IEEE Trans. Rel. **66**(4), 1085–1100 (2017)
15. Sevakula, R.K., Verma, N.K.: Assessing generalization ability of majority vote point classifiers. IEEE Trans. Neural Netw. Learn. Syst. **28**(12), 2985–2997 (2017)
16. Verma, N.K., Roy, A.: Self-optimal clustering technique using optimized threshold function. IEEE Syst. J. **99**, 1–14 (2013)
17. Sevakula, R.K., Verma, N.K.: Support vector machine for large databases as classifier. In: International Conference on Swarm, Evolutionary, and Memetic Computing, Bhubaneswar, India, pp. 303–313 (2012)
18. Verma, N.K., Singh, S., Gupta, J.K., Sevakula, R.K., Dixit, S., Salour, A.: Smartphone application for fault recognition. In: 6th International Conference on Sensing Technology, Kolkata, India, pp. 1–6 (2012)
19. Verma, N.K., Singh, J.V., Gupta, M., Sevakula, R.K., Dixit, S.: Windows mobile and tablet app for acoustic signature machine health monitoring. In: International Conference on Industrial and Information Systems, Gwalior, India, pp. 1–6 (2014)

Chapter 7
Fault Diagnosis System for Air Compressor Using Palmtop

Abstract This chapter provides details of health monitoring of the air compressor using Palmtop. Until now, it has been observed that the data acquisition and fault recognition systems both are separately handled. To reduce the complexity and combine both processes into one system, Palmtop is used. The aim of this chapter is to develop a quick fault diagnosis system for identifying and detecting the position and type of faults in the air compressor using Palmtop. The quick fault diagnosis system developed using Palmtop can be termed as an extension of the work done in Chap. 6.

7.1 Introduction

The main objective of a fault diagnosis system is to identify faults in the air compressor. The process of a fault diagnosis framework is mentioned in Chap. 1. The inbuilt microphone on Palmtop serves as the Analog Input (AI) sensor for acquiring the data from the air compressor. The data was collected from single-stage reciprocating air compressor. Seven faults are considered for the analysis as mentioned in Chap. 2. Features are extracted from the data acquired through Palmtop for characterization of faults. From these extracted features, relevant set of features are selected using existing feature selection methods. Feature selection outputs are arranged in training and testing sets for different conditions of the air compressor, i.e., healthy faults, Leakage Inlet Valve (LIV) faulty condition, Leakage Outlet Valve (LOV) faulty condition, Non-Returning Valve (NRV) faulty condition, bearing faulty condition on the opposite flywheel side, and bearing faulty condition on the flywheel side. Finally, Support Vector Machine (SVM)-based classification technique is used for the classification of faults.

7.2 Experimental Setup: Hardware and Seeded Faults

The data acquisition of a single-stage reciprocating air compressor for sound is done using the inbuilt microphone in Palmtop. Figure 7.1 shows an image of Sony Vaio

N. K. Verma and A. Salour, *Intelligent Condition Based Monitoring*, Studies in Systems, Decision and Control 256, https://doi.org/10.1007/978-981-15-0512-6_7

Fig. 7.1 Sony Vaio Palmtop. Image Courtesy: IDEA LAB, IIT Kanpur

Palmtop. The monaural microphone is a single channel sound recording source and the monaural sound is also known as monophonic sound or high-fidelity sound. The monophonic sound systems encode one single stream of sound which usually uses only one speaker. Monophonic sound is the most basic format of sound output. The microphone is used to record a sound at a distance while minimizing the surrounding noise.

7.2.1 Palmtop Specifications

- Operating System: Genuine Windows Vista® Business
- Processor Technology: Intel® Centrino® Processor Technology
- Processor Name: Intel® Core™ Solo Processor U1500 1.33 GHz
- Hard Disk: 40 GB
- Main Memory: 1 GB DDR2 SDRAM
- Built-in Monaural Microphone

7.2.2 Faults Seeded in Air Compressor

For the purpose of analysis, faults were deliberately introduced in the air compressor expect to give us an insight to the factors contributing strongly towards fault diagnosis. The faults introduced in the air compressor are the following:

- LOV fault
- LIV fault
- NIV fault
- Rider belt fault
- Bearing fault of opposite flywheel side

7.2.3 Bearing Fault of Flywheel Side Experimental Setup

The complete steps involved in the fault diagnosis system using Palmtop are given below and discussed in further subsections.

Step 1. Data Acquisition (DAQ) using inbuilt microphone of Palmtop
Step 2. Data pre-processing
Step 3. Extracting features from the pre-processed data
Step 4. Selection of salient features
Step 5. Classification using SVM
Step 6. Fault prediction using SVM

7.3 Data Acquisition Using Palmtop

As illustrated in Fig. 7.2, the data is collected using inbuilt microphone on Palmtop, which serves as an analog sensor deployed near the air compressor in various healthy and faulty conditions. Analog input subsystems convert real-world analog signals from a sensor into bits that can be read by our computer. These subsystems are typically multichannel devices offering 12 or 16 bits of resolution. The data collection is done with the help of DAQ toolbox in MATLAB that provides access to analog input devices through an analog input object. Various steps involved in DAQ using Palmtop through MATLAB are explained below.

1. **Create a device object**—create the analog input object AI for a sound card.
   MATLAB Command: AI = analoginput('winsound');

Fig. 7.2 Palmtop data acquisition system. Image Courtesy: IDEA Lab, IIT Kanpur

2. **Add channels—**add one channel to AI.
   MATLAB Command: chan = addchannel(AI,1);
3. **Configure property values**—assign values to the basic setup properties and create the blocksize and actual rate (Fs) for variables to be used for subsequent analysis. The actual sampling rate is retrieved because it might be set by the engine to a value that differs from the specified value.
   MATLAB Command: duration = 1; % acquisition for 1 s
   set(AI,'SampleRate',50000);
   ActualRate = get(AI,'SampleRate');
   set(AI,'SamplesPerTrigger',duration*ActualRate)
   set(AI,'TriggerType','Manual')
   blocksize = get(AI, 'SamplesPerTrigger');
   Fs = ActualRate;
4. **Acquire data**—start AI then issue a manual trigger and extract data from the engine. Before the trigger is issued, you should begin inputting data from the tuning fork into the sound card.
   MATLAB Command: start(AI)
   trigger(AI)
   data = getdata(AI)
5. **Clean up—**when AI is no longer needed, remove it from memory and from the MATLAB workspace as well.
   MATLAB Command: delete(AI)
   clear AI

As mentioned earlier, one must initially find the sensitive position of the machine to place the sensor. From sensitive position analysis discussed in Chap. 2, it is found that positions on top of the piston head side of air compressor are having high peak, RMS, standard deviation, and variance values as compared to the other three sides of the air compressor i.e., NRV side, opposite NRV side, and opposite flywheel side. The external unidirectional microphone in each faulty condition and single sensitive position on top of the piston head has been shown in Figs. 7.3 and 7.4, respectively. The total data recorded for sensitive position analysis is equal to: 50 k samples/sec × 5 sec (time duration for one reading) × 6 (total number of pressure slots for each reading) × 3 (number of readings for each sensitive position) ×

Fig. 7.3 An external microphone. Image Courtesy: IDEA Lab, IIT Kanpur

Fig. 7.4 Sensitive positions on top of the piston head [1]

6 (total number of sensitive positions on top of the piston head). Sensitive positions determined are tabulated in Table 7.1.

The details of the raw data are given below:

- Sampling frequency = 50 kHz
- Total number of samples = 250 K samples
- Total time of recording = 5 s
- Range of frequencies = Up to 25 kHz

For the air compressor pressure range of 0–150 lb/in, total six readings from the pressure slots 10, 20, 40, 75, 110, and 150 psi as shown in Fig. 7.5 were taken

Table 7.1 Final sensitive positions on top of the piston head

S. No.	Air compressor condition	Sensitive position
1	Healthy condition	Position 2
2	LIV fault	Position 1
3	LOV fault	Position 4
4	NRV fault	Position 1
5	Rider belt fault	Position 1
6	Bearing fault on opposite flywheel side	Position 1
7	Bearing fault on flywheel side	Position 1

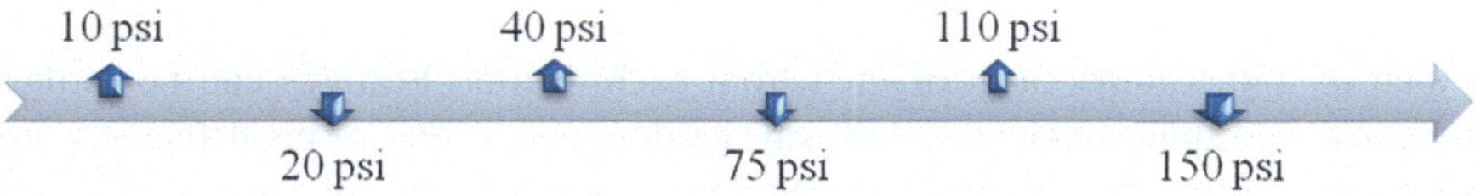

Fig. 7.5 DAQ at different pressure slots. Image Courtesy: IDEA Lab, IIT Kanpur

Table 7.2 Datasets in each condition of the air compressor

Air compressor condition	Palmtop position on top of the piston head	Data collected	Readings over 0–150 psi
Healthy condition	Position 2	30 samples	180
LIV fault	Position 1	30 samples	180
LOV fault	Position 4	30 samples	180
NRV fault	Position 1	30 samples	180
Rider belt fault	Position 1	30 samples	180
Bearing fault on opposite flywheel side	Position 1	30 samples	180
Bearing fault on flywheel side	Position 1	30 samples	180
Total		210 samples	1260

using Palmtop microphone placed at sensitive positions. Thirty datasets are taken at each pressure slot for all seven conditions (one healthy and six faulty conditions) of the air compressor. The number of times data acquired through Palmtop from air compressor at healthy and various faulty conditions are given in Table 7.2.

The next subsection presents details about the remaining modules of fault diagnosis framework.

7.4 Data Pre-processing

Once the raw data is acquired, it is required to pre-process the data before analysis. The pre-processing techniques followed are the same as described in Chap. 3. In summary, the pre-processing includes clipping, normalization, smoothing, and filtering. Initially, the recorded data is filtered using a Butterworth low pass filter which reduces high-frequency content in the sample. Next clipping is performed where out of 5 s best 1 s of the signal is selected based on standard deviation which could be used for further processing. The selected signal of 1 s is then smoothened using a moving average filter and then normalized using 0–1 normalization technique. Thus, the pre-processed data is ready for further processing.

7.5 Feature Extraction

A total of 286 features are extracted from each dataset. Features can be further bifurcated into time, frequency, and wavelet domain. A total of eight features are extracted via time domain analysis. Similarly, eight features are extracted via frequency domain and 270 features were extracted in wavelet domain. The explanation of feature extraction has already been discussed in Chap. 4 and results are given in Sect. 7.7. The entire features extracted from time, frequency, and wavelet domain are listed in Table 7.3.

Table 7.3 Description of features in various domains

Dataset features description		
Time domain (8 features)	F_1	Absolute mean
	F_2	Maximum peak
	F_3	Root mean square
	F_4	Variance
	F_5	Kurtosis
	F_6	Crest factor
	F_7	Shape factor
	F_8	Skewness
Frequency domain (8 features)	F_9	Bin 1/Total spectral energy
	F_{10}	Bin 2/Total spectral energy
	F_{11}	Bin 3/Total spectral energy
	F_{12}	Bin 4/Total spectral energy
	F_{13}	Bin 5/Total spectral energy
	F_{14}	Bin 6/Total spectral energy
	F_{15}	Bin 7/Total spectral energy
	F_{16}	Bin 8/Total spectral energy
Continuous Wavelet Transform (CWT) (7 features)	F_{17}	Standard deviation
	F_{18}	Entropy
	F_{19}	Kurtosis
	F_{20}	Skewness
	F_{21}	Standard variance
	F_{22}	Sum of peak
	F_{23}	Zero crossing rate
Discrete Wavelet Transform (DWT) (9 features)	F_{24}–F_{26}	Variance of coefficients at level-1, 2 and 3
	F_{27}–F_{29}	Mean of coefficients at level-1, 2 and 3
	F_{30}–F_{32}	Variance of auto-correlated coefficients at level-4, 5 and 6
Wavelet Packet Transform (WPT) (254 features)	F_{33}–F_{286}	Wavelet packet node energy

7.6 Feature Selection

The next step is the feature selection where 286 extracted features are ranked using feature selection methods mentioned below. Top-ranked features are selected for training and testing with multiclass SVM-based feature classification techniques. The above-mentioned feature selection techniques and their algorithms mentioned below are described in Chap. 5.

- Principle Component Analysis (PCA)
- Mutual Information Feature Selector (MIFS)
- Minimum Redundancy Maximum Relevance (mRMR)
- Normalized Mutual Information Feature Selection (NMIFS)
- Mutual Information Feature Selection (MIFS_U)
- Bhattacharyya Distance (BD)
- Independent Component Analysis (ICA)

7.7 Classification

The classification of data is done using SVM-based techniques. Following four multiclass SVM classification techniques are used for implementing training and testing of the model. The classification techniques mentioned below are described in Chap. 6.

- One-Against-One (OAO)
- One-Against-All (OAA)
- Fuzzy Decision Function (FDF)
- Decision Directed Acyclic Graph (DDAG)

7.8 Fault Recognition Model Development

The objective of fault recognition is to take stock of the problem for a fault, answer the questions regarding what we think the problem is, what information should be collected, and assessed to confirm the understanding of the problem. The main outcome of fault recognition is that it filters out the characteristics of a faulty condition.

7.8.1 DAQ to Feature Extraction

Table 7.4 shows the data acquired from the air compressor under healthy and various faulty conditions, along with the respective most sensitive positions. The total

Table 7.4 DAQ under various conditions of air compressor

Air compressor condition	Palmtop position on top of the piston head	Data collected	Readings over 0–150 psi
Healthy condition	Position 2	30 samples	180
LIV fault	Position 1	30 samples	180
LOV fault	Position 4	30 samples	180
NRV fault	Position 1	30 samples	180
Rider belt fault	Position 1	30 samples	180
Bearing fault on opposite flywheel side	Position 1	30 samples	180
Bearing fault on flywheel side	Position 1	30 samples	180
Total		210 samples	1260

samples collected from these positions are 210 (30 samples from each condition). The observations are made for a pressure range of 0–150 psi.

After implementing feature extraction on the complete pre-processed data, the final Condition-Based Monitoring (CBM) dataset contains 1260 data samples and 286 features. Among these 286 features;

- Eight are taken for Time domain
- Eight are taken for Frequency domain
- Seven are taken for CWT
- Nine are taken for DWT
- 254 are taken for WPT

In the final CBM dataset, one more column is added which represents its corresponding class. "Class" refers to that state of air compressor during which the data is recorded. This is illustrated in Table 7.5.

Table 7.5 Class and its description

Class	Description
1	Healthy condition
2	LIV
3	LOV
4	NRV
5	Rider belt fault
6	Bearing fault on opposite flywheel side
7	Bearing fault on flywheel side

7.8.2 Training and Testing

The CBM dataset is trained with seven feature selection methods. Features are selected using each feature selection method. All SVM methods were trained by half of the selected feature dataset, with all classes in enough percentage. The remaining half dataset is utilized for finding the classification accuracy of each SVM method. SVM methods based on Gaussian kernel functions have been applied. The kernel parameter "σ" and the regularization parameter "λ" are optimized by minimizing the error rate.

The procedure of training and testing involves feature selection and feature classification as shown in Fig. 7.6 that finally yields training results which help in drawing implications whether the tested data falls under the category of healthy or faulty. It also gives the extent to which the data is associated with a condition. Training and testing datasets are assigned to one of the multiclass SVM methods to calculate the accuracy for different combinations of kernel parameter sigma (σ) and the regularization parameter lambda (λ). The considered values of these parameters are as follows:

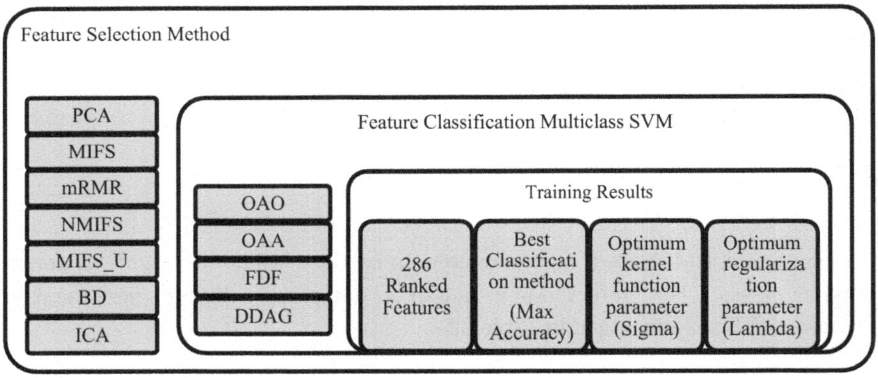

Fig. 7.6 Feature selection and feature classification [2]

We run 13 different samples of sigma (σ) as mentioned above, while setting lambda (λ) to infinity, thus calculating the classification accuracy for each sigma. The training set is again run for 13 different samples of lambda at that value of sigma for which the classification accuracy is highest, thus calculating the classification accuracy for each lambda (λ). Finally, one can find the optimum lambda (λ) and sigma (σ) values which can be denoted as

- σ_{optimum}: the optimum Gaussian kernel parameter sigma
- λ_{optimum}: the optimum regularization parameter lambda

7.9 Results and Discussions

7.9.1 Data Pre-processing Results

See Figs. 7.7, 7.8, 7.9, 7.10, 7.11, 7.12, and 7.13.

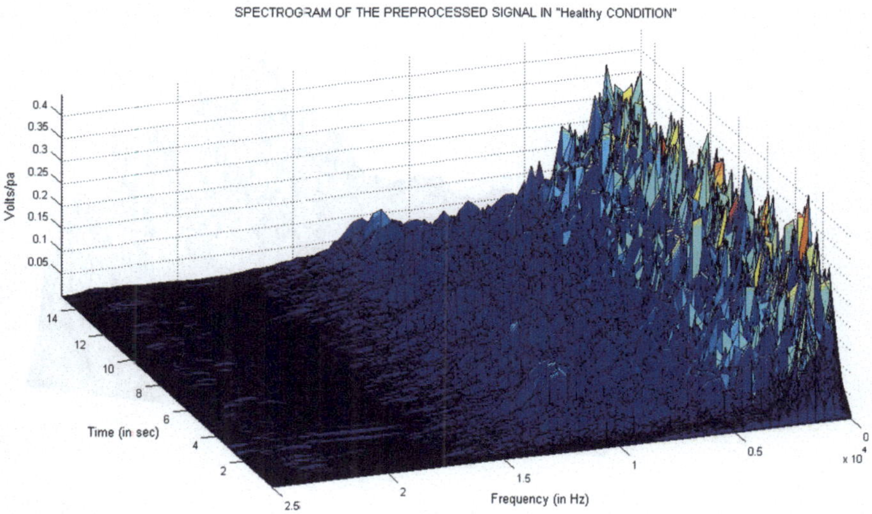

Fig. 7.7 Spectrogram of the pre-processed signal in "Healthy Condition" [2]

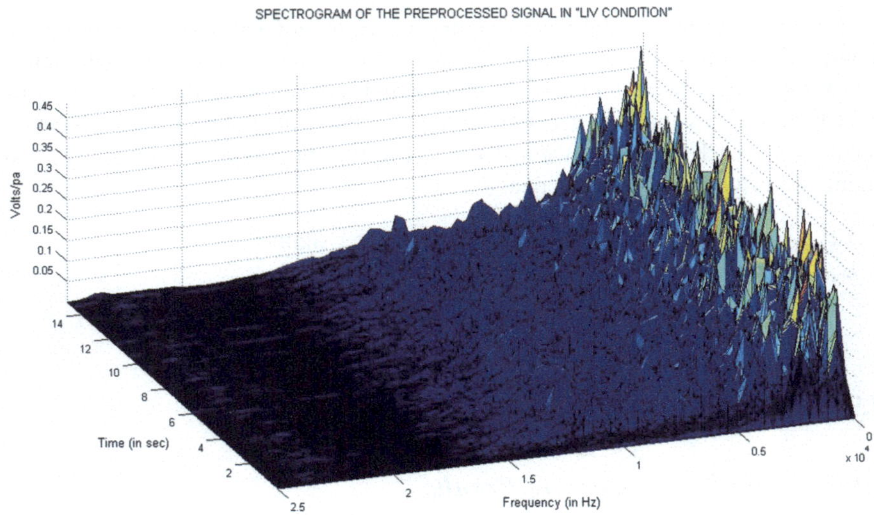

Fig. 7.8 Spectrogram of the pre-processed signal in "LIV Condition" [2]

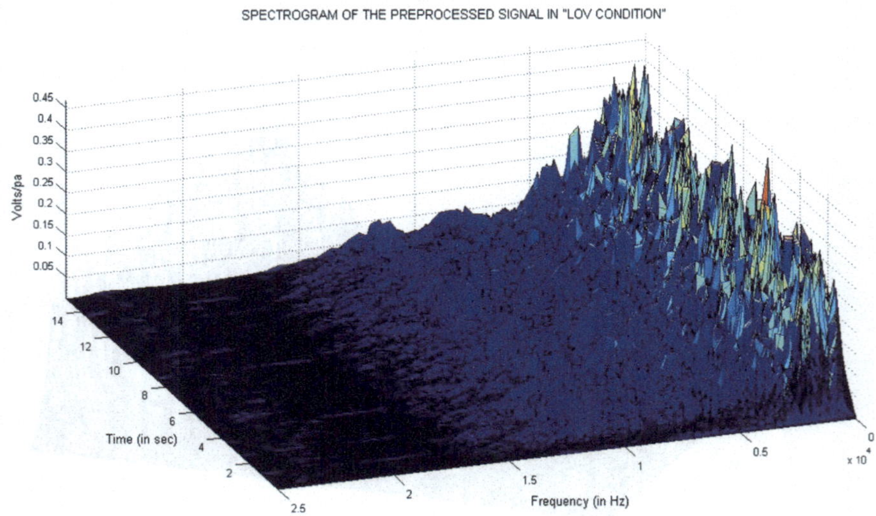

Fig. 7.9 Spectrogram of the pre-processed signal in "LOV Condition" [2]

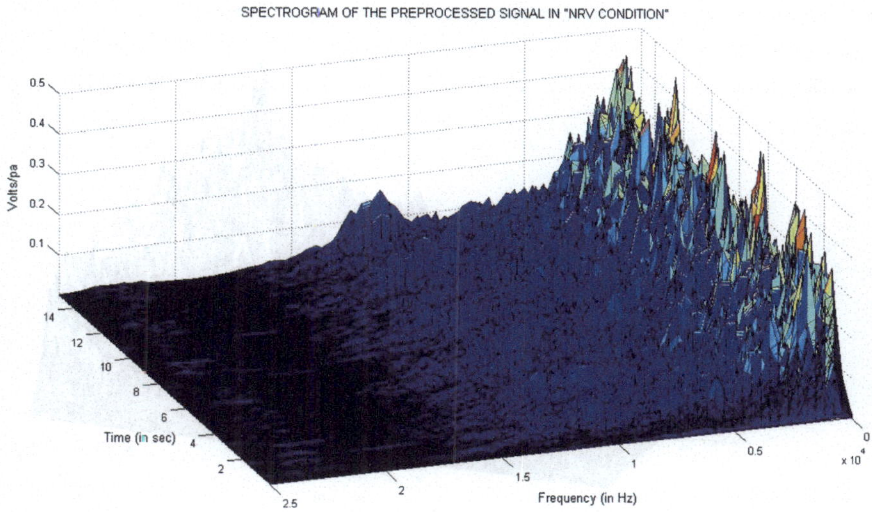

Fig. 7.10 Spectrogram of the pre-processed signal in "NRV Condition" [2]

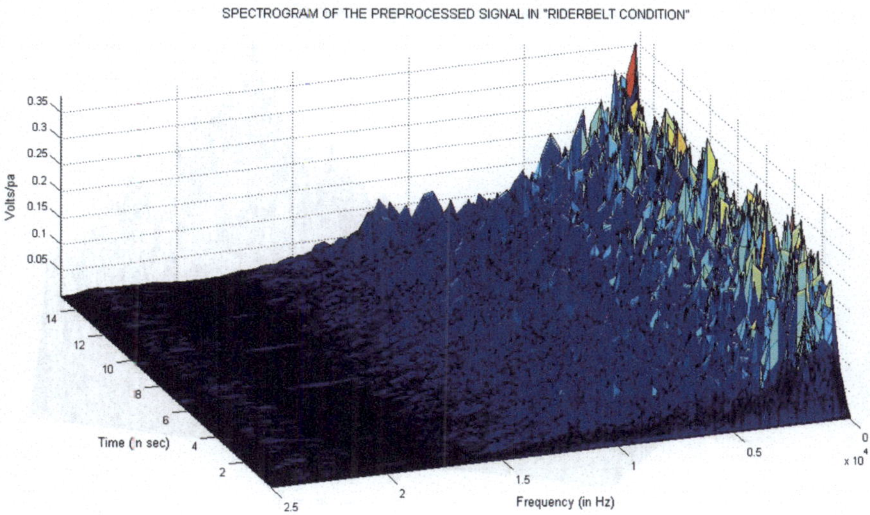

Fig. 7.11 Spectrogram of the pre-processed signal in "Riderbelt Condition" [2]

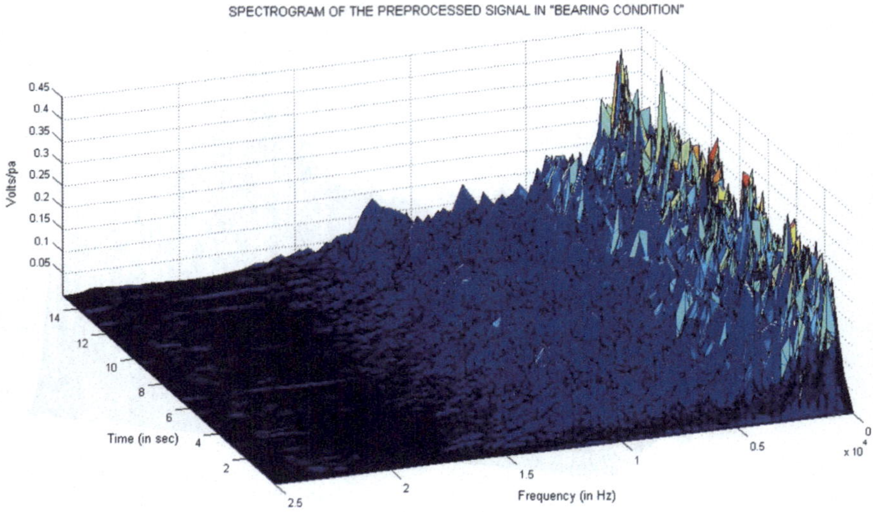

Fig. 7.12 Spectrogram of the pre-processed signal in "Bearing Condition" [2]

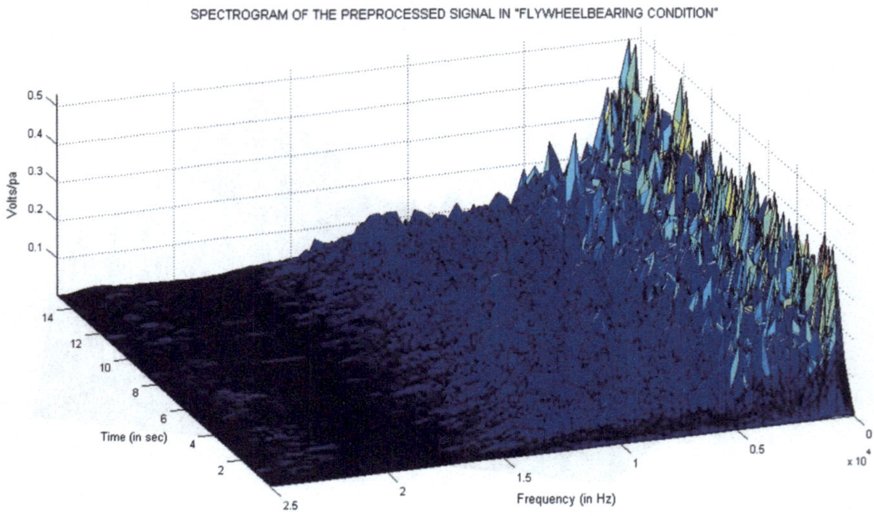

Fig. 7.13 Spectrogram of the pre-processed signal in "Flywheel Bearing Condition" [2]

7.9.2 Feature Extraction Results

- **CBM sheet of extended features: time domain**

			PALMTOP MICROPHONE							
READINING NO.	PRESSURE (PSI)	SR. NO	TIME DOMAIN							
			F1	F2	F3	F4	F5	F6	F7	F8
HealthyREADING1	10psi	1	0.005321	0.023236	0.006541	4.28E-05	2.500765	3.552574	5.54E+16	0.064859
	20psi	2	0.00538	0.02274	0.006622	4.38E-05	2.554659	3.434189	3.5E+17	0.081428
	40psi	3	0.005507	0.023503	0.00685	4.69E-05	2.690538	3.43088	1.57E+17	0.12356
	75psi	4	0.0056	0.02383	0.00695	4.83E-05	2.750166	3.428782	3.49E+16	0.169164
	110psi	5	0.005683	0.024298	0.007044	4.96E-05	2.708667	3.449362	1.51E+17	0.112198
	150psi	6	0.005416	0.021665	0.00663	4.4E-05	2.546612	3.267625	1.17E+17	0.110067
HealthyREADING2	10psi	7	0.00527	0.02141	0.006515	4.24E-05	2.562807	3.286463	7.53E+16	0.117106
	20psi	8	0.005192	0.022148	0.006344	4.03E-05	2.433174	3.490928	1.05E+17	0.093524
	40psi	9	0.00537	0.021757	0.006624	4.39E-05	2.565784	3.284554	5.77E+16	0.128594
	75psi	10	0.005405	0.022328	0.006686	4.47E-05	2.641144	3.33967	6.99E+16	0.182547
	110psi	11	0.005477	0.024	0.006705	4.5E-05	2.580465	3.579358	3.59E+16	0.130138
	150psi	12	0.005622	0.023379	0.00696	4.84E-05	2.727362	3.359022	4.02E+17	0.083809
HealthyREADING3	10psi	13	0.005069	0.018433	0.006199	3.84E-05	2.420647	2.973487	1.18E+17	0.10702
	20psi	14	0.00531	0.021245	0.006524	4.26E-05	2.493246	3.256728	3.25E+16	0.090119
	40psi	15	0.005007	0.017728	0.006081	3.7E-05	2.342564	2.915343	4.62E+16	0.104866
	75psi	16	0.005413	0.022252	0.006659	4.44E-05	2.631497	3.341353	7.87E+16	0.219
	110psi	17	0.005371	0.024167	0.006608	4.37E-05	2.655073	3.65729	9.18E+16	0.122953
	150psi	18	0.005535	0.024355	0.006789	4.61E-05	2.612889	3.587485	2.12E+17	0.136788
HealthyREADING4	10psi	19	0.005308	0.020637	0.006538	4.27E-05	2.522241	3.156659	8.03E+16	0.074165
	20psi	20	0.005113	0.020595	0.006287	3.95E-05	2.465693	3.276126	5.16E+16	0.053389
	40psi	21	0.005123	0.021047	0.006213	3.86E-05	2.369452	3.387438	4.13E+16	0.128082
	75psi	22	0.005176	0.019487	0.006305	3.98E-05	2.467254	3.090612	6.15E+16	0.169289
	110psi	23	0.005312	0.02212	0.006547	4.29E-05	2.616361	3.37864	6.84E+16	0.157952
	150psi	24	0.005499	0.021099	0.006737	4.54E-05	2.555783	3.131993	8.8E+16	0.100413
HealthyREADING5	10psi	25	0.005156	0.019252	0.00633	4.01E-05	2.446634	3.041537	3.42E+16	0.077992
	20psi	26	0.005231	0.020809	0.006445	4.15E-05	2.529322	3.228818	8.49E+16	0.088437
	40psi	27	0.005397	0.022631	0.006607	4.37E-05	2.46374	3.42517	7.6E+16	0.113443
	75psi	28	0.005525	0.022711	0.006827	4.66E-05	2.636969	3.326818	9.99E+16	0.11758
	110psi	29	0.005436	0.024525	0.006673	4.45E-05	2.603963	3.675187	3.46E+16	0.111326
	150psi	30	0.0054	0.02208	0.006616	4.38E-05	2.569671	3.337167	4.13E+16	0.12713
HealthyREADING6	10psi	31	0.004891	0.017007	0.005927	3.51E-05	2.316327	2.86918	5.26E+16	0.125517
	20psi	32	0.005002	0.020063	0.006125	3.75E-05	2.445453	3.275588	2.79E+17	0.067347
	40psi	33	0.004985	0.018088	0.006059	3.67E-05	2.340883	2.985474	8.3E+16	0.092405
	75psi	34	0.005377	0.022029	0.006629	4.39E-05	2.605626	3.323162	4.83E+16	0.177244
	110psi	35	0.005351	0.023179	0.00658	4.33E-05	2.616061	3.522338	1.34E+17	0.181735
	150psi	36	0.005427	0.022417	0.006645	4.42E-05	2.552407	3.373424	1.02E+17	0.115203
	10psi	37	0.004858	0.017254	0.005911	3.49E-05	2.34483	2.918963	3.22E+16	0.096918
	20psi	38	0.005141	0.018909	0.006282	3.95E-05	2.404498	3.009758	4.49E+16	0.073856

- **CBM sheet of extended features: frequency domain**

			PALMTOP MICROPHONE							
READINING NO.	PRESSURE (PSI)	SR. NO	FREQUENCY DOMAIN							
			F9	F10	F11	F12	F13	F14	F15	F16
HealthyREADING1	10psi	1	0.36276	0.249683	0.15537	0.119378	0.07303	0.018702	0.006992	0.014086
	20psi	2	0.417055	0.236929	0.142589	0.119572	0.050208	0.015151	0.006997	0.011499
	40psi	3	0.40013	0.244821	0.132667	0.116069	0.073082	0.014521	0.006945	0.011764
	75psi	4	0.394066	0.244016	0.152798	0.11224	0.060824	0.017786	0.005916	0.012353
	110psi	5	0.334815	0.258384	0.141706	0.144331	0.078928	0.02238	0.008505	0.01095
	150psi	6	0.356878	0.262287	0.158037	0.113137	0.074398	0.017805	0.008178	0.00928
HealthyREADING2	10psi	7	0.377405	0.254148	0.148414	0.118787	0.065739	0.018375	0.007191	0.009941
	20psi	8	0.371507	0.252536	0.159105	0.119923	0.063609	0.017654	0.006237	0.00943
	40psi	9	0.469998	0.196748	0.142254	0.10514	0.058016	0.014541	0.004716	0.008588
	75psi	10	0.405655	0.216977	0.16531	0.120757	0.058183	0.016775	0.005871	0.010472
	110psi	11	0.416608	0.231502	0.137915	0.116032	0.068643	0.014442	0.005529	0.009329
	150psi	12	0.396426	0.253844	0.143105	0.107163	0.067536	0.015871	0.005735	0.010319
HealthyREADING3	10psi	13	0.38924	0.256011	0.173163	0.09913	0.053544	0.012794	0.005681	0.010437
	20psi	14	0.361234	0.279503	0.151824	0.105187	0.066485	0.016198	0.006992	0.012577
	40psi	15	0.352195	0.248676	0.174372	0.125862	0.062442	0.019941	0.005426	0.011085
	75psi	16	0.382605	0.267078	0.136999	0.111194	0.075105	0.012468	0.005608	0.008943
	110psi	17	0.388393	0.246238	0.12967	0.136294	0.064186	0.017565	0.006857	0.010797
	150psi	18	0.348417	0.245439	0.156635	0.135786	0.078455	0.018419	0.007506	0.009344
HealthyREADING4	10psi	19	0.46042	0.227504	0.129297	0.105161	0.051523	0.011053	0.004852	0.01019
	20psi	20	0.497429	0.223327	0.117881	0.089703	0.043187	0.013493	0.005319	0.009661
	40psi	21	0.35824	0.270565	0.158912	0.105829	0.068103	0.018698	0.005747	0.013906
	75psi	22	0.355156	0.24452	0.149743	0.130583	0.083329	0.017366	0.006059	0.013244
	110psi	23	0.335597	0.280293	0.146416	0.136458	0.064523	0.01754	0.00504	0.014133
	150psi	24	0.419066	0.235808	0.164735	0.097918	0.055211	0.013856	0.003907	0.0095
HealthyREADING5	10psi	25	0.354867	0.300557	0.142208	0.106024	0.057703	0.018239	0.007493	0.012909
	20psi	26	0.409769	0.212141	0.183624	0.106324	0.057711	0.017094	0.004838	0.0085
	40psi	27	0.435453	0.227235	0.131962	0.112798	0.061246	0.016006	0.005787	0.009513
	75psi	28	0.435697	0.245657	0.121405	0.102151	0.064639	0.014608	0.004953	0.010891
	110psi	29	0.364627	0.253142	0.149235	0.113979	0.087081	0.01751	0.004943	0.009482
	150psi	30	0.414146	0.225695	0.135024	0.128124	0.064176	0.015989	0.006327	0.010519
HealthyREADING6	10psi	31	0.382079	0.257861	0.156178	0.10371	0.062679	0.018827	0.006938	0.011728
	20psi	32	0.433751	0.238448	0.134049	0.104714	0.059268	0.014509	0.004126	0.011135
	40psi	33	0.378995	0.249643	0.160401	0.124286	0.054113	0.014885	0.006449	0.011227
	75psi	34	0.387267	0.222414	0.157031	0.139179	0.0592	0.017331	0.006367	0.01121
	110psi	35	0.382184	0.239963	0.142156	0.132494	0.067655	0.017626	0.005066	0.012857
	150psi	36	0.375806	0.26413	0.156719	0.109927	0.061646	0.014693	0.005943	0.011136
	10psi	37	0.397892	0.269284	0.134296	0.112799	0.056814	0.016074	0.004772	0.008069
	20psi	38	0.403405	0.253411	0.151188	0.102664	0.057044	0.017115	0.004839	0.010333

- **CBM sheet of extended features: CWT**

			PALMTOP MICROPHONE						
READINING NO.	PRESSURE (PSI)	SR. NO	CONTINUOUS WAVELET TRANSFORM						
			F17	F18	F19	F20	F21	F22	F23
HealthyREADING1	10psi	1	0.008006	1.734336	4.014488	-0.32179	6.41E-05	-0.00461	2458
	20psi	2	0.008184	1.75338	4.083862	-0.23775	6.7E-05	-0.0014	2375
	40psi	3	0.008269	1.752444	4.534605	-0.30591	6.84E-05	-0.00022	2340
	75psi	4	0.008632	1.794516	4.386702	-0.28054	7.45E-05	0.003342	2349
	110psi	5	0.008244	1.74899	4.726315	-0.18722	6.8E-05	0.002814	2268
	150psi	6	0.008058	1.745315	4.194994	-0.18733	6.49E-05	0.005011	2294
HealthyREADING2	10psi	7	0.008774	1.825407	3.571184	-0.2692	7.7E-05	-0.00067	2586
	20psi	8	0.008254	1.774756	3.490946	-0.21997	6.81E-05	-0.00199	2461
	40psi	9	0.008617	1.802807	3.861469	-0.28064	7.42E-05	-0.00224	2438
	75psi	10	0.008367	1.774998	4.181746	-0.29876	7E-05	-0.00118	2345
	110psi	11	0.008236	1.757315	4.413326	-0.22219	6.78E-05	0.005052	2286
	150psi	12	0.008904	1.833791	4.164515	-0.19253	7.93E-05	0.001072	2360
HealthyREADING3	10psi	13	0.007958	1.740072	3.561469	-0.26613	6.33E-05	-0.00058	2447
	20psi	14	0.008436	1.783603	3.761643	-0.29141	7.12E-05	-0.00197	2401
	40psi	15	0.007583	1.701758	3.674572	-0.27144	5.75E-05	-5.5E-05	2245
	75psi	16	0.008204	1.756049	4.267229	-0.28565	6.73E-05	0.002272	2289
	110psi	17	0.008088	1.747838	4.199133	-0.24122	6.54E-05	0.000303	2354
	150psi	18	0.008463	1.790785	4.06786	-0.23466	7.16E-05	-0.00185	2358
HealthyREADING4	10psi	19	0.00857	1.80112	3.701349	-0.24159	7.34E-05	0.001277	2459
	20psi	20	0.00782	1.722219	3.627037	-0.20012	6.11E-05	-0.00154	2335
	40psi	21	0.00828	1.788477	3.546564	-0.28352	6.86E-05	-0.00062	2418
	75psi	22	0.0075	1.678804	4.137797	-0.39253	5.63E-05	-0.00453	2254
	110psi	23	0.0082	1.761017	3.970933	-0.27128	6.72E-05	0.000947	2423
	150psi	24	0.008476	1.792084	4.047644	-0.1972	7.18E-05	0.004382	2290
HealthyREADING5	10psi	25	0.007886	1.730402	3.708033	-0.28332	6.22E-05	-0.00229	2389
	20psi	26	0.008296	1.778425	3.644543	-0.26357	6.88E-05	0.00097	2464
	40psi	27	0.008445	1.794209	3.812511	-0.27112	7.13E-05	-0.00073	2352
	75psi	28	0.008122	1.739793	4.398323	-0.25571	6.6E-05	-0.00163	2309
	110psi	29	0.008239	1.758168	4.255464	-0.26609	6.79E-05	0.001448	2272
	150psi	30	0.008262	1.776161	3.829193	-0.23053	6.83E-05	-0.00118	2328
HealthyREADING6	10psi	31	0.007322	1.669452	3.662824	-0.35492	5.36E-05	-0.00157	2345
	20psi	32	0.00772	1.712436	3.830045	-0.30108	5.96E-05	0.00457	2364
	40psi	33	0.007644	1.709372	3.588653	-0.26791	5.84E-05	-0.00367	2298
	75psi	34	0.007943	1.7271	4.263141	-0.35244	6.31E-05	-0.00418	2342
	110psi	35	0.007826	1.717939	4.267487	-0.33678	6.12E-05	0.000163	2263
	150psi	36	0.008321	1.780121	4.005408	-0.22986	6.92E-05	0.002249	2305
	10psi	37	0.007362	1.672603	3.506488	-0.27599	5.42E-05	0.000388	2410
	20psi	38	0.007858	1.726874	3.760124	-0.27599	6.17E-05	-0.00117	2365

7.9.3 Classification Results

The classification results are presented for various number of selected features from feature selection techniques. Different proportions of CBM dataset are used for training and testing of each set of selected features.

Training Results With 5 Selected Features With five selected features using a training and testing sets of 50% each in Table 7.6, it is observed that the combination of MIFS feature selection and OAA feature classification gives overall maximum accuracy of 73.3334%.

With five selected features using a training set of 60% and a testing set of 40% in Table 7.7, it is observed that the combination of mRMR feature selection and DDAG feature classification gives overall maximum accuracy of 75.5952%.

With five selected features using a training set of 70% and a testing set of 30% in Table 7.8, it is observed that the combination of mRMR feature selection and DDAG feature classification gives overall maximum accuracy of 75.3968%.

With five selected features using a training set of 80% and a testing set of 20% in Table 7.9, it is observed that the combination of MIFS-U feature selection and DDAG feature classification gives overall maximum accuracy of 73.2919%.

With five selected features using a training set of 90% and a testing set of 10% in Table 7.10, it is observed that the combination of mRMR feature selection and DDAG feature classification, along with yet another combination of MIFS feature selection and DDAG feature classification, both give overall maximum accuracy of 75.3246%.

Table 7.6 Results with training set = 50% and testing set = 50%

Training results							
Multiclass SVM	Training set = 50%, Testing set = 50 %						
	PCA	MIFS	mRMR	MIFS-U	NMIFS	BD	ICA
OAO	27.7778	50.2778	50.8333	48.0556	48.0556	33.8889	42.5
OAA	25.2778	73.3334	68.3333	70.8334	60.2778	48.0556	51.1667
FDF	27.5	69.1667	66.9444	63.0556	58.8889	44.1667	47.3334
DDAG	28.3334	72.2222	71.3889	70.3334	61.3889	49.4445	56.2778

Table 7.7 Results with training set = 60% and testing set = 40%

Training results							
Multiclass SVM	Training set = 60 %, Testing set = 40 %						
	PCA	MIFS	mRMR	MIFS-U	NMIFS	BD	ICA
OAO	44.3452	47.9166	48.8095	50.8928	39.8809	45.8333	20.2381
OAA	50.5952	69.0476	72.0238	73.5119	56.8452	61.0119	21.7261
FDF	48.2142	63.6904	69.3452	67.8571	61.3095	59.2261	23.2142
DDAG	56.5476	71.4285	75.5952	73.5119	65.4761	66.3690	18.1547

Table 7.8 Results with training set = 70% and testing set = 30%

Training Results							
Multiclass SVM	Training set = 70 %, Testing set = 30 %						
	PCA	MIFS	mRMR	MIFS-U	NMIFS	BD	ICA
OAO	33.7301	46.4285	48.8095	49.2063	36.5079	48.0158	26.1904
OAA	32.1428	64.6825	67.0634	70.6349	58.7301	57.5396	28.9682
FDF	26.1904	67.8571	69.4444	65.4761	60.3174	57.1428	27.3809
DDAG	38.0952	72.6190	75.3968	71.0317	64.2857	66.6666	19.8412

Table 7.9 Results with training set = 80% and testing set = 20%

Training results							
Multiclass SVM	Training set = 80 %, Testing set = 20 %						
	PCA	MIFS	mRMR	MIFS-U	NMIFS	BD	ICA
OAO	42.8571	47.8260	49.0683	49.0683	44.7205	43.4782	21.7391
OAA	42.2360	64.5962	65.8385	68.9441	54.6583	62.1118	24.2236
FDF	32.9192	63.9751	70.8074	67.7018	59.6273	60.2484	23.6024
DDAG	52.1739	72.6708	72.6708	73.2919	63.9751	68.3229	22.9813

Table 7.10 Results with training set = 90% and testing set = 10%

Training results							
Multiclass SVM	Training set = 90 %, Testing set = 10 %						
	PCA	MIFS	mRMR	MIFS-U	NMIFS	BD	ICA
OAO	50.6493	54.5454	51.9480	49.3506	37.6623	46.7532	24.6753
OAA	44.1558	64.9350	64.9350	71.4285	49.3506	54.5454	24.6753
FDF	37.6623	71.4285	64.9350	68.8311	49.3506	55.8441	27.2727
DDAG	57.1428	75.3246	75.3246	72.7272	53.2467	70.1298	24.6753

Figure 7.14 illustrates the training and testing results for five selected features: (a) with 50% training set and 50% testing set, (b) with 60% training set and 40% testing set, (c) with 70% training set and 30% testing set, (d) with 80% training set and 20% testing set, and (e) with 90% training set and 10% testing set.

Training Results With 10 Selected Features: With ten selected features using a training set of 50% and a testing set of 50% in Table 7.11, it is observed that the combination of mRMR feature selection and DDAG feature classification gives overall maximum accuracy of 75.5556%.

With ten selected features using a training set of 60% and a testing set of 40% in Table 7.12, it is observed that the combination of MIFS feature selection and DDAG feature classification gives overall maximum accuracy of 75.8928%.

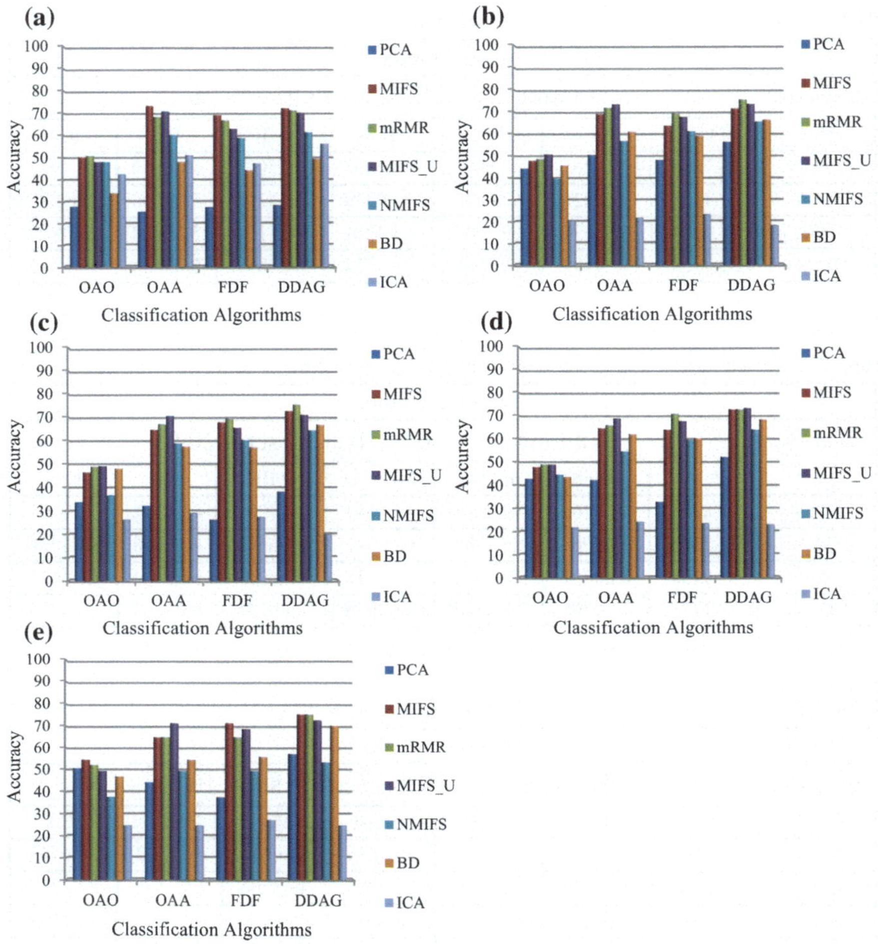

Fig. 7.14 Classification results with five selected features: (a) 50% training set and 50% testing set (b) 60% training set and 40% testing set (c) 70% training set and 30% testing set (d) 80% training set and 20% testing set and (e) 90% training set and 10% testing set [2]

Table 7.11 Results with training set = 50% and testing set = 50%

Training results							
Multiclass SVM	Training set = 50 %, Testing set = 50 %						
	PCA	MIFS	mRMR	MIFS-U	NMIFS	BD	ICA
OAO	27.2222	52.5	52.2223	53.3334	54.1667	49.7222	45.7223
OAA	27.2222	68.0558	71.6667	69.7223	66.1112	63.8889	61.1111
FDF	27.7778	59.1667	67.2222	63.0556	61.9445	59.4445	56.3337
DDAG	26.9444	72.7778	75.5556	71.1112	73.0556	70	65.1667

Table 7.12 Results with training set = 60% and testing set = 40%

Training results							
Multiclass SVM	Training set = 60 %, Testing set = 40 %						
	PCA	MIFS	mRMR	MIFS-U	NMIFS	BD	ICA
OAO	48.5119	53.86905	52.97619	48.80952	42.55952	50.59524	22.32143
OAA	53.57143	69.34524	70.83333	61.0119	49.40476	64.58333	22.61905
FDF	46.13095	63.69048	69.04762	60.41667	43.15476	58.03571	24.10714
DDAG	59.82143	75.89286	74.10714	71.72619	58.33333	70.53571	23.5119

With ten selected features using a training set of 70% and a testing set of 30% in Table 7.13, it is observed that the combination of mRMR feature selection and DDAG feature classification gives overall maximum accuracy of 76.5873%.

With ten selected features using a training set of 80% and a testing set of 20% in Table 7.14, it is observed that the combination of MIFS feature selection and DDAG feature classification gives overall maximum accuracy of 83.22981%.

With ten selected features using a training set of 90% and a testing set of 10% in Table 7.15, it is observed that the combination of mRMR feature selection and DDAG feature classification gives overall maximum accuracy of 77.9221%.

Figure 7.15 illustrates the training and testing results for ten selected features: (a) with 50% training set and 50% testing set, (b) with 60% training set and 40%

Table 7.13 Results with training set = 70% and testing set = 30%

Training results							
Multiclass SVM	Training set = 70 %, Testing set = 30 %						
	PCA	MIFS	mRMR	MIFS-U	NMIFS	BD	ICA
OAO	50.39683	53.57143	50.39683	49.60317	44.04762	47.61905	20.2381
OAA	46.8254	68.65079	75	63.09524	49.60317	52.77778	21.8254
FDF	34.12698	64.68254	70.2381	62.30159	44.04762	60.31746	23.4127
DDAG	53.57143	76.19048	76.5873	73.4127	59.92063	71.03175	22.22222

Table 7.14 Results with training set = 80% and testing set = 20%

Training results							
Multiclass SVM	Training set = 80 %, Testing set = 20 %						
	PCA	MIFS	mRMR	MIFS-U	NMIFS	BD	ICA
OAO	45.96273	56.52174	52.17391	50.93168	47.82609	54.03727	40.37267
OAA	49.06832	74.53416	67.08075	56.52174	49.68944	62.73292	32.91925
FDF	44.09938	73.91304	68.9441	68.32298	44.09938	56.52174	27.95031
DDAG	60.24845	83.22981	78.26087	75.15528	66.45963	69.56522	32.29814

Table 7.15 Results with training set = 90% and testing set = 10%

Training results							
Multiclass SVM	Training set = 90 %, Testing set = 10 %						
	PCA	MIFS	mRMR	MIFS-U	NMIFS	BD	ICA
OAO	53.2467	55.84416	48.0519	53.24675	55.8441	46.7533	28.5714
OAA	63.6364	76.62338	75.3246	58.44156	76.6234	67.5325	28.5714
FDF	45.4546	63.63636	76.6234	53.24675	63.6363	55.8442	27.2727
DDAG	68.8312	77.92208	77.9221	70.12987	77.9221	72.7273	25.9741

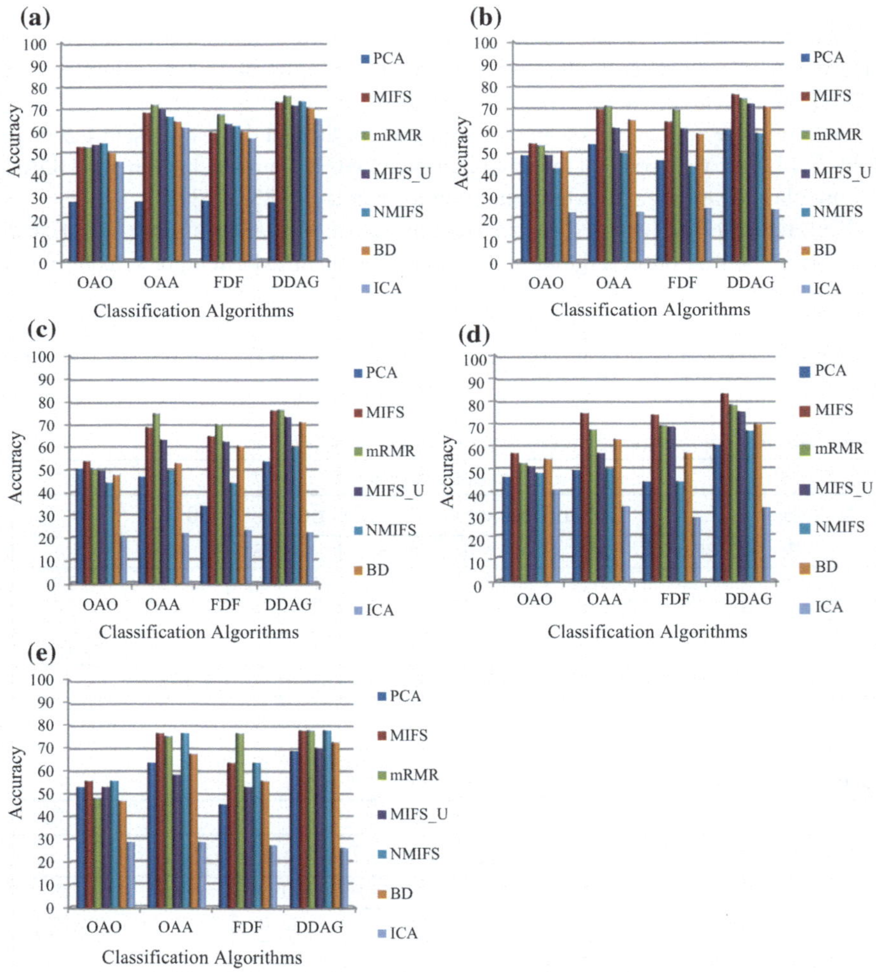

Fig. 7.15 Classification results with ten selected features: (a) 50% training set and 50% testing set (b) 60% training set and 40% testing set (c) 70% training set and 30% testing set (d) 80% training set and 20% testing set and (e) 90% training set and 10% testing set [2]

testing set, (c) with 70% training set and 30% testing set, (d) with 80% training set and 20% testing set, and (e) with 90% training set and 10% testing set.

Training Results With 23 Selected Features: With 23 selected features using a training set of 50% and a testing set of 50% in Table 7.16, it is observed that the combination of MIFS feature selection and DDAG feature classification gives overall maximum accuracy of 75.8333%.

With 23 selected features using a training set of 60% and a testing set of 40% in Table 7.17, it is observed that the combination of MIFS feature selection and DDAG feature classification gives overall maximum accuracy of 75.89286%.

With 23 selected features using a training set of 70% and a testing set of 30% in Table 7.18, it is observed that the combination of mRMR feature selection and DDAG feature classification gives overall maximum accuracy of 75.8333%.

Table 7.16 Results with training set = 50% and testing set = 50%

Training results							
Multiclass SVM	Training set = 50 %, Testing set = 50 %						
	PCA	MIFS	mRMR	MIFS-U	NMIFS	BD	ICA
OAO	48.3333	55.2778	58.6111	55.5556	54.4445	54.7222	52.6333
OAA	53.3333	70.8333	71.1111	69.1667	67.2223	63.6112	61.1667
FDF	38.8889	64.7222	59.7223	58.6111	52.2223	55.2778	54.2777
DDAG	55	75.8333	74.1667	72.5	68.8889	72.2222	69.7223

Table 7.17 Results with training set = 60% and testing set = 40%

Training results							
Multiclass SVM	Training set = 60 %, Testing set = 40 %						
	PCA	MIFS	mRMR	MIFS-U	NMIFS	BD	ICA
OAO	48.5119	53.86905	52.97619	48.80952	42.55952	50.59524	22.32143
OAA	53.57143	69.34524	70.83333	61.0119	49.40476	64.58333	22.61905
FDF	46.13095	63.69048	69.04762	60.41667	43.15476	58.03571	24.10714
DDAG	59.82143	75.89286	74.10714	71.72619	58.33333	70.53571	23.5119

Table 7.18 Results with training set = 70% and testing set = 30%

Training results							
Multiclass SVM	Training set = 70 %, Testing set = 30 %						
	PCA	MIFS	mRMR	MIFS-U	NMIFS	BD	ICA
OAO	50.39683	53.57143	50.39683	49.60317	44.04762	47.61905	20.2381
OAA	46.8254	68.65079	75	63.09524	49.60317	52.77778	21.8254
FDF	34.12698	64.68254	70.2381	62.30159	44.04762	60.31746	23.4127
DDAG	53.57143	76.19048	76.5873	73.4127	59.92063	71.03175	22.22222

Table 7.19 Results with training set = 80% and testing set = 20%

Training results							
Multiclass SVM	Training set = 80 %, Testing set = 20 %						
	PCA	MIFS	mRMR	MIFS-U	NMIFS	BD	ICA
OAO	45.96273	56.52174	52.17391	50.93168	47.82609	54.03727	40.37267
OAA	49.06832	74.53416	67.08075	56.52174	49.68944	62.73292	32.91925
FDF	44.09938	73.91304	68.9441	68.32298	44.09938	56.52174	27.95031
DDAG	60.24845	83.22981	78.26087	75.15528	66.45963	69.56522	32.29814

Table 7.20 Results with training set = 90% and testing set = 10%

Training results							
Multiclass SVM	Training set = 90 %, Testing set = 10 %						
	PCA	MIFS	mRMR	MIFS-U	NMIFS	BD	ICA
OAO	53.2467	55.84416	48.0519	53.24675	55.8441	46.7533	28.5714
OAA	63.6364	76.62338	75.3246	58.44156	76.6234	67.5325	28.5714
FDF	45.4546	63.63636	76.6234	53.24675	63.6363	55.8442	27.2727
DDAG	68.8312	77.92208	77.9221	70.12987	77.9221	72.7273	25.9741

With 23 selected features using a training set of 80% and a testing set of 20% in Table 7.19, it is observed that the combination of MIFS feature selection and DDAG feature classification gives overall maximum accuracy of 83.22981%.

With 23 selected features using a training set of 90% and a testing set of 10% in Table 7.20, it is observed that the combination of mRMR feature selection and DDAG feature classification gives overall maximum accuracy of 77.92208%.

Figure 7.16 illustrates the training and testing results for 23 selected features: (a) with 50% training set and 50% testing set, (b) with 60% training set and 40% testing set, (c) with 70% training set and 30% testing set, (d) with 80% training set and 20% testing set, and (e) with 90% training set and 10% testing set.

Training Results With 40 Selected Features:

With 40 selected features using a training set of 50% and a testing set of 50% in Table 7.21, it is observed that the combination of MIFS-U feature selection and DDAG feature classification gives overall maximum accuracy of 75%.

With 40 selected features using a training set of 60% and a testing set of 40% in Table 7.22, it is observed that the combination of MIFS feature selection and DDAG feature classification gives overall maximum accuracy of 78.8690%.

With 40 selected features using a training set of 70% and a testing set of 30% in Table 7.23, it is observed that the combination of MIFS feature selection and DDAG feature classification gives overall maximum accuracy of 77.3809%.

With 40 selected features using a training set of 80% and a testing set of 20% in Table 7.24, it is observed that the combination of MIFS feature selection and DDAG feature classification gives overall maximum accuracy of 77.6397%.

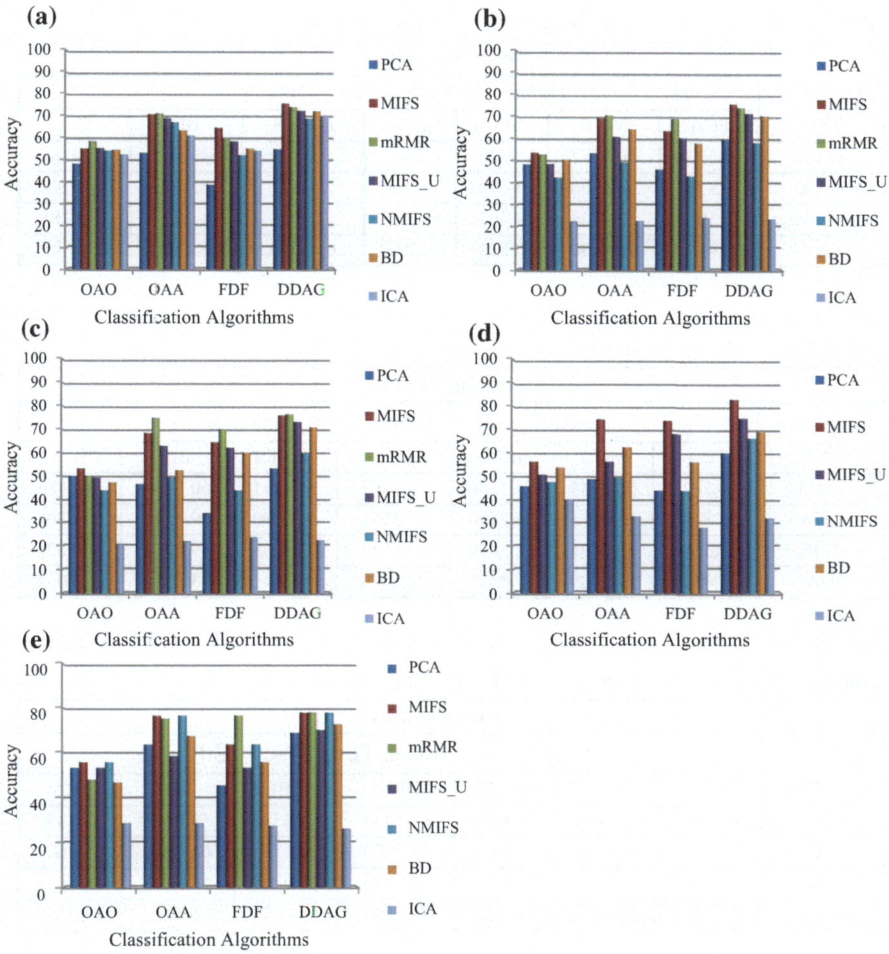

Fig. 7.16 Classification results with 23 selected features: (a) 50% training set and 50% testing set (b) 60% training set and 40% testing set (c) 70% training set and 30% testing set (d) 80% training set and 20% testing set and (e) 90% training set and 10% testing set [2]

Table 7.21 Results with training set = 50% and testing set = 50%

Training results							
Multiclass SVM	Training set = 50 %, Testing set = 50 %						
	PCA	MIFS	mRMR	MIFS-U	NMIFS	BD	ICA
OAO	45.0	56.1112	56.1112	59.4445	51.9444	51.3889	50.7778
OAA	62.7778	64.7223	65.5556	68.0556	62.2222	70.5556	61.0556
FDF	50.8333	48.0556	52.5	51.3889	45.5556	56.9445	54.4445
DDAG	66.3889	71.9445	73.3334	75	63.3333	73.8889	68.1112

Table 7.22 Results with training set = 60% and testing set = 40%

Training results							
Multiclass SVM	Training set = 60 %, Testing set = 40 %						
	PCA	MIFS	mRMR	MIFS-U	NMIFS	BD	ICA
OAO	59.5238	60.7142	56.25	50	50	54.4642	35.4166
OAA	59.8214	66.6666	63.6904	66.3690	51.4881	58.9285	28.2738
FDF	49.1071	57.4404	54.4642	55.3571	43.75	45.5357	27.6785
DDAG	72.6190	78.8690	70.8333	74.7023	61.3095	68.1547	39.8809

Table 7.23 Results with training set = 70% and testing set = 30%

Training results							
Multiclass SVM	Training set = 70 %, Testing set = 30 %						
	PCA	MIFS	mRMR	MIFS-U	NMIFS	BD	ICA
OAO	51.5873	58.3333	53.1746	58.7301	51.9841	50.3968	36.1111
OAA	57.5396	62.3015	66.2698	62.6984	47.6190	59.9206	37.6984
FDF	54.3650	48.8095	52.3809	55.5555	32.9365	49.6031	30.9523
DDAG	73.4127	77.3809	71.0317	72.6190	60.7142	73.0158	49.6031

Table 7.24 Results with training set = 80% and testing set = 20%

Training results							
Multiclass SVM	Training set = 80 %, Testing set = 20 %						
	PCA	MIFS	mRMR	MIFS-U	NMIFS	BD	ICA
OAO	48.4472	60.2484	53.4161	60.2484	54.6583	59.6273	48.4472
OAA	57.1428	60.8695	69.5652	62.7329	52.1739	64.5962	45.9627
FDF	52.7950	48.4472	52.7950	50.9316	36.0248	48.4472	34.1614
DDAG	75.7764	77.6397	70.1863	75.7764	59.6273	76.3975	65.8385

With 40 selected features using a training set of 90% and a testing set of 10% in Table 7.25, it is observed that the combination of MIFS feature selection and DDAG feature classification gives overall maximum accuracy of 87.0129%. Figure

Table 7.25 Results with training set = 90% and testing set = 10%

Training results							
Multiclass SVM	Training set = 90 %, Testing set = 10 %						
	PCA	MIFS	mRMR	MIFS-U	NMIFS	BD	ICA
OAO	31.1688	59.7402	61.0389	61.0389	51.9480	57.1428	41.5584
OAA	46.7532	63.6363	61.0389	62.3376	44.1558	48.0519	36.3636
FDF	41.5584	70.1298	48.0519	53.2467	31.1688	50.6493	31.1688
DDAG	63.6363	87.0129	77.9220	77.9220	66.2337	75.3246	45.4545

7.17 illustrates the training and testing results for 40 selected features: (a) with 50% training set and 50% testing set, (b) with 60% training set and 40% testing set, (c) with 70% training set and 30% testing set, (d) with 80% training set and 20% testing set, and (e) with 90% training set and 10% testing set.

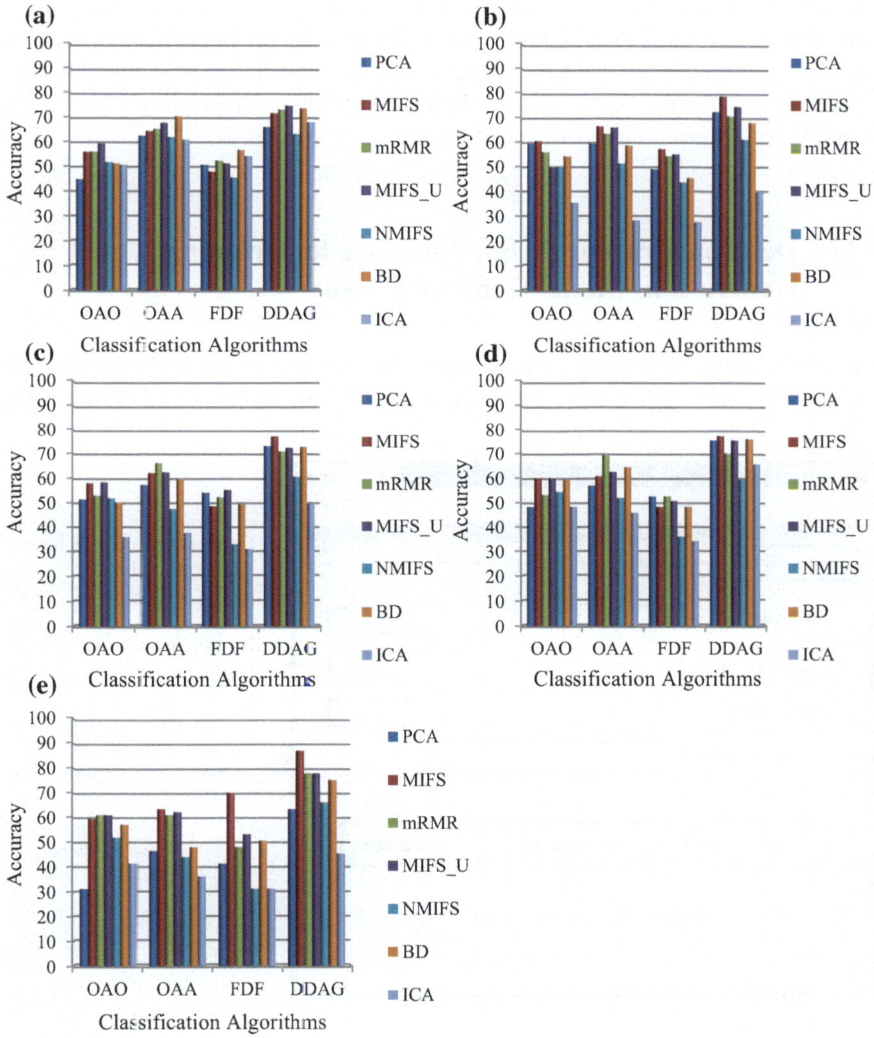

Fig. 7.17 Classification results with 40 selected features: (a) 50% training set and 50% testing set (b) 60% training set and 40% testing set (c) 70% training set and 30% testing set (d) 80% training set and 20% testing set and (e) 90% training set and 10% testing set [2]

7.10 Conclusions

The quick fault diagnosis system using a Palmtop makes life simple for accomplishing online health monitoring of the air compressor and quick detection of any fault in the compressor. With Palmtop inbuilt microphone working as a DAQ tool and with a compact feature extraction set, the results presented in this chapter are more inclined towards DDAG multiclass SVM method for implementing a quick fault diagnosis algorithm using the Palmtop. The implementation of standard feature selection methods like PCA, ICA, or Mutual Information (MI) methods for feature selection would give much better results.

7.11 Desktop Graphical User Interface for Training Model Generation: Model Generator Tool

The main purpose of this subsection is to introduce the MATLAB application, which can be used for the generation of training model and classification of the input data.

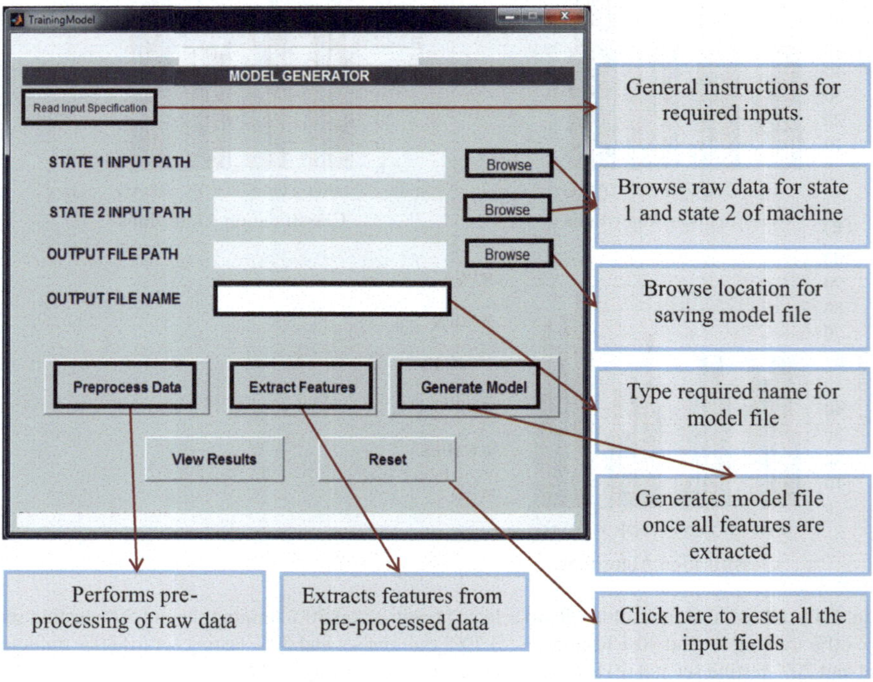

Fig. 7.18 GUI of model generator tool [2]

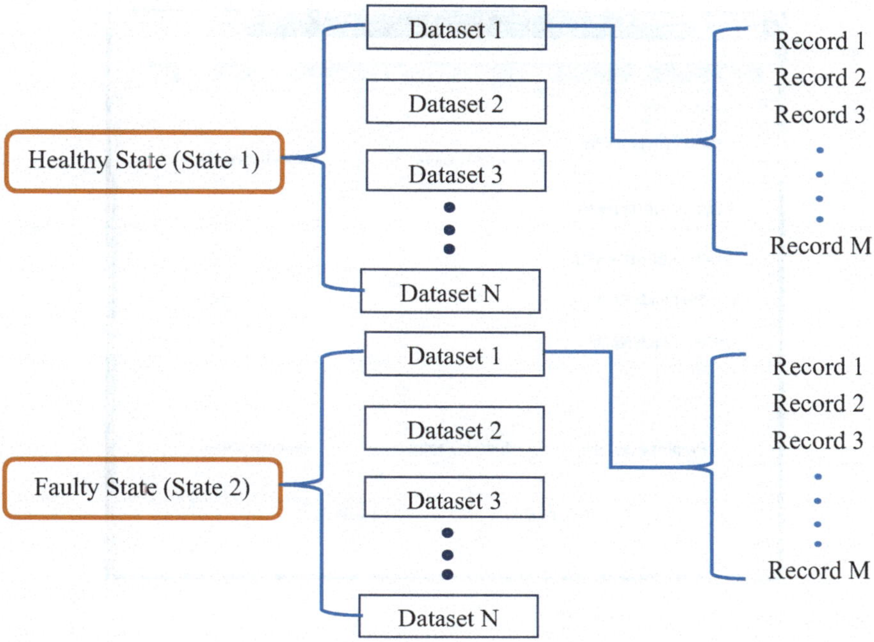

Fig. 7.19 Data organization [2]

The main window of the tool along its description for data organization is shown in Figs. 7.18 and 7.19.

7.11.1 Execution Steps for Model Generator Tool on Desktop Personal Computer (PC)

Step 1. As user clicks of "TrainingModel.exe," the following window appears (Refer Fig. 7.20).

Step 2. On clicking browse button, the current directory containing data folders opens. From this location, user can select the respective folder containing data files. Expected form of data organization is shown in Fig. 7.21.

Step 3. If the user selects the "Healthy" folder, the path of the selected folder appears in the text box adjacent to the browse button as shown in Fig. 7.22.

Step 4. Similarly, browse the "STATE 2" folder, and once the folder gets selected, its path will also be displayed in GUI as shown in Fig. 7.23.

Step 5. Similarly, browse the "Output Folder" where results of model generator tool will get stored. Once this folder gets selected, its path also appears in the adjacent box as shown in Fig. 7.24.

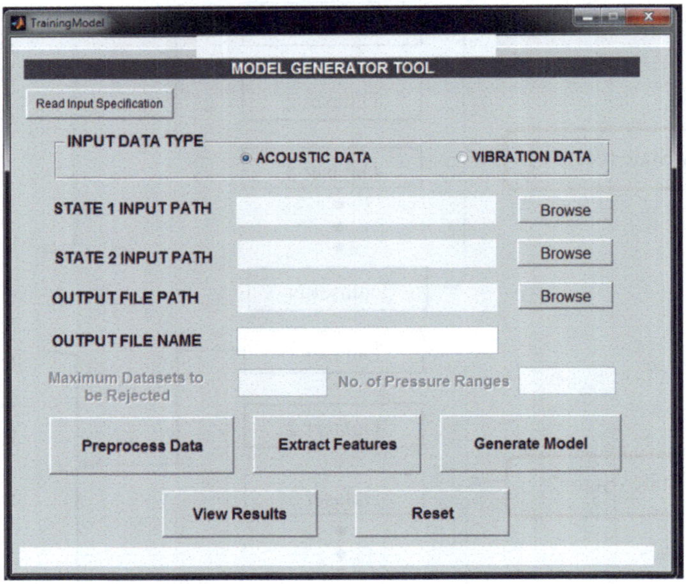

Fig. 7.20 Model generator GUI [2]

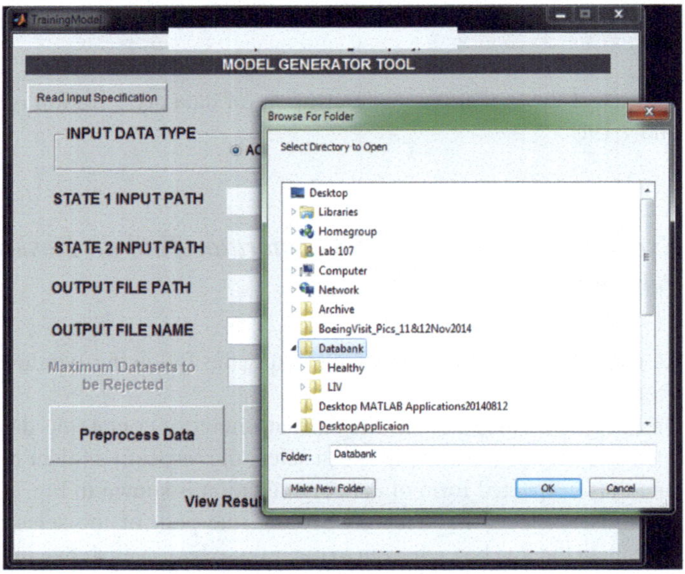

Fig. 7.21 Browsing option on interface [2]

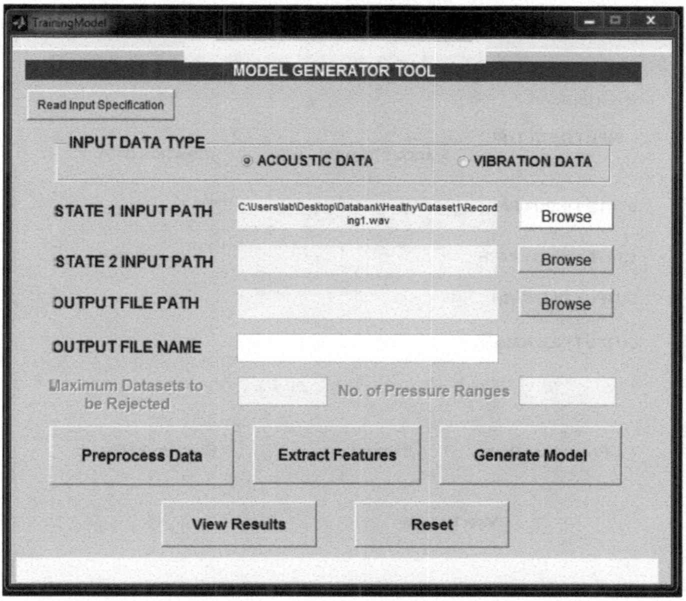

Fig. 7.22 Interface displaying "Healthy Dataset" has been selected [2]

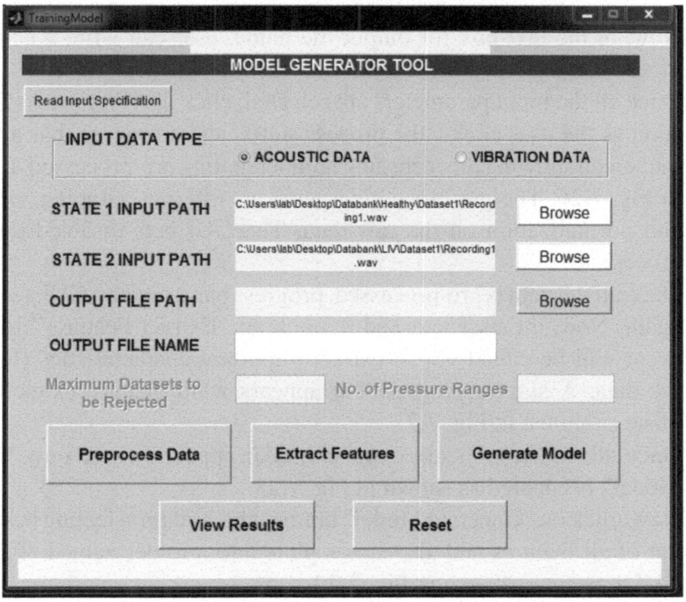

Fig. 7.23 Interface displaying "Faulty Dataset" has been selected [2]

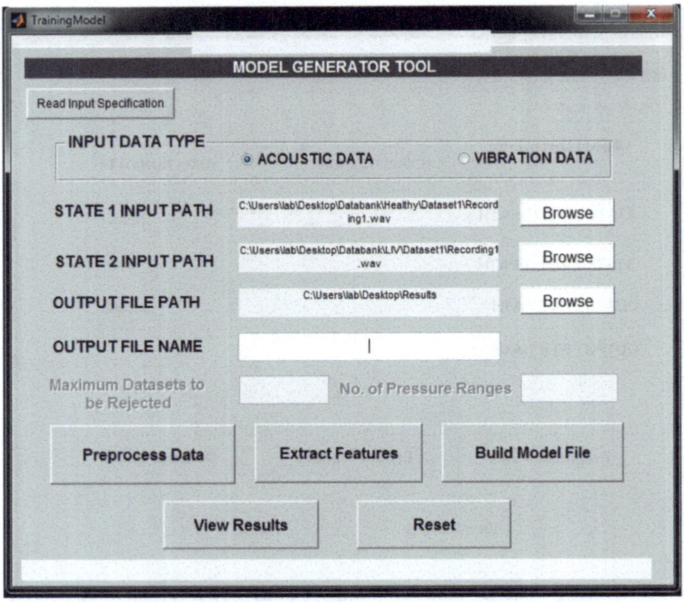

Fig. 7.24 Interface displaying "Output Folder" has been selected [2]

Step 6. Now, in the text box for output file name, user can write a name of the model file, e.g.,"healthy_liv_model" as shown in Fig. 7.25.

Step 7. Once all the input parameters are selected, click on "Preprocess Data". As soon as the user clicks, the process starts, and a progress bar appears on the screen showing the recording number getting pre-processed. It is shown in Fig. 7.26. The pre-processing performs filtering, clipping, smoothing, and normalization on the raw data. The GUI gets disabled during this process.

Step 8. Once all the data is pre-processed, progress bar stops and GUI gets enabled again. Now, the user is asked to click on "Extract Features" button. An event will be called which extract important characteristics (feature) of the data. A similar progress bar appears while feature extraction takes place as shown in Fig. 7.27.

Step 9. Once all the features are extracted, GUI appears where only "Generate Model" is enabled as shown in Fig. 7.28.

Step 10. Now, click on "Generate Model" button which starts selecting best features out of all features and also stores plots into a folder named "Plots". The GUI appears as shown in Fig. 7.29.

Step 11. Once this process is complete, a dialog box appears on the GUI as shown in Fig. 7.30. A text file is generated consisting of five rows. The first row contains the indices of good features. The second and fourth rows contain the mean of these features while third and fifth rows contain the standard deviation of theses features for State 1 and State 2.

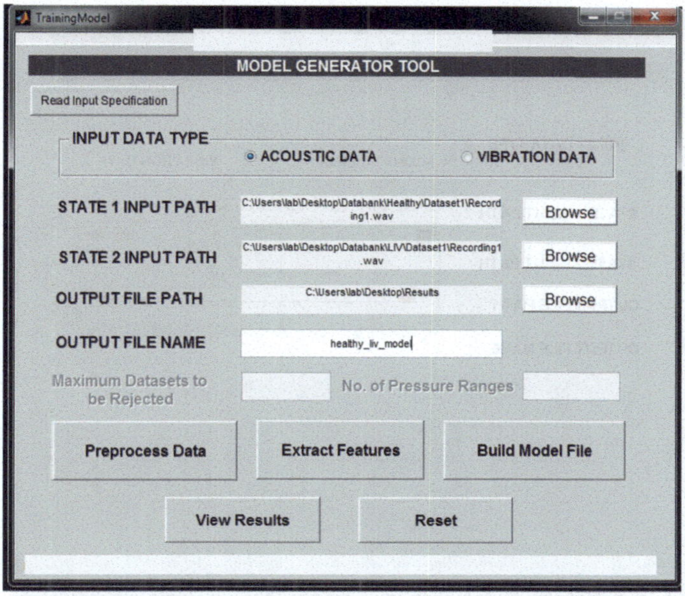

Fig. 7.25 Interface displaying "healthy_liv_model" name has been written [2]

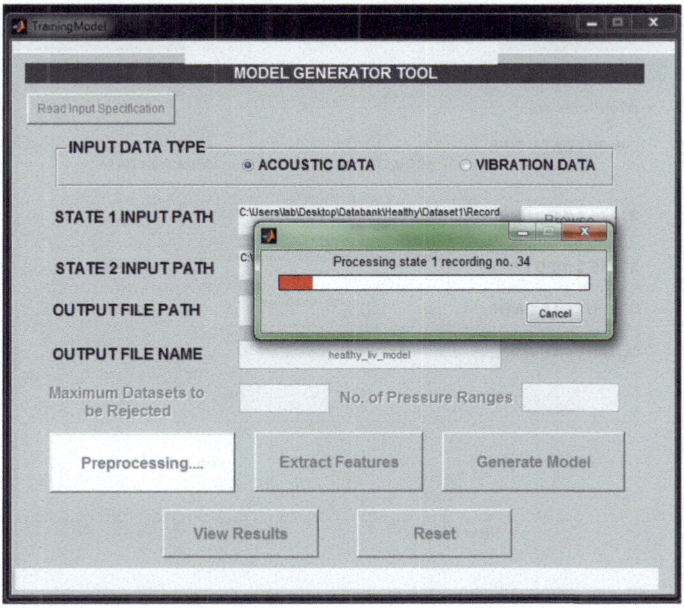

Fig. 7.26 Interface displaying "Recording 34" is being pre-processed [2]

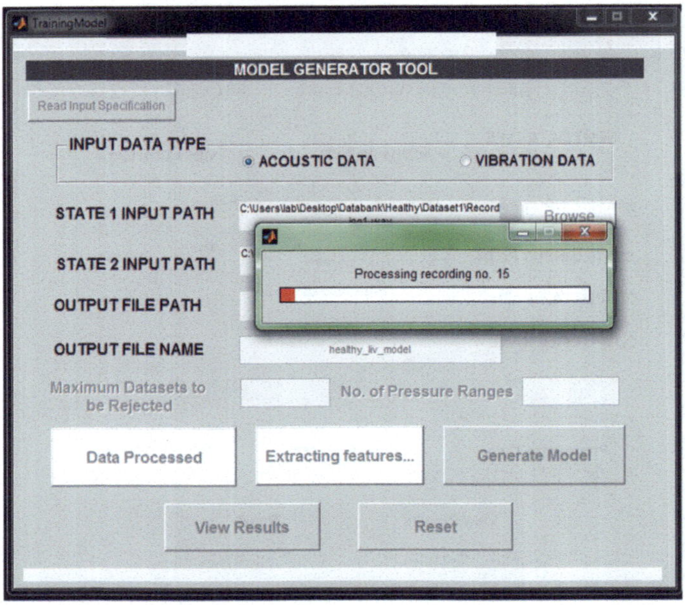

Fig. 7.27 Interface displaying that features are being extracted from "Recording 15" [2]

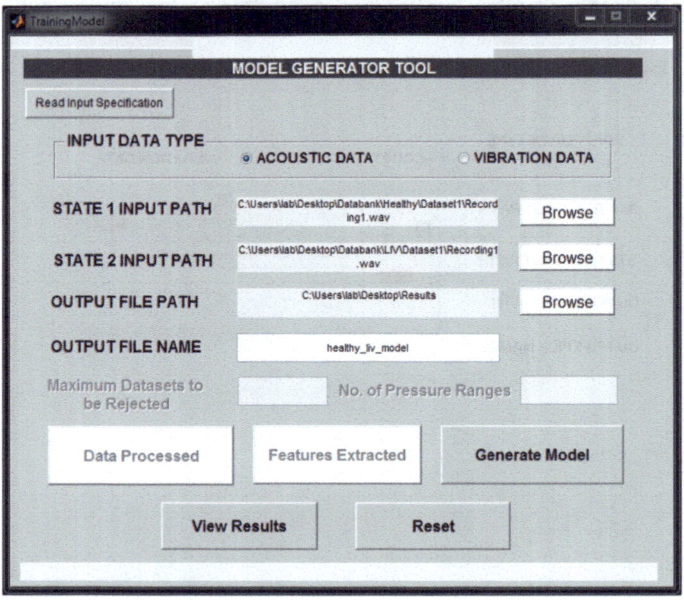

Fig. 7.28 Interface displaying that all features have been extracted [2]

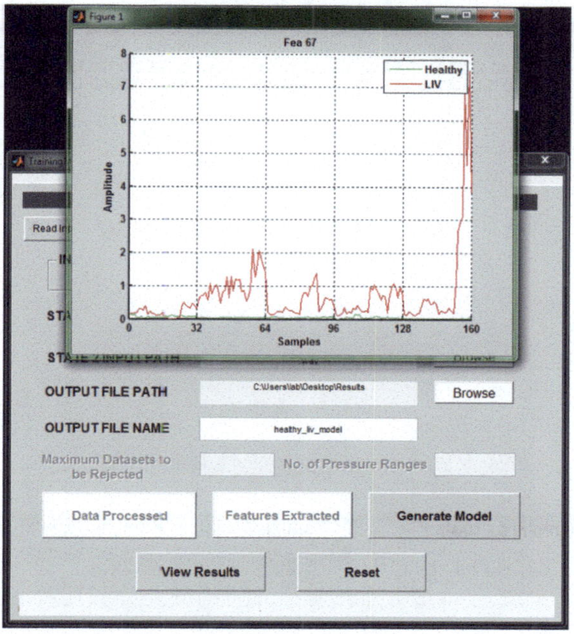

Fig. 7.29 Interface displaying plots of best selected feature [2]

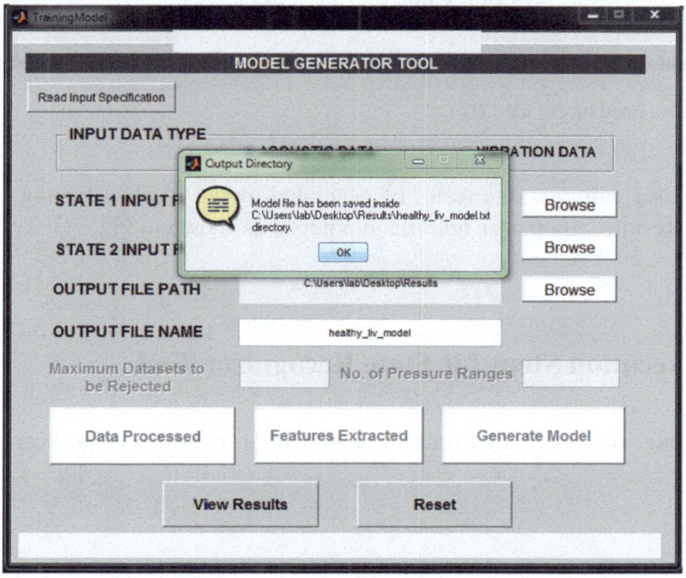

Fig. 7.30 Dialogue box for model file generation [2]

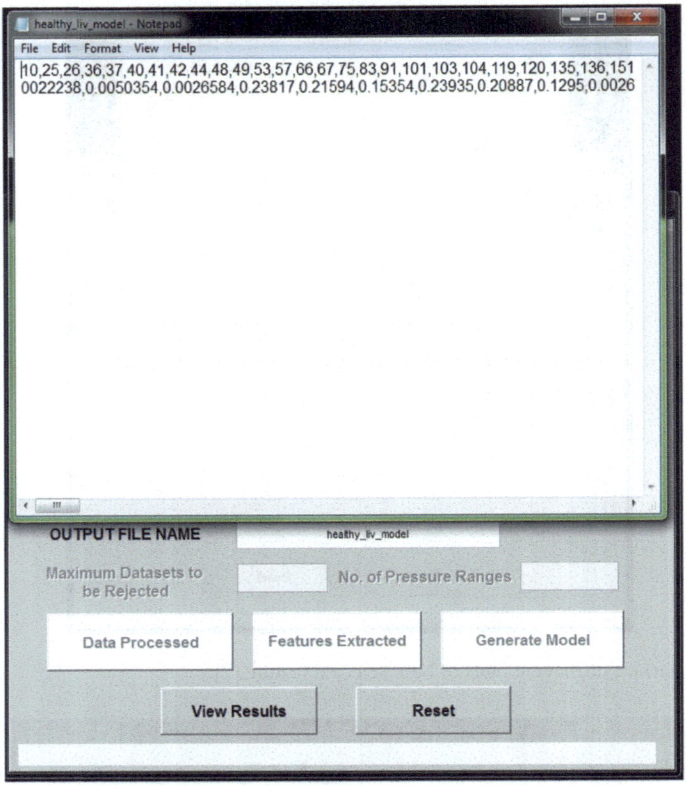

Fig. 7.31 Generated model file [2]

Step 12. User can view the result of generated model file by clicking on "View Results". It opens a text file in notepad as shown in Fig. 7.31.

7.12 Execution Steps for State Recognizer Tool

The main page of state recognizer tool along with its description has been shown in Fig. 7.32.

7.12.1 Execution of State Recognizer Tool on Desktop PC

Step 1 As user clicks on "Classification.exe", the following window appears as shown in Fig. 7.33.

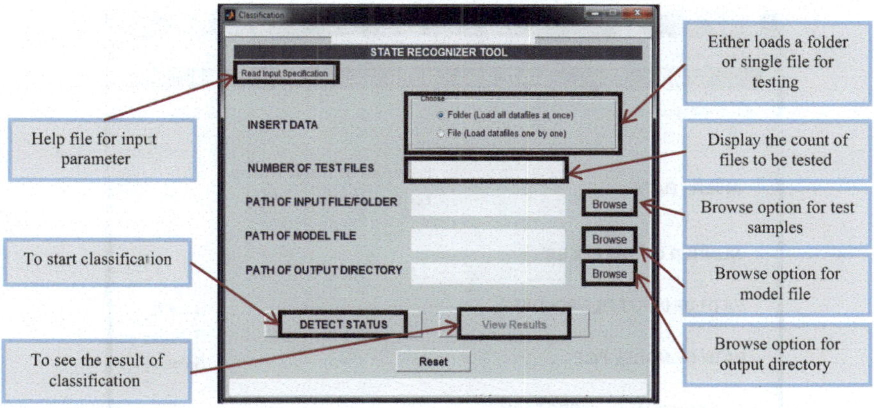

Fig. 7.32 Specification details for classification GUI [2]

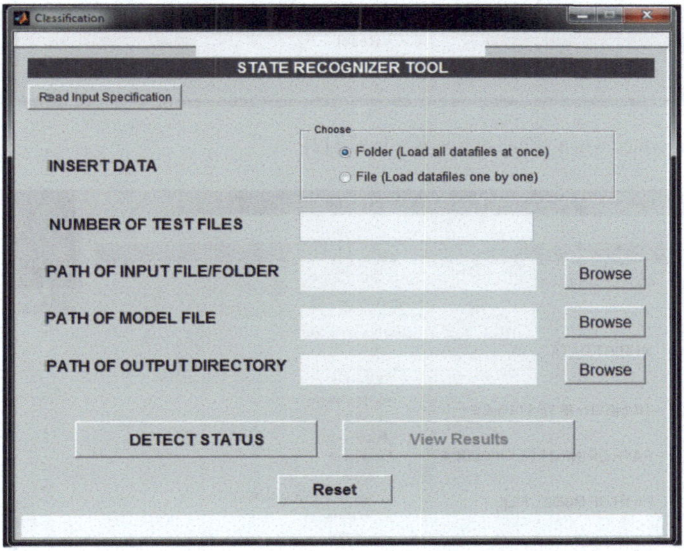

Fig. 7.33 State recognizer GUI [2]

Step 2. The test data can be chosen in two ways either a file or a folder. As per the user choice, one option gets selected. If user chooses "File" option, then the number of test files to be loaded must be entered in text box adjacent to "Number of Test Files" label as shown in Fig. 7.34. For "Folder" option, it should remain blank.

Step 3. Here, on selecting the folder option in the next step, browse the folder in the next step. It opens the current directory as shown in Fig. 7.35 from where user can select a test sample folder.

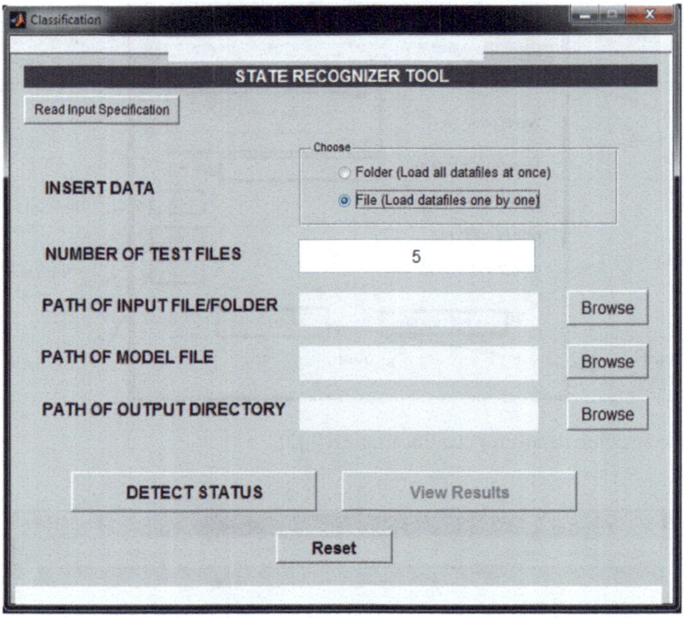

Fig. 7.34 Interface when "File" option is selected [2]

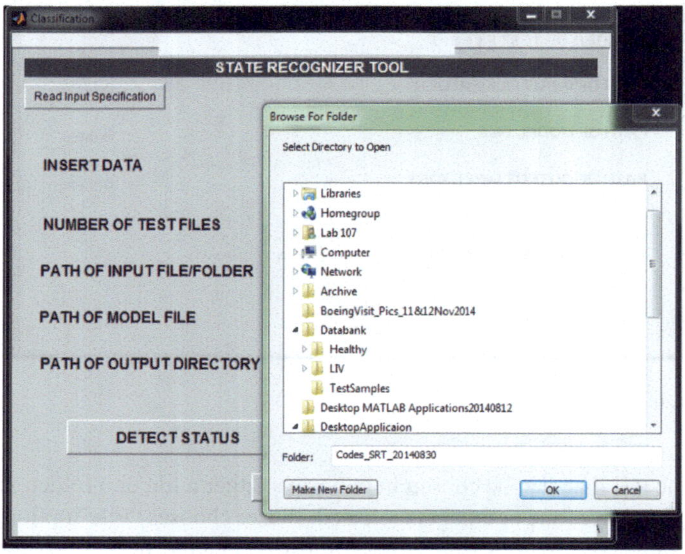

Fig. 7.35 Interface displaying the directories [2]

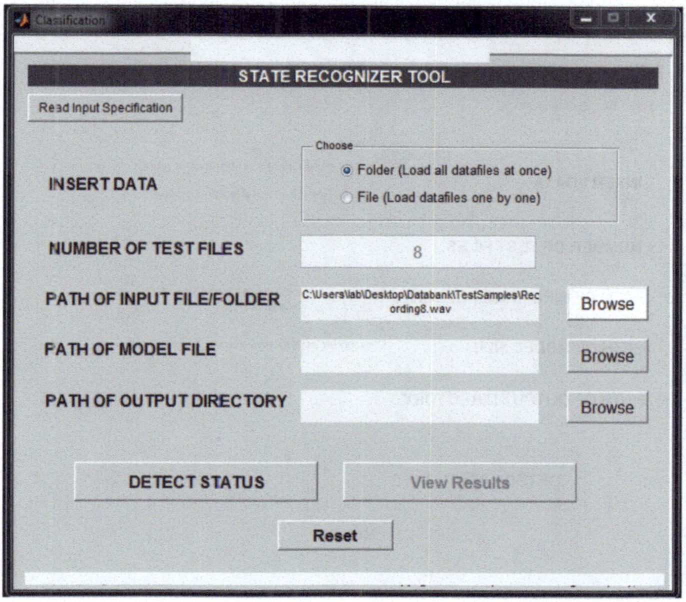

Fig. 7.36 Interface displaying the name of selected test folder [2]

Step 4. Once a folder is selected, its name appears in the adjacent text box "Test Samples" as shown in Fig. 7.36. In "Number of Test Files" text box, the number of files inside the folders will appear. For this case, it is "8".

Step 5. Now, browse the model file. The selected file name will appear in the text box as shown in Fig. 7.37.

Step 6. In the next step, user can browse the location of output directory where results of analysis will get stored in a text file "StateAnalysis_Result". This is shown in Fig. 7.38.

Step 7. Now, after all the input parameters have been selected, click on "Detect Status" button. As soon as this button gets clicked, a progress bar appears as shown in Fig. 7.39.

Step 8. Once all the recording gets processed, progress bar closes, and by clicking on "view result" button, user can open the result file as shown in Fig. 7.40. In the result file, the name of the test file appears then the result of classification. For example, here, Recording1 belongs to State 1 while Recording2 belongs to State 2. The results also get stored in the system for later use.

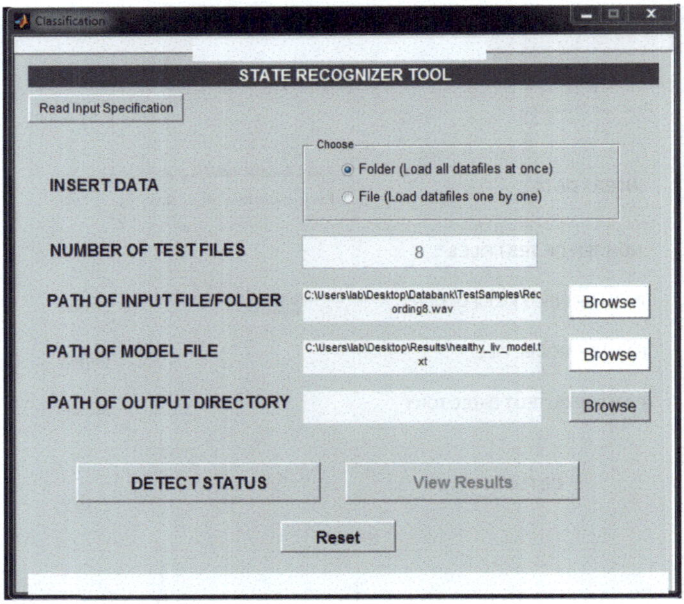

Fig. 7.37 Interface displaying the name of selected model file [2]

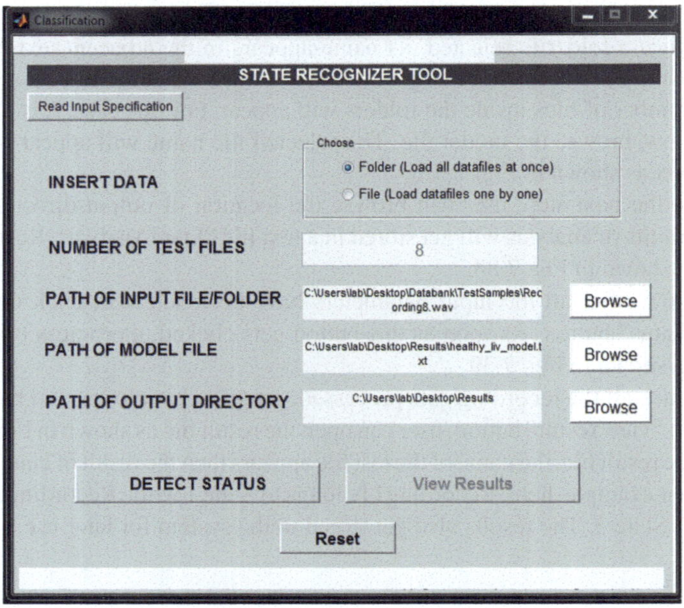

Fig. 7.38 Interface displaying name of output folder [2]

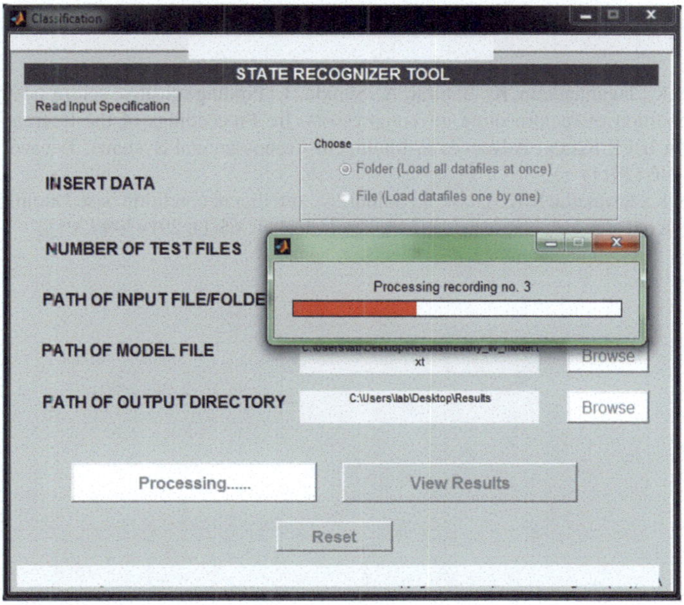

Fig. 7.39 Progress bar showing that recording no. 3 is getting processed [2]

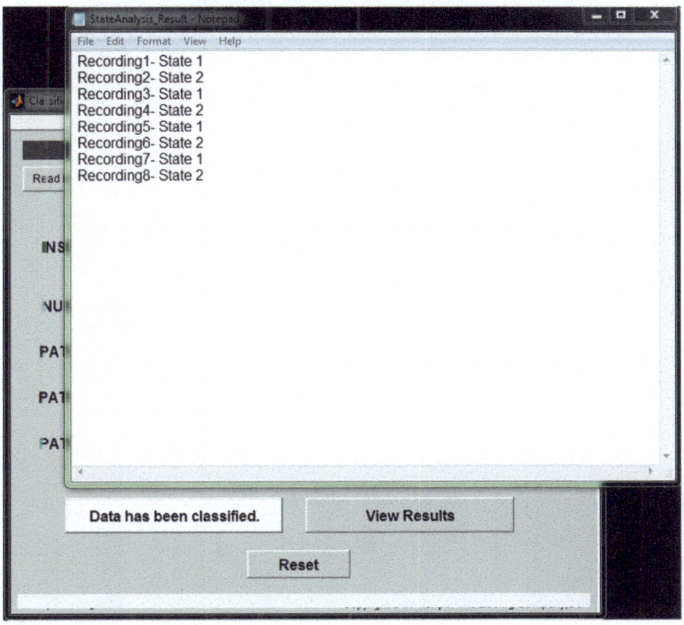

Fig. 7.40 Classification results of all recordings [2]

References

1. Verma, N.K., Jagannatham, K., Bahirat, A., Shukla, T.: Finding sensitive sensor positions under faulty condition of reciprocating air compressors. In: Proceedings of the International Conference on IEEE Recent Advances in Intelligent Computational Systems, Trivandrum, India, pp. 242–246 (2011)
2. Verma, N.K., Sevakula, R.K., Dixit, S., Salour, A.: Intelligent condition based monitoring using acoustic signals for air compressors. IEEE Trans. Reliab. **65**(1), 291–309 (2016)

Chapter 8
Improved Fault Detection Model

Abstract This chapter details about the improvement in different modules of the fault diagnosis framework. During data acquisition, the data is acquired in cyclic order instead of capturing in one shot. For feature extraction purpose, a new set of 343 features were introduced for detailed analysis. The next step is about to improve the feature selection where we proposed a novel feature selection method based on the graphical analysis. This method is based on how nicely a feature can differentiate between two states based on their feature's plots for multiple recordings. In the classification phase, each feature votes for a class and the final class is decided based on the majority.

8.1 Introduction

Visual inspection revealed that the fault identification is not consistent over time due to changes in compressor's acoustics (Even for the same state, same model may not work, i.e., healthy state may be recognized once but the same model may not work perfectly over a period). After performing a lot of experiments on air compressor, we observed that the air compressor, over a long period of time, gradually changes its acoustic characteristics, even while lying in the same state. Hence, we need a more Generalized Model for Fault Detection (GMFD). Earlier, while using SVM classifier model, we trained our model with one large dataset in which total 900 data samples were taken and for each machine condition, 300 recordings were taken as shown in Fig. 8.1. All of this was done in one cycle.

Cycle refers to the change in compressor's states from one state to another, i.e., from Healthy state to LIV state and then to LOV state. Formerly, it was assumed that signature of machine does not change with time. Later, we realized that this is not the case. To verify the above fact, we took acoustic and vibration recordings over a long period of time while changing the compressor states randomly and then returning to the original state. The features of these recordings were then studied as shown in Fig. 8.2.

N. K. Verma and A. Salour, *Intelligent Condition Based Monitoring*, Studies in Systems, Decision and Control 256, https://doi.org/10.1007/978-981-15-0512-6_8

Fig. 8.1 Data acquisition cycle [1]

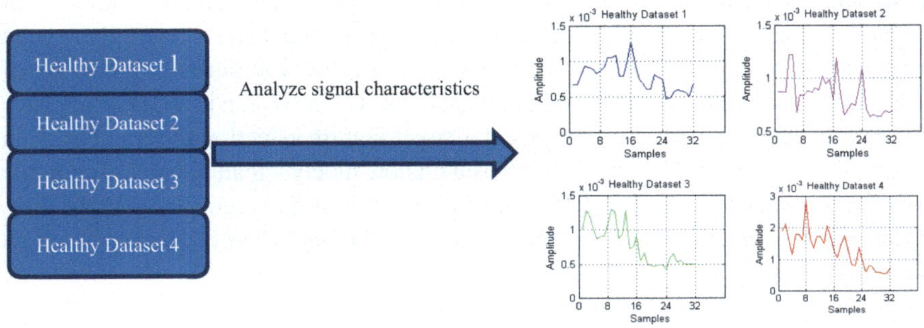

Fig. 8.2 Plots of healthy dataset [2]

Each acoustic recording dataset consists of 32 readings (4 recordings per 1 kg/cm^2 range). Similarly, each vibration recording dataset consists of 40 recordings (5 recordings per 1 kg/cm^2 range). In Fig. 8.3, plots of autocorrelation feature for four different healthy datasets are shown. As we can see, it is difficult to find any consistent trend in the plots. Similarly, Figs. 8.4 and 8.5 show autocorrelation feature for four different LOV and LIV datasets, respectively.

8.2 Improved Fault Detection Model (IFDM) Methodology

The process flow of generalized classifier model is shown in Fig. 8.6.

In order to develop a robust and reliable model, following things need to be considered

- Identify those signal characteristics which are having less variation with respect to time, i.e., consistent over time.
- Fault detection process should be able to handle external disturbances/environmental uncertainties.

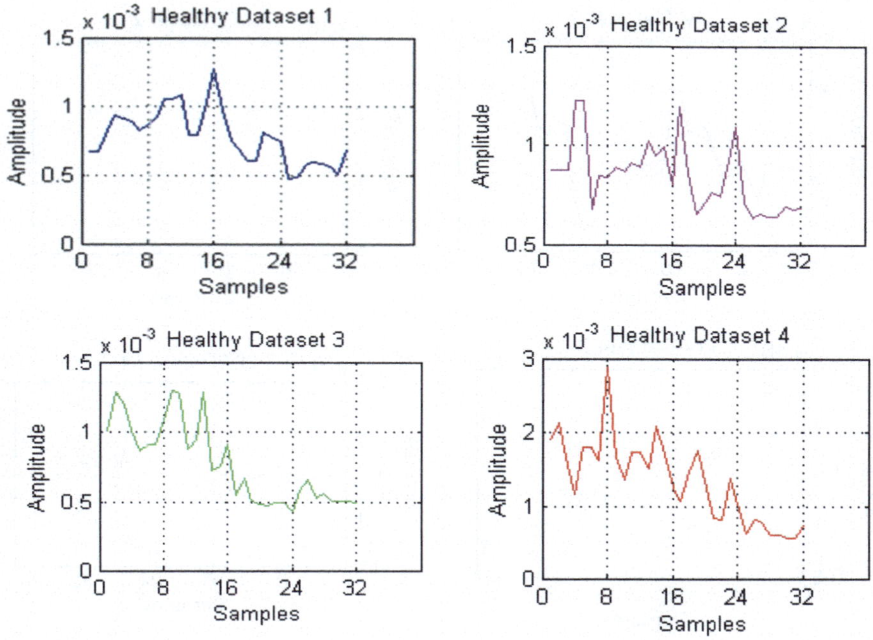

Fig. 8.3 Different healthy dataset plots [2]

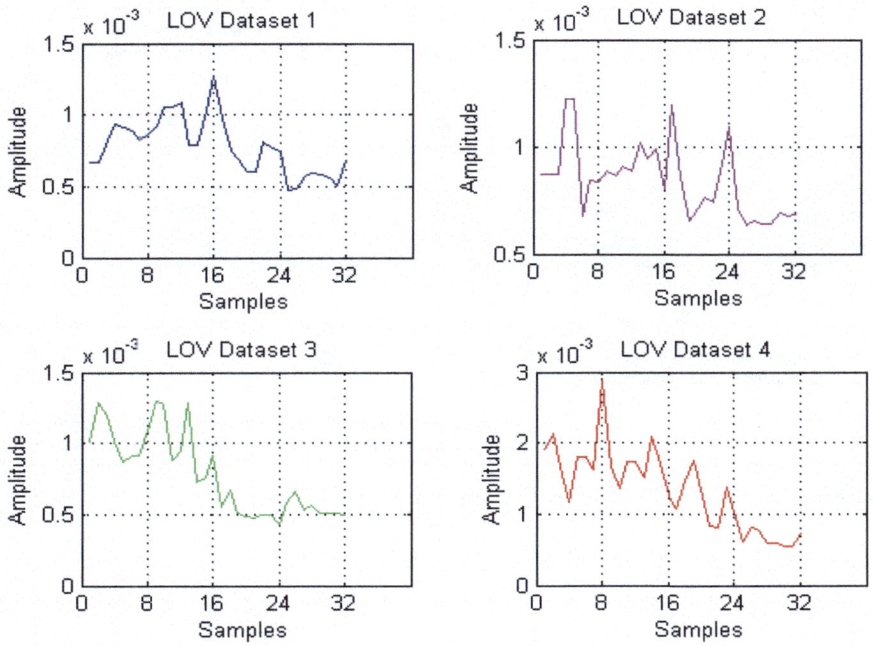

Fig. 8.4 Different LOV dataset plots [2]

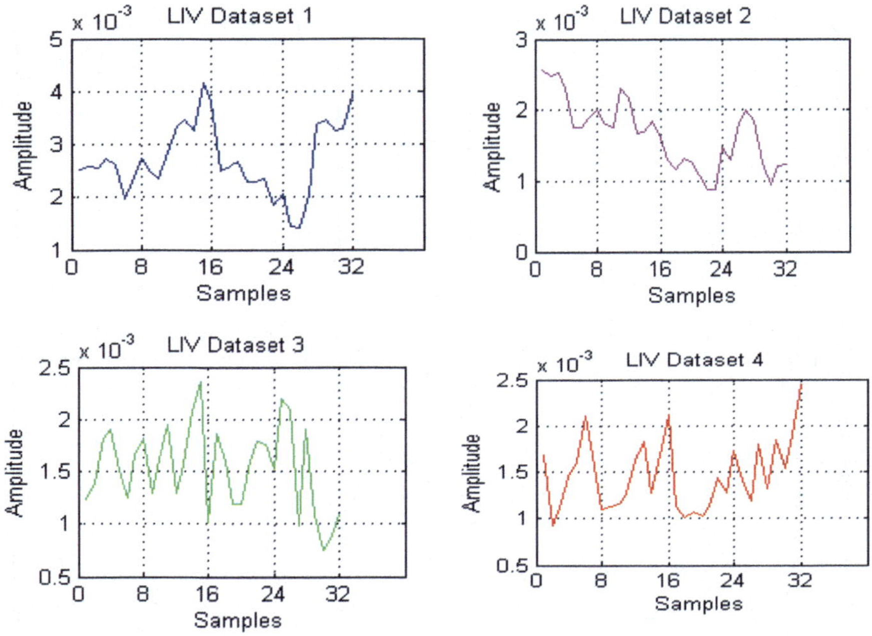

Fig. 8.5 Different LIV dataset plots [2]

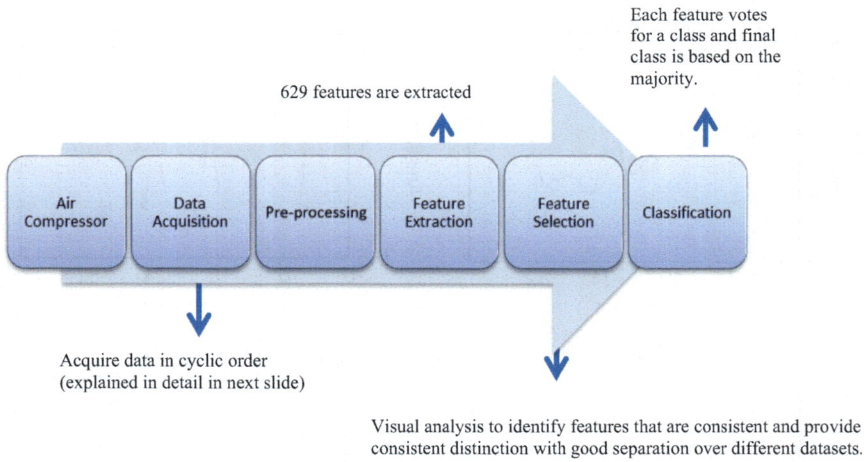

Fig. 8.6 Flow chart for IFDM. Image Courtesy: IDEA LAB, IIT Kanpur

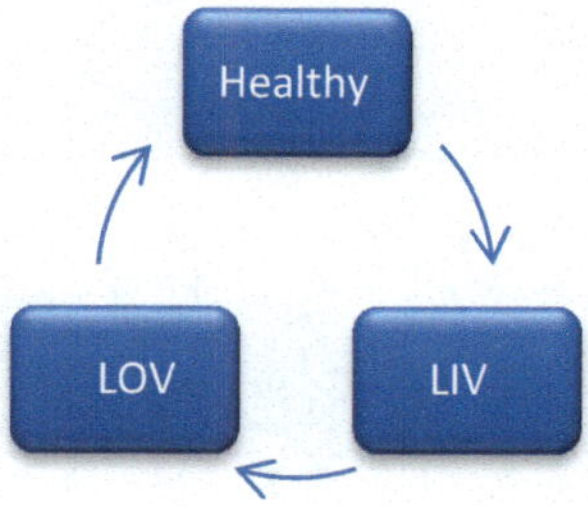

Fig. 8.7 Data acquisition in cyclic order [1]

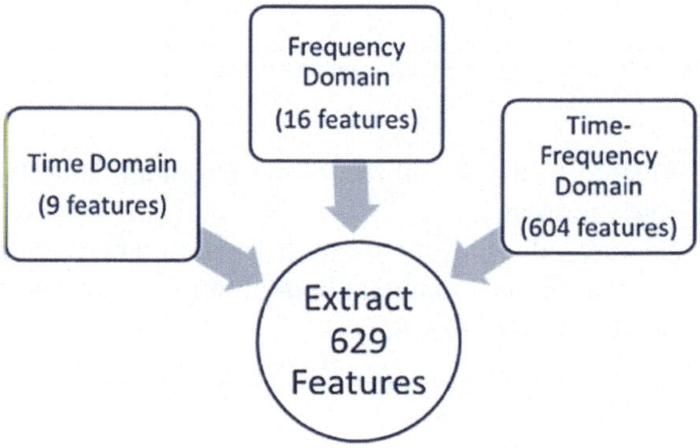

Fig. 8.8 Feature extraction in different domains [4]

8.3 IFDM: Data Acquisition

We took eight different training datasets, i.e., a total of 320 vibration recordings and 256 acoustic recordings for each state, namely healthy, LIV, and LOV over a long period of time. The main idea was to capture trends in data which do not change with time (Fig. 8.7).

8.4 IFDM: Feature Extraction

Including the newly introduced features, we will extract 629 features [3] from the pre-processed data (Fig. 8.8).

8.5 IFDM: Feature Selection

Out of total 629 features, only a subset of relevant features was taken for further analysis. A feature will be selected only if it satisfies the following:

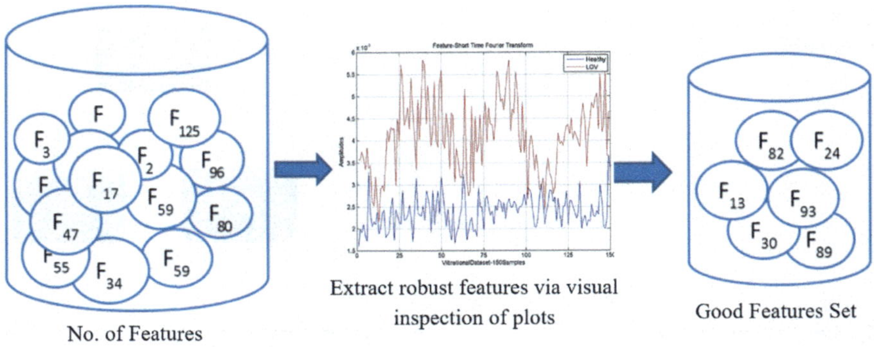

Fig. 8.9 Selecting best features using visual analysis [2]

- Feature values for all recordings (from all datasets) of the same class must have low variation.
- Feature values must have high separation between two separate classes.
- A feature must maintain both aforementioned properties consistently for a long period of time (Fig. 8.9).

8.5.1 Feature Selection Based on Separation

As shown in Fig. 8.10, Kurtosis feature is taken and plotted. Circle represents the separation between two different states of machine. It is observed that Kurtosis feature can discriminate four datasets out of eight but this separation was not consistent over a period.

8.5.2 Feature Selection Based on Consistency

Here, Wigner–Ville distribution samples are taken and plotted in Fig. 8.11. It can be seen from the figure that feature values are consistent throughout the model with low variation within class but no variation between classes, though the feature maintains consistency but does not provide separation. Hence, consistency without separation will not result in a good feature.

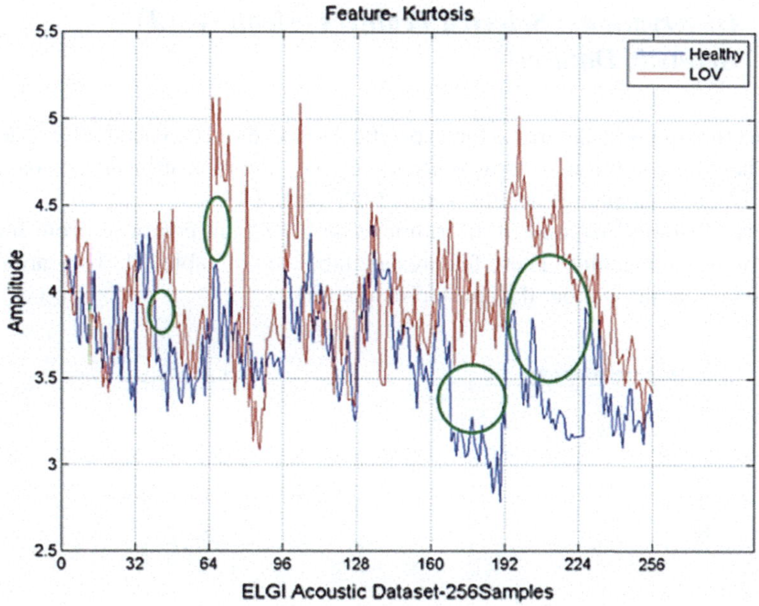

Fig. 8.10 Selecting best features based on separation [2]

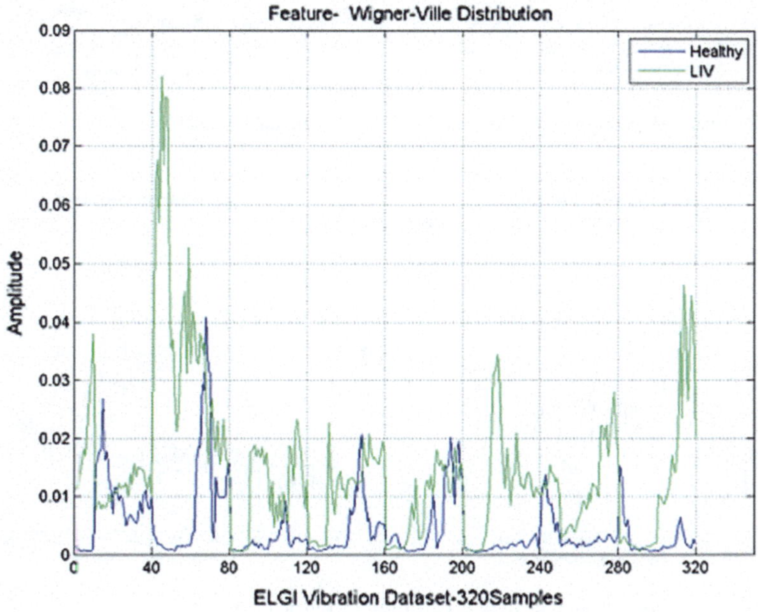

Fig. 8.11 Selecting best features based on consistency [2]

8.5.3 *Observations: Selected Features (Healthy–LIV Acoustic Dataset)*

The selection of good features is fault specific. Hence, plots for each faulty condition versus healthy condition are drawn separately, i.e., the plot of healthy versus LIV (Fig. 8.12), and the plot of healthy versus LOV (Fig. 8.13).

Twelve features are used to discriminate the LIV fault condition from healthy condition of the machine. These features are tabulated in Table 8.1. They are found to be consistent throughout the entire pressure range.

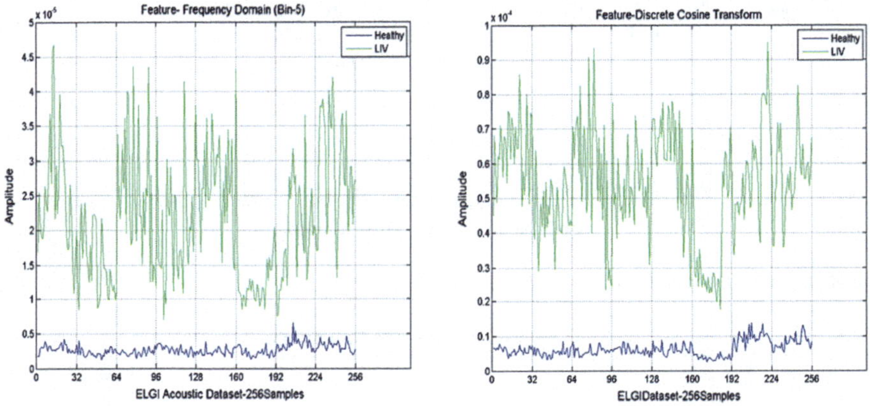

Fig. 8.12 Visual inspection of healthy versus LIV (acoustic dataset) [2]

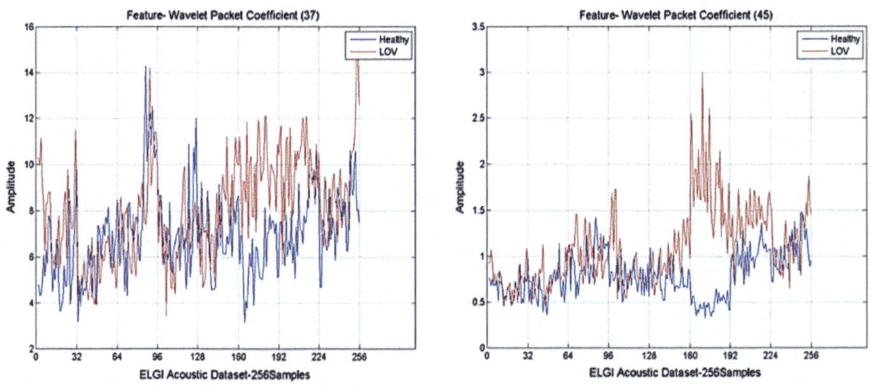

Fig. 8.13 Visual inspection of healthy versus LOV (acoustic dataset) [2]

Table 8.1 Features from healthy–LIV acoustic dataset

Acoustic dataset (healthy–LIV)	
Frequency domain feature	F_{13}
Discrete wavelet transform	F_{24}, F_{25}
Wavelet packet transform	F_{34}, F_{36}, F_{38}, F_{41}, F_{45}, F_{51}, F_{87}, F_{89}, F_{112}

Table 8.2 Features from healthy–LOV acoustic dataset

Acoustic dataset (healthy-LOV)	
Wavelet packet transform	F_{70}, F_{78}, F_{96}, F_{109}, F_{133}, F_{161}, F_{190}, F_{220}, F_{236}, F_{252}, F_{254}

8.5.4 Observations: Selected Features (Healthy–LOV Acoustic Dataset)

It was found that some features from Wavelet Packet Transform (WPT) were able to distinguish the healthy condition from LOV faulty condition of the machine. The selected features based on visual inspection for healthy versus LOV fault condition based on acoustic data are tabulated in Table 8.2.

8.5.5 Observations: Selected Features (Vibration Dataset)

It was found that feature values of the vibration data were generally consistent with respect to the pressure range (in fact only at certain ranges), i.e., recordings taken for certain pressure range can be expected to be consistent and otherwise not. This is because feature values change as the air compressor's pressure increases. In fact, it is noticed that consistency and separation for different states are best only for certain ranges. Therefore, while finding best features, the vibration dataset is analysed not only at feature level but also at pressure level. To perform such pressure-wise feature level analysis, recording was taken as shown in Table 8.3.

Table 8.3 Different pressures for vibration dataset

Pressure range (kg/cm^2)	No. of recordings
2–4	10
4–6	10
6–8	10
8–10	10

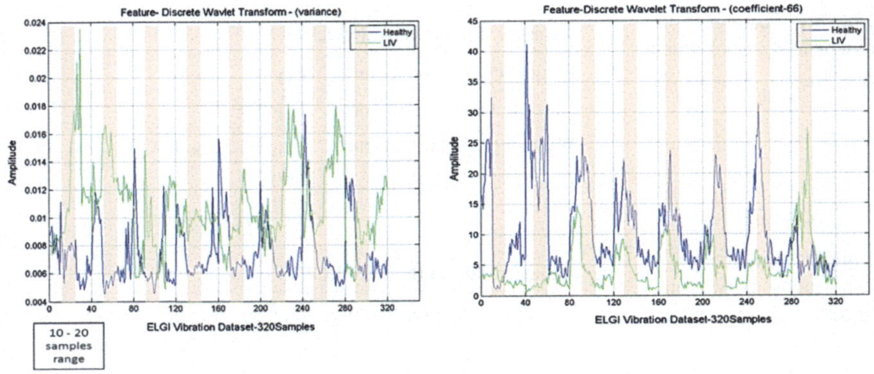

Fig. 8.14 Visual inspection of healthy versus LIV (vibration dataset) [2]

Table 8.4 Features from healthy–LIV vibration dataset

Vibration dataset (healthy–LIV)	
Discrete wavelet transform	F_{29}
Wavelet packet transform	F_{65}, F_{69}, F_{99}, F_{107}, F_{167}, F_{168}, F_{183}, F_{184}, F_{288}, F_{336}

8.5.5.1 Observations: Selected Features (Healthy–LIV Vibration Dataset)

Figure 8.14 shows some of the features obtained from WPT which can be considered as good features. By visual inspection, we found that 11 features were able to discriminate the LIV fault condition from healthy condition. The selected features are tabulated in Table 8.4.

8.5.5.2 Observations: Selected Features (Healthy–LOV Vibration Dataset)

Figure 8.15 shows the plots for some of the features obtained from WPT can be considered as good features. By visual inspection, we found 12 features which were able to discriminate the LOV fault condition from healthy condition of the machine. The selected features are tabulated in Table 8.5.

8.6 IFDM: Classification—Training Phase

In the training phase, good features from all sample recordings are collected and their mean and standard deviation (s.d.) are calculated for each class as shown in

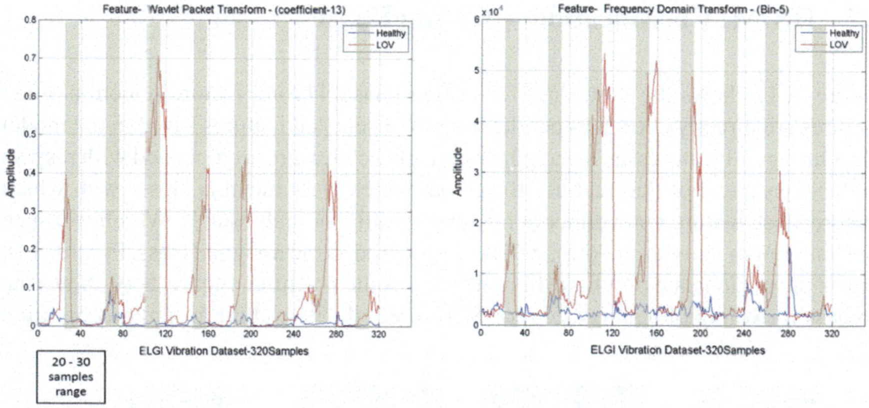

Fig. 8.15 Visual inspection of healthy versus LOV (vibration dataset) [2]

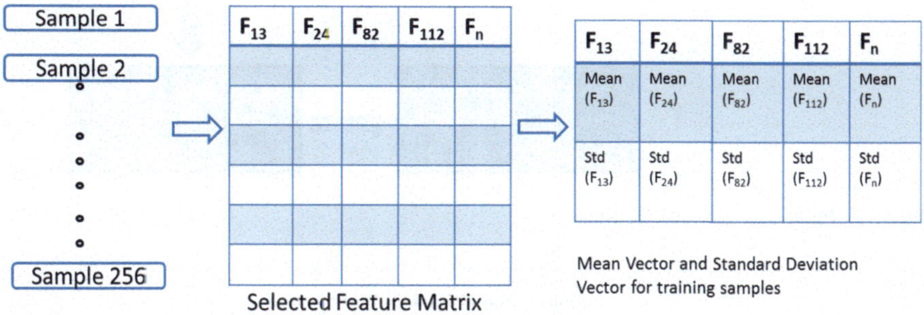

Fig. 8.16 Feature selection process [2]

Fig. 8.16. The mean and standard deviation of selected features for a given class are the characteristics of that class. The indices of good features, their mean, and s.d. of each class form the model files for classification.

Table 8.5 Features from healthy–LOV vibration dataset

Vibration dataset (healthy–LOV)	
Frequency domain features	F_{11}, F_{12}
Wavelet packet transform	F_{30}, F_{31}, F_{36}, F_{38}, F_{42}, F_{44}, F_{45}, F_{46}, F_{51}, F_{53}, F_{57}, F_{59}, F_{60}, F_{67}, F_{75}, F_{83}, F_{87}, F_{88}, F_{90}, F_{103}, F_{119}, F_{120}, F_{135}, F_{136}, F_{151}, F_{157}, F_{207}, F_{208}, F_{239}, F_{240}, F_{271}, F_{274}

8.7 IFDM: Classification—Testing Phase

Figure 8.17 shows the flow chart for testing phase. During feature selection, good features are directly chosen as per the indices of selected features saved in the model file. Once good features are selected, z-score of each feature is calculated with respect to the class mean of that feature. Based on, which class mean is closer with which feature; each feature votes either in favour of healthy or faulty states. Whichever state gets maximum number of votes, will be considered as the assigned class. Figure 8.18 shows the mean value plot for different states of the machine. Figure 8.19 shows the process for voting of features. Figure 8.20 shows the flow chart for fault recognition

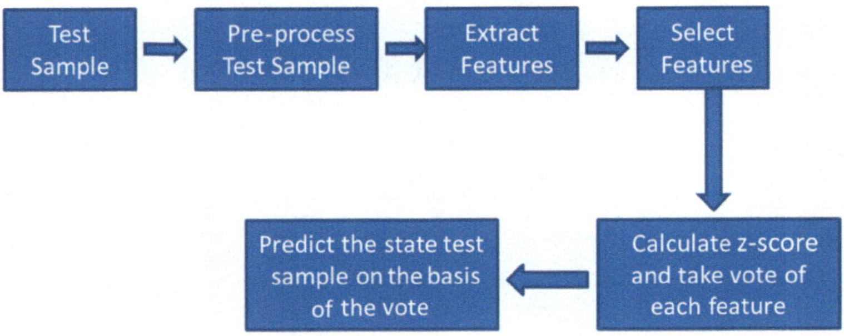

Fig. 8.17 Flow chart for testing phase

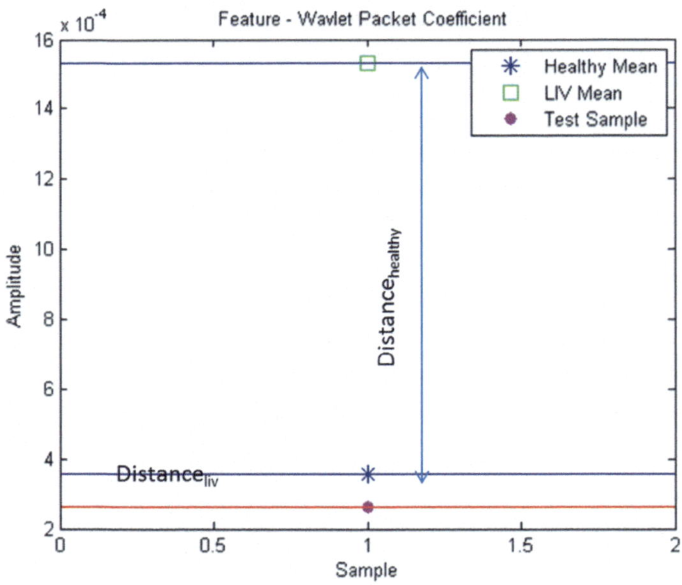

Fig. 8.18 Mean value plot for different states of the machine [2]

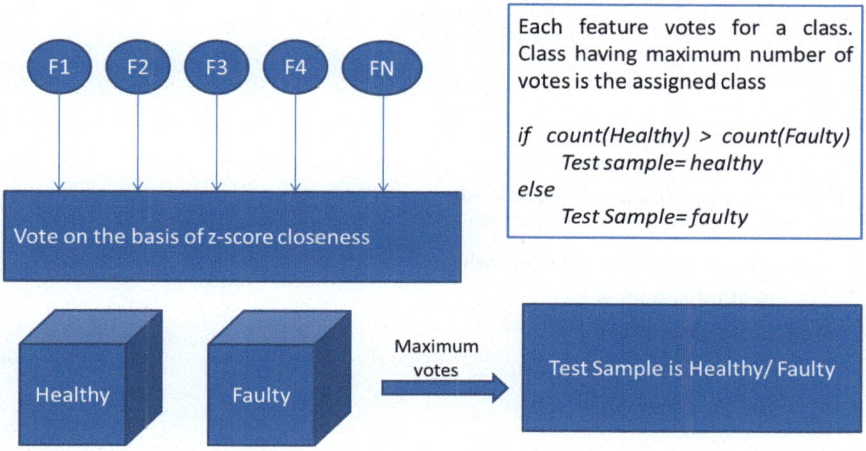

Fig. 8.19 Voting of features [5]

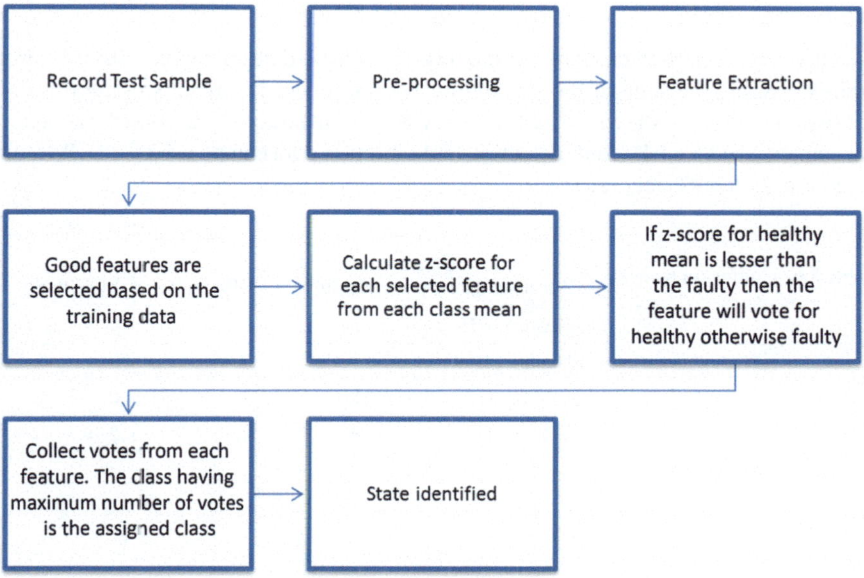

Fig. 8.20 Flow chart for fault recognition process in IFDM [2]

process in IFDM. The z-score for each feature is evaluated as

$$z \text{ score} = \frac{x - \mu}{\sigma}$$

where x is the raw data, μ is the mean, and σ is the standard deviation.

IFDM Testing Phase—Pseudo-code:

```
If
        Distancehealthy > Distanceliv
        healthy_ count ++
elseif Distancehealthy <
        Distanceliv
        faulty_ count ++
```

8.8 Real-Time Results and Conclusions—IFDM

We checked the performance of our models on acoustic dataset and vibration dataset while changing the state of the air compressor ten times. The results in the form of confusion matrix are shown in Tables 8.6 and 8.7. It can be concluded that the acoustic data can be used for LIV fault detection and vibration data can be used for LOV fault detection with 90% accuracy.

Table 8.6 Confusion matrix for acoustic dataset

Acoustic dataset	Healthy	LIV fault	LOV fault
Healthy–LIV	9	1	0
Healthy–LOV	10	0	0
LIV fault	1	9	0
LOV fault	8	0	2

Table 8.7 Confusion matrix for vibration dataset

Vibration dataset	Healthy	LIV fault	LOV fault
Healthy–LIV	9	1	0
Healthy–LOV	10	0	0
LIV fault	7	3	0
LOV fault	1	0	9

References

1. Verma, N.K., Sevakula, R.K., Dixit, S., Salour, A.: Intelligent condition based monitoring using acoustic signals for air compressors. IEEE Trans. Reliab. **65**(1), 291–309 (2016)
2. Verma, N.K., Sevakula, R.K., Thirukovalluru, R.: Pattern analysis framework with graphical indices for condition-based monitoring. IEEE Trans. Reliab. **66**(4), 1085–1100 (2017)
3. Verma, N.K., Sevakula, R.K., Goel, S.: Study of transforms for their comparison. In: International Conference on Industrial and Information Systems, Gwalior, India, pp. 1–6 (2014)
4. Verma, N.K., Agrawal, A.K., Sevakula, R.K., Prakash, D., Salour, A.: Improved signal preprocessing techniques for machine fault diagnosis. In: 2013 IEEE 8th International Conference on Industrial and Information Systems, pp. 403–408 (2013)
5. Sevakula, R.K., Verma, N.K.: Assessing generalization ability of majority vote point classifiers. IEEE Trans. Neural Netw. Learn. Syst. **28**(2), 2985–2997 (2017)

Chapter 9
Fault Diagnosis System Using Smartphone

Abstract This chapter presents an Android application of a general-purpose fault diagnosis system developed earlier for desktop and Palmtop. The basic methodology of fault recognition using Android is similar to the methods discussed in Chap. 8. This chapter first describes the data mining model used for the fault diagnosis purpose and then presents summarized theories of all the modules covered in the previous chapter. A detailed description about the development and usage of smartphone with Android application for fault diagnosis is provided in this chapter.

9.1 Introduction

With the increasing pervasiveness and processing power, smartphones have proven to be a very useful tool. In the context of monitoring, a lot of work can be shifted to a single device which, otherwise, requires a separate computer and data acquisition tools. This certainly saves skilled manpower required for processing and collection of data, such that even an unskilled worker can use a smartphone and get the results. This motivates to build a smartphone-based application [1] which can facilitate all the operations, i.e., from data acquisition to training of model. Traditionally, the machines are checked manually for faults, by the mechanics having expertise in the field. This way of diagnosing machine's health condition is fair but always has some drawbacks. Some of these drawbacks are stated as:

- The development of manual fault diagnosis process is a time-consuming process.
- As the operator's opinion can be biased, the decision-making process is highly subjective. This means the opinions about the machine's condition vary from one operator to the other based on their perception levels.
- The ability of making a decision by an individual depends on the mental condition and fatigue.

The above mentioned drawbacks encouraged us to pursue the "Intelligent Condition Based Monitoring" of machines [2, 3] which will automate the entire procedure and will be accepted by all such that one can not only overcome these drawbacks but will also perform fault recognition efficiently.

N. K. Verma and A. Salour, *Intelligent Condition Based Monitoring*, Studies in Systems, Decision and Control 256, https://doi.org/10.1007/978-981-15-0512-6_9

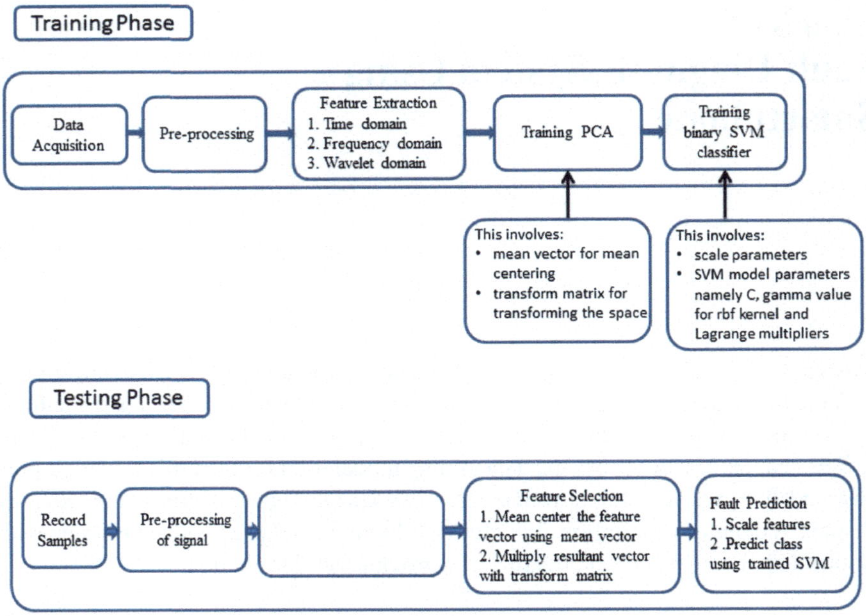

Fig. 9.1 Data mining model for training and testing phases [1]

The primary idea is to develop an easily available and accessible platform which allows the identification of machine health condition in real-time. Further, this chapter details the methodology and smartphone-based application used for development of fault diagnosis [4].

9.2 Data Mining Model for Fault Recognition

The flow diagram of intelligent fault diagnosis system using data mining model [5] for fault recognition is shown in Fig. 9.1.

9.3 Data Acquisition (DAQ)

As mentioned in the fault diagnosis framework, the first step of fault diagnosis is to measure the machine characteristics for analysis of machine condition. It is known as data acquisition. The data acquisition is done by recording acoustic data with the inbuilt microphone of smartphone at 44.1 kHz sampling rate in ".wav" format (Fig. 9.2). Once the recording is done in ".wav" format which is a lossless format of 16 bit 2's compliment coding, the files can be directly read to obtain raw input sampled data [6–8]. The recording is done at sensitive positions only.

Fig. 9.2 Data acquisition with smartphone [1]

Before starting the DAQ process, one must identify significant positions of the machine and around the surface of air compressor from where microphone is able to pick maximum information for healthy and faulty conditions [9] . For air compressor, 24 sensor positions were selected. The positions were analysed based on five attributes, namely maximum peak, rms, crest factor, absolute mean, and standard deviation. The mode of positions having maximum of each of the given attributes is selected as the final sensitive position for given health condition of the machine. Different conditions of machine may give different sensitive positions. Figure 9.3 shows the flowchart for finding sensitive positions.

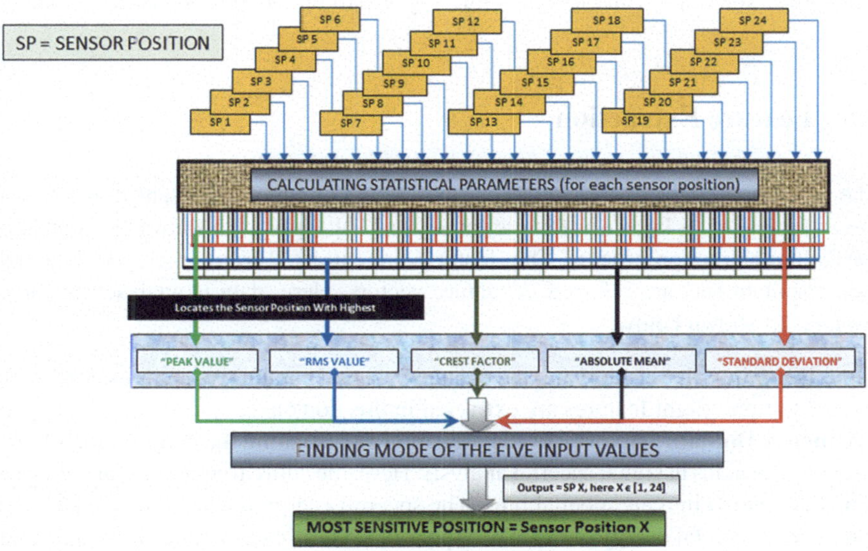

Fig. 9.3 Flowchart for finding sensitive positions [9]

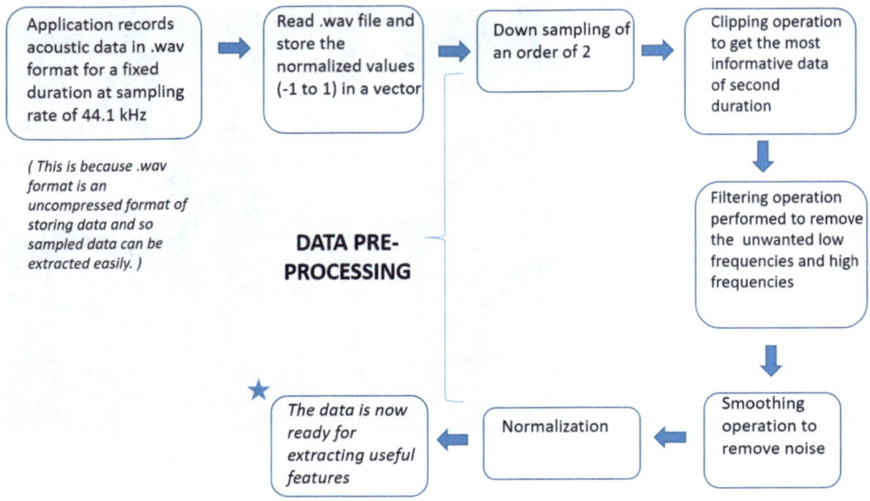

Fig. 9.4 Flowchart for data pre-processing [2]

9.4 Pre-processing

Once the data is collected, the pre-processing is performed. Figure 9.4 shows the flowchart for data pre-processing. It improves the quality of data as explained in Chap. 3. Data pre-processing also transforms one form of data into a format that can be easily and effectively processed. Here, we perform downsampling, clipping, normalization, smoothing, and low pass filtering as part of data pre-processing module.

9.5 Feature Extraction

The process of extracting characteristics of a signal, which represent it at a much lower dimension, is known as feature extraction [10]. Figure 9.5 shows the flowchart for feature extraction process. The features are extracted from processed data and their resultant sets are referred as feature vectors. Here, data is analysed in three domains as shown below:

Time Domain— In this domain, the change in signal parameters are presented with respect to time. Eight features are extracted in this domain.
Frequency Domain—It involves identification and isolating the contribution of different frequencies in signal spectral analysis. Here, the entire frequency band of input signal is divided into eight equal bins. The spectral energy in each bin is calculated and divided by total energy of the entire spectrum. These energy ratios form the eight features of frequency domain.

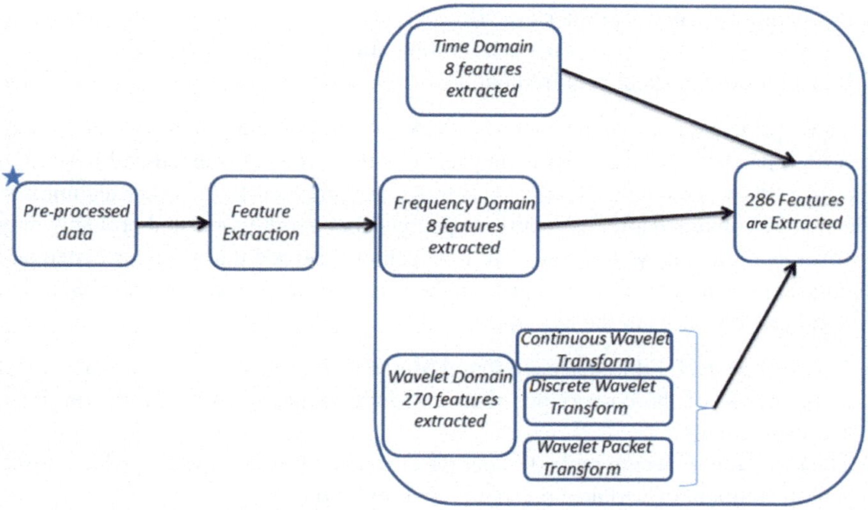

Fig. 9.5 Flowchart for feature extraction [2]

Wavelet Domain—It involves analysis of signal's frequency characteristics at different times which is possible if the signal is represented in time–frequency spectrum. We have used three types of transforms for analysis of data in wavelet domain. These transforms are:

- **Continuous Wavelet Transform**—Firstly, the input signal is convolved with Morlet wavelet. This provides Morlet wavelet coefficients. Several statistical parameters are calculated using these coefficients.
- **Discrete Wavelet Transform**—The signal is passed through a series of high and low pass filters which provide detail and approximation coefficients, respectively. Again, the decomposition of approximation coefficients is done using both the filters, and it is repeated till *n*th level. These coefficients are further used for calculation of features.
- **Wavelet Packet Transform (WPT)**—In WPT, decomposition is applied to both—approximation and detail coefficients.

9.6 Feature Selection

Feature selection is the technique of selecting a subset of relevant features for building robust learning models. By removing irrelevant and redundant features from the data, there is improvement in the performance of learning models. This is because of the following:

- Removing the curse of dimensionality
- Speeding up the learning process as computational complexity reduces
- Reduced storage space requirement

Here, principal component analysis (PCA), a very popular dimensionality reduction tool, has been used for extracting features that are uncorrelated and have high variance. PCA is simple, effective, and faster as compared to the other techniques. The features obtained after extraction stage may not be orthogonal and probably may have some redundancy. The objective is to achieve a space that is uncorrelated and maintains the variance. For this purpose, the following steps have been performed to obtain feature vectors in the new space .

- Basis functions for this space are found from the eigenvectors (known as principal components) of the co-variance matrix of input dataset, which are uncorrelated and orthogonal.
- The basic functions are ranked as per the corresponding eigenvalues which represent the amount of variance present in that direction.
- For feature selection, top-ranked principal components have been chosen.

Through PCA, we have selected 23 features as we found from the previous experiments that it provides good results for the same.

9.7 Classification

Classification is the problem of learning characteristics of different classes/categories of data and then predicting to which category the test sample belongs to. Fault diagnosis is also a classification problem where fault classes need to be identified by analysing collected features. In this research, classification has been done using support vector machine (SVM). SVM is a binary classifier. It tries to find a linear separating hyperplane that separates the two classes with maximal margin. Margin is defined as the minimum distance between the hyperplane and the points of both classes. For the classes which are nonlinearly separable, a linear hyperplane is found in a higher dimension without mapping the data to a higher dimension. This is done with the help of kernels. The projection of hyperplane in current feature space is mostly nonlinear in nature.

For experimentation, we perform check on three states of air compressor for which data has been collected: Healthy condition, LIV faulty condition, and LOV faulty condition. We have used RBF as kernel function for SVM. As we are interested in checking whether a specific fault exists or not, we use SVM binary classifier to check if that fault exists in the compressor or not. So, in a generalized case, if we have N possible faults in an air compressor, then we need N SVM models, the ith SVM model is trained with all the training data of ith fault labelled as positive and data belonging to all other faults along with healthy class is labelled as negative. Radial basis function has been used as kernel function in this model.

9.8 Android Application

An application interface is a vital component of applied research work. It ensures that the framework/model/algorithm developed during research can be used in real-time by any user. To develop the interface, some tools [11–13] and platforms are required [14, 15]. The tools used for our application development are as follows:

Eclipse Classic 3.7.2—It is an open-source software development platform which comprises of an integrated development environment and an extensible plugin system, suitable for developing Java-based applications.
Android SDK—It provides tools and libraries needed for building, testing, and debugging Android applications. It includes a virtual mobile device emulator that runs on PC for simulating the application interface. The tools provided by Android SDK are:

- Android Virtual Device
- Android Debug Bridge
- Dalvik Virtual Machine
- Android Emulator

Android Developer Tools (ADT)—Basically, it consists of a set of components (plug-ins) which extend the Eclipse IDE with Android development capabilities. It provides code editor features, SDK tool integration, and the graphical layout editor.
SQLite Database Browser 2.0 b1— It is an open-source database embedded in Android. The Android applications are written in Java programming language. The Android SDK tools compile the code—along with any data and resource files into an Android package of an archive file with a .apk suffix. All the codes are in a single .apk file in the form of one application. It is the file that Android-powered devices use to install the application.

Each Android application may consist of one or more activities. Activity is basically a container where user interface can be placed. Our application [16] consists of five activity pages. They act with each other to build a complete fault diagnosis system. A user provides inputs to the activity pages as instructed to perform fault diagnosis. Every activity page can be divided into two parts, namely the front end and the back end. A front end is the interface through which user interacts with application directly. It is basically responsible for collection of inputs from user. The back end of any application is where all tasks for setting up the front-end user interface and information processing is performed. Data processing, querying from databases, and performing checks related to the applications like secure digital (SD) card availability are all part of the back end. In the next section, detailed information about each activity page is provided.

9.8.1 *Activity Page 1 (Model Selection Activity)*

The purpose of this activity page is to select the model of air compressor on which diagnosis must be performed. Then, user can enter a unique compressor id so that the current air compressor can be identified. In the front end of this activity page, four components appear to choose air compressor type as shown in Fig. 9.6.

A dialog box where type of air compressor can be chosen from list of compressors mentioned in the selection dialog box. The compressor list is automatically pulled from a database which is stored in the SD card [17]. A dialogue box is a small window which appears (pop-up) in front of the current activity page as shown in Fig. 9.6(b). After it is displayed, user can interact only with this dialogue box and the activity page in the background becomes inactive until a choice is made by the user. User can select the type or model of compressor from the list of compressors provided. A text field for "Compressor Unique ID" is also present through which user is asked to provide the Compressor Unique ID so that the compressor can be identified in the reports to be sent to the maintenance server. This field cannot be left empty. A "*Reset*" button is also present for resetting both the fields, namely air compressor selection field and Compressor Unique ID field. If user wants to revert, both values will be cleared and set to null. A "*Proceed*" button, when clicked by user, directs to the second activity page of this application and all the information collected in this activity page is transferred to the next page.

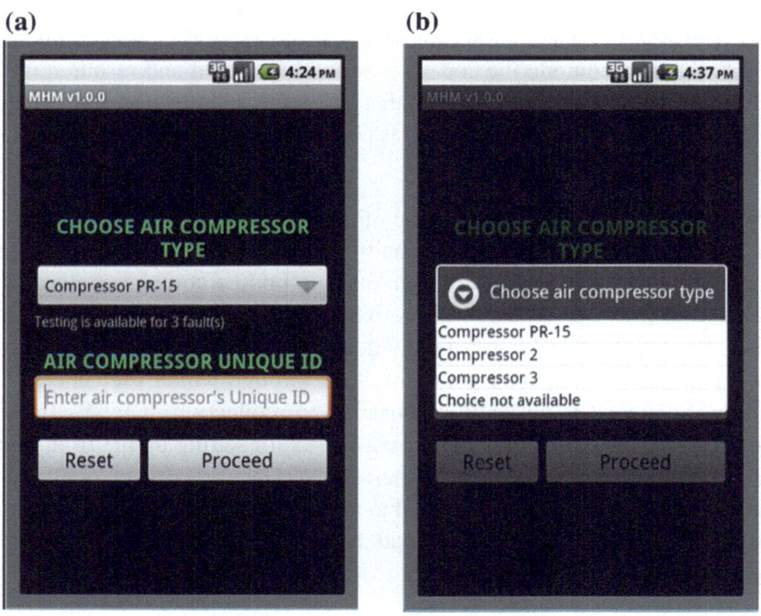

Fig. 9.6 **a** User interface for taking input from user, **b** dialogue box to choose the compressor type [1]

In the back end of this activity page, we have two choices for storing large amount of data including the apps install folder or external storage. Using SD card for storage has certain advantages such as—many devices have limited internal storage for apps, storage capacity of the SD card can be increased as per the need. It also provides freedom to update the resource files for the application very easily. Whenever a new air compressor is introduced in the market and the manufacturer wants fault detection of that air compressor to be included in the application, the details of those compressors can be included simply by adding all the required files in the downloadable file to resource server and the updated resource files can be downloaded. So, we have chosen external storage director of the smartphone to store application resources. Every time when the application runs, it is necessary to check whether these resources are present in the SD card. All resources are saved in a directory named as "CBM_ResourceFiles" inside the SD card. The following requirement checks are performed at the back end of this activity page:

(a) **SD Card Availability**—The application checks if SD card is available. If the SD card is not available, or the application is unable to access it due to any reason like the SD card is being accessed in PC using data cable, the user is notified with the alert dialogue message as shown in Fig. 9.7(a). The application will continue once SD card is mounted in the device. The user can mount the SD card here itself when the alert dialogue is displayed and can continue by

(a) (b)

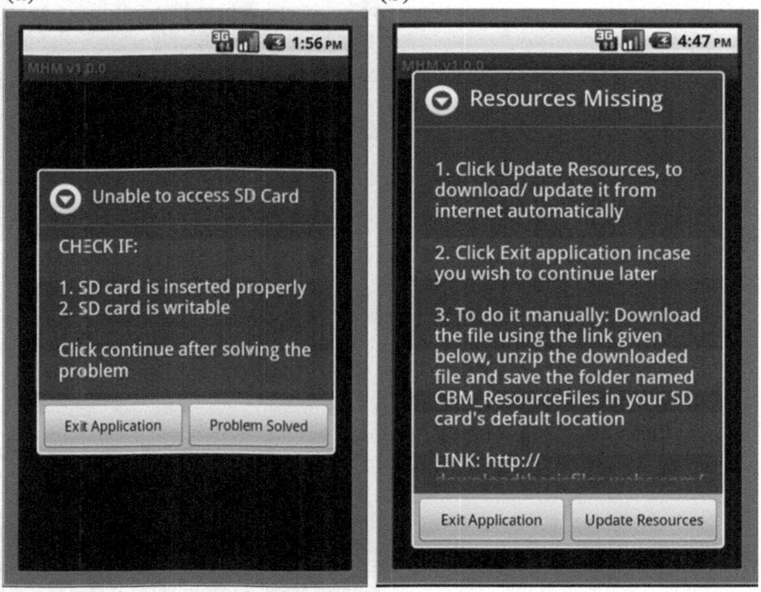

Fig. 9.7 **a** Dialogue box to show unavailability of SD card, **b** dialogue box to show resource folder is missing [1]

clicking the *Problem Solved* button to continue or the user can choose to exit the application.

(b) **Application Resource Folder**—The check for application resource folder is performed after it is confirmed that the SD card is available and accessible. Here, the application checks for the resource folder required by the application in the SD card. In case it is not available, the user is prompted with message as shown in Fig. 9.7(b) to download it from the server or exit application.

(c) **Internet Availability**—When *Update Resources* button is clicked, application checks for the Internet. In case, if the Internet is not available, the application exits by displaying the message "*Internet connection is not available*" else the download of resource file begins while displaying a progress bar, as shown in Fig. 9.8.

(d) **Internet Authentication**—When the user clicks on *Update Resources*, one more possibility may be that no file gets downloaded or the file getting downloaded is corrupted because the available Internet is working without a proxy authentication. In this case, it is required to connect the Internet by entering proxy details without which download/upload processes from the Internet are impossible. Android does not provide any function to detect such a case. The steps to handle this case are given below.

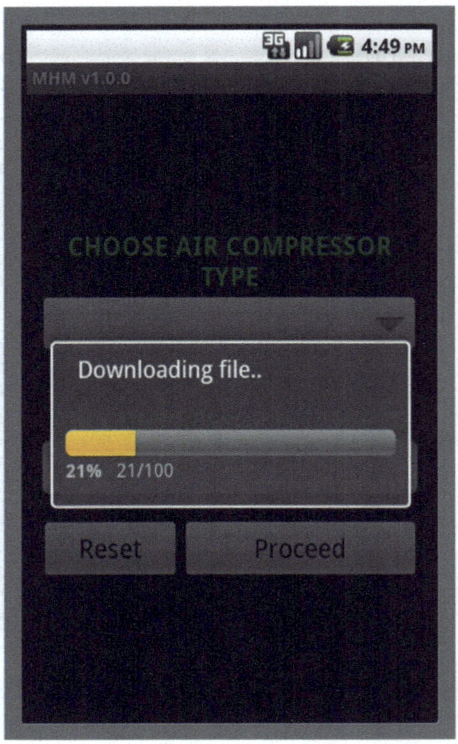

Fig. 9.8 Dialogue box showing that the resource file is being downloaded [1]

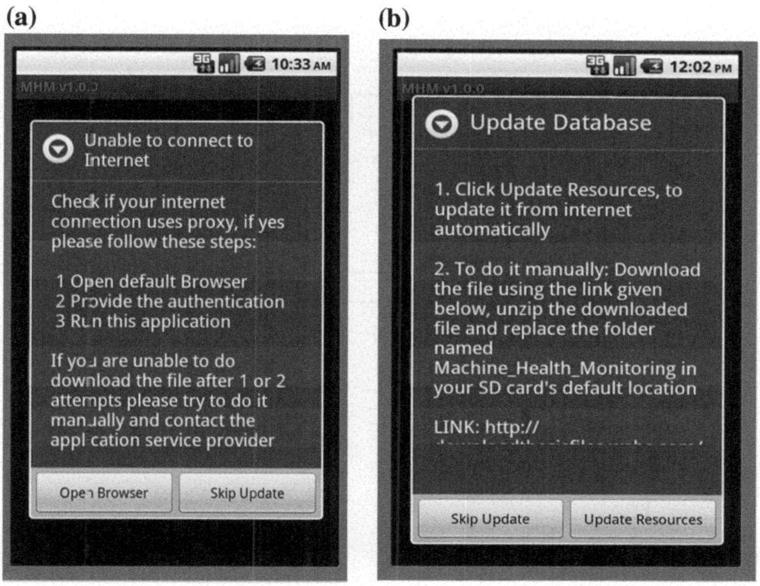

Fig. 9.9 **a** Dialogue box for internet authentication error, **b** dialogue box to update resources [1]

i. When the file is being downloaded from the Internet, it tries to extract a file named as *CompressorDetails.db* into a temporary storage location where the downloaded files are stored.
ii. If the downloaded file is corrupt, it will not contain the *CompressorDetails.db* and hence no such file will be extracted in the temporary storage location.
iii. The application performs a check if the *CompressorDetails.db* file was extracted or not. If the file exists, the downloaded file is extracted into the desired folder of SD card.
iv. In case the application is unable to find the file, a dialogue box will be displayed as shown in Fig. 9.9. Once all the above checks are proper, the application pulls data from *CompressorDetails.db* database and updates the Compressor Type selection box from where the user can select the necessary compressor for conducting the fault diagnosis.

An extra field named as *Choice Not Available* is also provided for the selection of compressor type. It has to chosen if the type compressor required is not available. Once this option is selected by user, the dialogue box is displayed as shown in Fig. 9.9.

When the user clicks on the button *Update Resources,* the application downloads the necessary file for update from the application server. *Skip Update* can be used if the user does not want to have any update. Once the user selects the compressor type, information is fetched from the database and displayed. After clicking on *Proceed* button, Activity page 2 appears. Figure 9.10 shows the back end of app activity page 1.

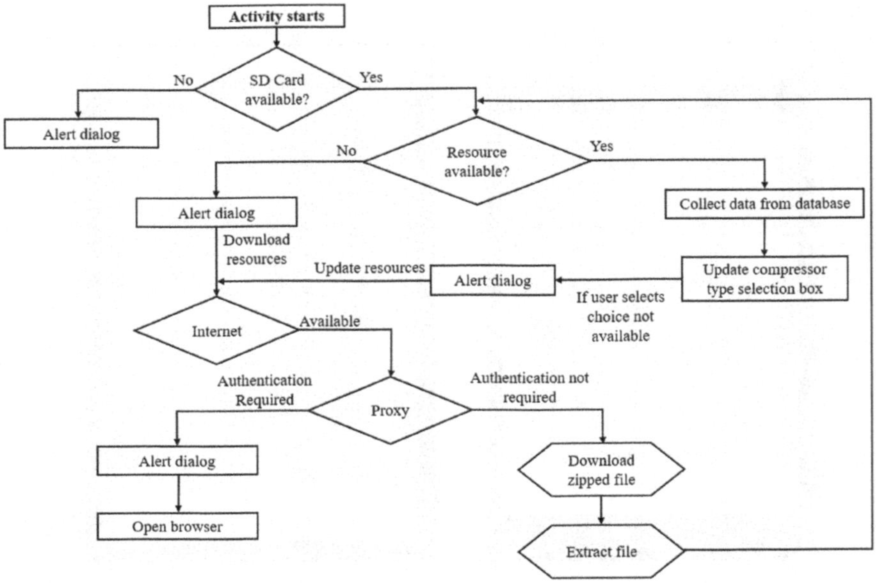

Fig. 9.10 Back end of app activity page 1 [1]

9.8.2 *Activity Page 2 (Confirmation Activity)*

This activity page is meant for confirming the user choices. In the front end, application displays all the machine details fetched from database, i.e., compressor id, type of compressor, and available number of faulty conditions as shown in Fig. 9.11. If the user wants to change the inputs that were entered in the first page, user can go back to the previous page by simply clicking the back button of smartphone. In the back end, application checks for a database file named "*CompressorDetails.db*" which contains information of the air compressor types and respective faults. This database file is created using open-source SQLite Database Browser 2.0. There is no back-end for this activity page.

9.8.3 *Activity Page 3 (Data Recording Activity)*

The purpose of this activity page is to record acoustic data corresponding to a fault. While recording data from a smartphone, it must be placed at a specific position. In the front end, this page consists of components needed to record data and display sensitive position locations and instruction details as shown in Fig. 9.12. The *view sensitive position* button is provided to display the sensitive position image corresponding to each fault as shown in Fig. 9.12. Once the user clicks the button to view sensitive position image, *Record data for fault* button gets enabled which then can

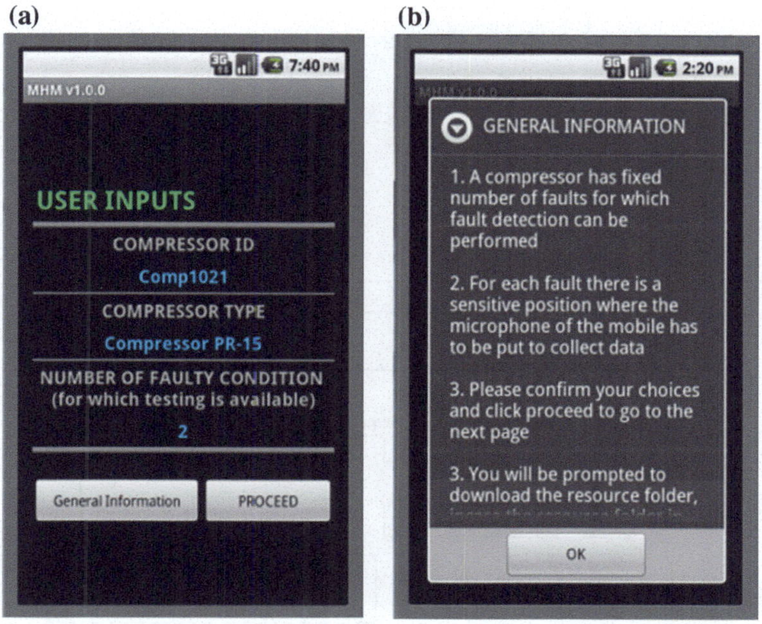

Fig. 9.11 **a** User interface of activity page 2, **b** dialogue box appears when general information is clicked [1]

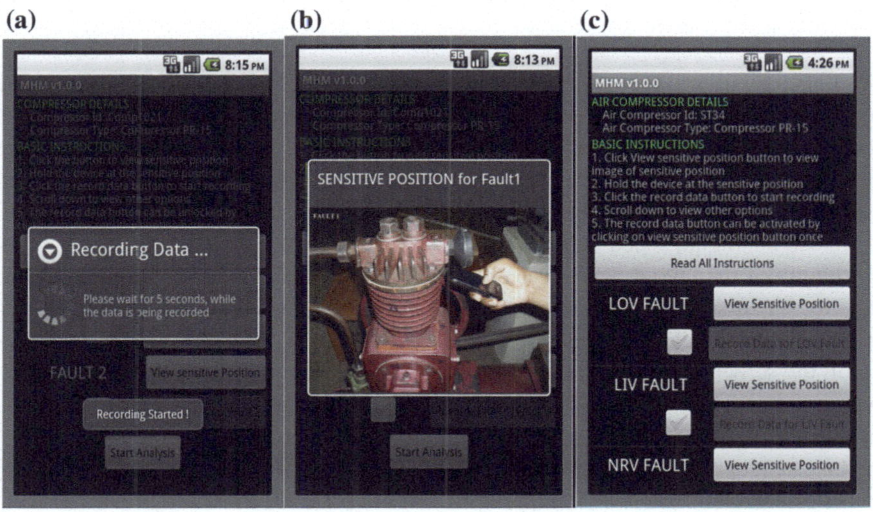

Fig. 9.12 **a** User interface page for app activity page 3, **b** dialogue box showing sensitive position location, **c** dialogue box appears while recording takes place [1]

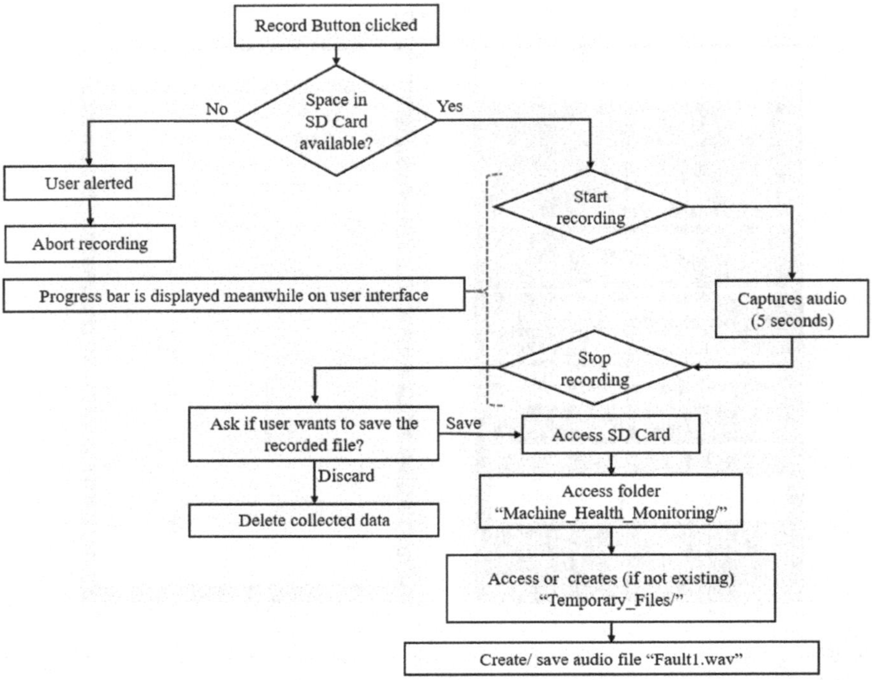

Fig. 9.13 Flowchart of "Record Button" of app activity page 3 [1]

be clicked by user to record data as shown in Fig. 9.12. The application gives option to perform check for fewer faults instead of all faults. This can be done by recording data for only those faults for which diagnosis is to be done. After recording, *Start Analysis* button can be clicked to begin the analysis. Figure 9.13 shows the flow chart of "record button" of app activity page 3.

In the back end, when activity page starts, application creates user interface as the length of the layout depends on the number of faults available for testing. This activity has a scrollable layout, which means options available in the application page can extend below the assigned screen size and to access these options user can scroll the screen. The concept used to create the activity page is of dynamic layout, whereas the page is generated by programming based on given parameters. The view of the activity page is decided by the number of faults available in the compressor. In this page, application gathers information of number of faults available for testing in the selected compressor from previous activity pages as shown in Fig. 9.13.

9.8.4 Activity Page 4 (Fault Diagnosis Activity)

Activity page 4 is designed for intimating the user that fault diagnosis process is going on. In the front end, activity page will display a progress spinner with dialogue box as shown in Fig. 9.14. The user can move on to another application by clicking home button. The application will automatically prompt with results when processing is completed.

In the back-end, as soon as the activity is called, current activity page will collect all the information from previous activity pages. In order to perform fault diagnosis, an iterative process is performed. The application creates an array where an individual fault is checked and the result is stored. At first, application checks if the recording file is present or not. In case if it is not present, then a message "data not collected" will be stored in the result array. If the file is present, application will check that fault specific model files are present or not. If the model files are not present, then the application will store a message "resources not updated" in the result array. In case the related model files are present, recorded audio data corresponding to each fault is passed through various stages of testing phase model, namely pre-processing, feature extraction, feature selection, and classification. Once the status of signal is determined, this page moves to the next page. The flowchart of processing algorithm for this page is shown in Fig. 9.15.

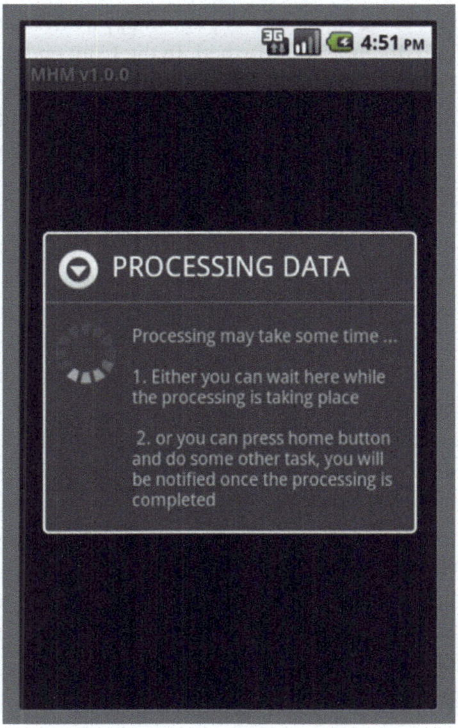

Fig. 9.14 User interface of app activity page 4 [1]

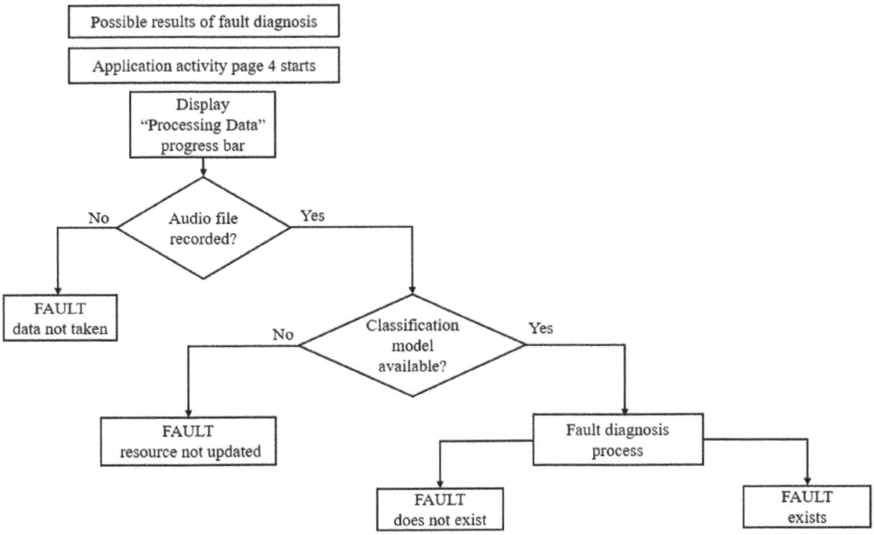

Fig. 9.15 Back-end flowchart of app activity page 4 [1]

9.8.5 Activity Page 5 (Report Generation Activity)

This is the final activity page of application and it is meant for displaying the results of fault diagnosis performed. In the front end, activity page displays "Test Results". The user can see detailed results of all faults tested for fault diagnosis where information of whether the fault exists or not is provided, as shown in Fig. 9.16. If no fault is detected, then "Tested Faults Do Not Exist" is displayed and if even one fault exists, then "Fault(s) Detected in Machine" is displayed. The details of individual fault checks can be seen by clicking on "Show detailed results" button as shown in Fig. 9.16. If the user has not recorded machine sound for a fault, "Data is not recorded" message will be displayed for the corresponding fault. The user can send test result to maintenance department either by sending a mail using "Send report" button or by sending a SMS using "Send SMS" button. This page also provides the user an option to perform another fault diagnosis test on machine without exiting the application. In the back end, as soon as the activity page is started, it starts collecting information of air compressor and results of the fault diagnosis from previous activity page. After receiving the results, application performs a check if any of the faults available for testing has been detected in air compressor. In case a fault is detected, the notification area will display a red-coloured text "Fault Detected in Machine". If no fault is detected then the application performs another check whether data for any of the faults was taken by the user. In case data for none of the faults was collected the text displayed in notification area will be "Data Not Collected for Any Fault". The Send Email button activates email intent activity and then sends email to the person as entered by the user.

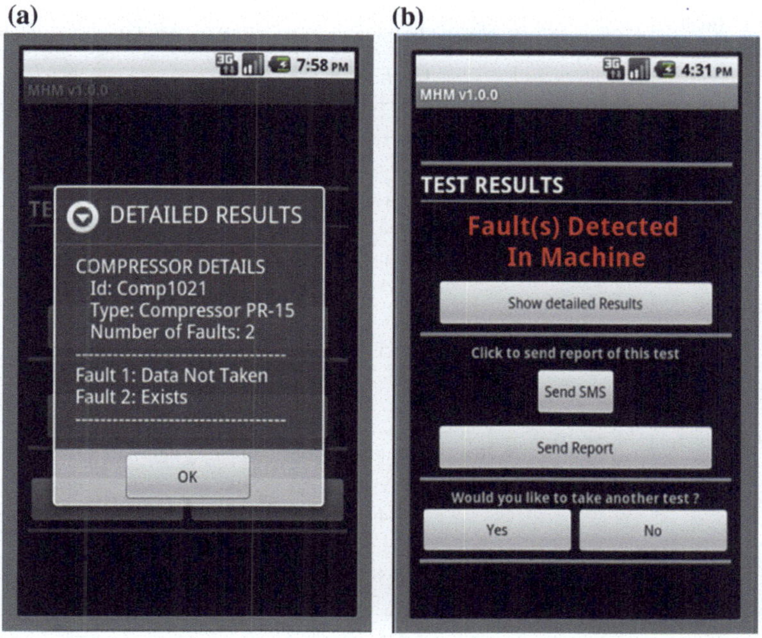

Fig. 9.16 **a** User interface of app activity page 5, **b** dialogue box appears when show detailed results is clicked [1]

9.9 Performance Evaluation

9.9.1 Fivefold Cross-Validation

At first, we collected 1500 data samples for healthy, LIV fault, and LOV fault conditions, and then fivefold cross-validation was performed. The dataset is divided into five sets, namely G1, G2, G3, G4, and G5. The accuracies have been calculated by taking different combinations of these data groups as training and testing sets as shown in Tables 9.1 and 9.2. Hence, 1200 data samples were taken as the training dataset each time and remaining 300 were taken as the testing dataset. The accuracy of a classification model is defined as the percentage of correct results given by the model over the total number of data samples.

9.9.2 Without Fivefold Cross-Validation

In this case, we took different number of datasets and divided them into two parts using *libsvm* function "*subset.py*" [18] into the ratio of 70% training dataset and 30% testing dataset. We worked with two types of kernel and all the training models were

Table 9.1 Fivefold cross-validation for LIV fault

S. No.	Training dataset	Testing dataset	Accuracy (%)
1	G1 + G2 + G3 + G4	G5	99.67
2	G1 + G3 + G4 + G5	G2	100
3	G1 + G2 + G4 + G5	G3	100
4	G1 + G2 + G3 + G5	G4	100
5	G2 + G3 + G4 + G5	G1	100

Table 9.2 Fivefold cross-validation for LOV fault

S. No.	Training dataset	Testing dataset	Accuracy (%)
1	G1 + G2 + G3 + G4	G5	90.00
2	G1 + G3 + G4 + G5	G2	94.67
3	G1 + G2 + G4 + G5	G3	97.33
4	G1 + G2 + G3 + G5	G4	94.33
5	G2 + G3 + G4 + G5	G1	92.33

built using RBF kernel as it gives good results for large number of samples. Table 9.3 shows the classification accuracy for LIV, LOV, and NRV faults. It also shows the results obtained with linear kernel and the RBF.

Table 9.3 Classification accuracy results

S. No.	No. of samples	Kernel	LIV (%)	LOV (%)	NRV (%)
1	800	RBF	99.375	94.375	92.5
2	800	RBF	100	93.125	92.5
3	800	RBF	100	96.25	91.825
4	375	RBF	98.66	100	–
5	375	Linear	98.66	100	–

9.10 Conclusions

The major issue in intelligent condition-based monitoring of air compressors was to perform live fault detection and put all the components required in data acquisition and processing into a single device. This chapter explains building a complete system on Android-based smartphone to perform condition-based monitoring. All the mentioned results were checked for offline data. The real-time fault detection was also done and up to one extent we achieved good results. Now, the future work involves real-time thorough testing. As of now, the developed ICBM has been tested only for two fault types. The system needs to be trained for more faults on offline and online data. The same can be extended for large number of faults and various models of the machines. The challenges still lie in hand to develop a full-fledged ICBM for air compressors and further improve the model to work in real-time with robustness.

References

1. Verma, N.K., Sarkar, S., Dixit, S., Sevakula, R.K., Salour, A.: Android app for intelligent CBM. In: 22nd IEEE Symposium on Industrial Electronics, Taipei, Taiwan, pp. 1–6 (2013)
2. Verma, N.K., Sevakula, R.K., Dixit, S., Salour, A.: Intelligent condition based monitoring using acoustic signals for air compressors. IEEE Trans. Reliab. **65**(1), 291–309 (2016)
3. Verma, N.K., Sevakula, R.K., Dixit, S., Salour, A.: Data driven approach for drill bit monitoring. IEEE Rel. Mag. 19–26 (2015)
4. Verma, N.K., Singh, J.V., Gupta, M., Dixit, S., Sevakula, R.K., Salour, A.: Windows mobile and tablet app for acoustic signature based machine health monitoring. In: 9th IEEE Conference on Industrial and Information Systems, Gwalior, India, pp. 1–6 (2014)
5. Verma, N.K., Singh, S., Gupta, J.K., Sevakula, R.K., Dixit, S., Salour, A.: Smartphone application for fault recognition. Int. J. Smart Sens. Intell. Syst. **6**(4), 1763–1782 (2013)
6. Wave PCM Soundfile Format: Available at https://ccrma.stanford.edu/courses/422/projects/WaveFormat/. Accessed June 2014
7. WasApiAudioComp Package: Available at https://github.com/nokiadeveloper/audio-recorder/tree/master/WasapiAudioComp. Accessed June 2014
8. NAudio Library: Available at http://naudio.codeplex.com/. Accessed May 2014
9. Verma, N.K., Kumar, P., Sevakula, R.K., Dixit, S., Salour, A.: Ranking of sensitive positions based on statistical parameters and cross correlation analysis. In: 6th IEEE International Conference on Sensing Technology, Kolkata, India, pp. 815–821 (2012)
10. Lambrou, T., Kudumakis, P., Speller, R., Sandler, M., Linney, A.: Classification of audio signals using statistical features on time and wavelet transform domains. In: Proceedings of IEEE International Conference Acoustics, Speech and Signal Processing, Washington, USA, **6**, 3621–3624 (1998)
11. Reserved File and URI Associations: http://msdn.microsoft.com/enus/library/windowsphone/develop/jj207065(v=vs.105).aspx. Accessed May 2014
12. Math.Net Package: Available at https://www.nuget.org/packages/MathNet.Numerics. Accessed May 2014
13. Microsoft Phone Controls Toolkit: Available at http://phone.codeplex.com/. Accessed May 2014
14. Commons Math Developers: Apache Commons Math, Release 2.2. Available: http://commons.apache.org/math. Accessed June 2014

15. Nuget Package Gallery: Available at https://www.nuget.org/. Accessed May 2014
16. Verma, N.K., Singh, S., Gupta, J.K., Sevakula, R.K., Dixit, S., Salour, A.: Smartphone application for fault recognition. In: 6th International Conference on Sensing Technology, Kolkata, India, pp. 1–6 (2012)
17. Windows Phone Power Tools 2.5.5: Available at http://wptools.codeplex.com/. Accessed May 2014
18. Chang, C.-C., Lin, C.-J.: LIBSVM: a library for support vector machines. ACM Trans. Intell. Syst. Tech. **2**(3), 27 (2011)

The manufacturer's authorised representative in the EU is Springer Nature Customer Service Centre GmbH, Europaplatz 3, 69115 Heidelberg, Germany. If you have any concerns regarding our products, please contact ProductSafety@springernature.com

Printed and bound by CPI Group (UK) Ltd, Croydon, CR0 4YY
30/09/2026
02225556-0004